Advanced Wireless Sensor Networks: Applications, Challenges and Research Trends

Advanced Wireless Sensor Networks: Applications, Challenges and Research Trends

Editors

Dionisis Kandris
Eleftherios Anastasiadis

MDPI

Basel • Beijing • Wuhan • Barcelona • Belgrade • Novi Sad • Cluj • Manchester

Editors
Dionisis Kandris
University of West Attica
Athens
Greece

Eleftherios Anastasiadis
University of West Attica
Athens
Greece

Editorial Office
MDPI AG
Grosspeteranlage 5
4052 Basel, Switzerland

This is a reprint of articles from the Special Issue published online in the open access journal *Electronics* (ISSN 2079-9292) (available at: https://www.mdpi.com/journal/electronics/special_issues/S8SDAKH5Z0).

For citation purposes, cite each article independently as indicated on the article page online and as indicated below:

Lastname, A.A.; Lastname, B.B. Article Title. *Journal Name* **Year**, *Volume Number*, Page Range.

ISBN 978-3-7258-1513-5 (Hbk)
ISBN 978-3-7258-1514-2 (PDF)
doi.org/10.3390/books978-3-7258-1514-2

Contents

About the Editors

Dionisis Kandris

Professor Dionisis Kandris is a graduate of Electrical and Computer Engineering Department at the University of Patras, Greece. He received his doctorate from the same academic institute in the Control of Wirelessly Interconnected Systems field. He also holds an M.Sc. in Manufacturing Systems Engineering from the University of Bradford, United Kingdom. After working in the industry, Dr. Kandris joined the University of West Attica, Greece, where he is currently a Professor at the Department of Electrical and Electronics Engineering and a member of the microSENSES Research Laboratory. So far, Professor Kandris has served as either a senior researcher or the project leader in many international research projects and has been the author of several scientific articles and books. Also, he is an active member of the Editorial Board of international scientific journals and various scientific committees. His current research interests focus on developing control algorithms for Wireless Sensor Networks and Automation Systems.

Eleftherios Anastasiadis

Dr. Eleftherios Anastasiadis received a BSc degree from the Department of Informatics and Telecommunications of the University of Athens, Greece, in 2011. He received an MSc and a PhD in Theoretical Computer Science from the Department of Computer Science at the University of Liverpool, UK, in 2012 and 2017, respectively. His PhD thesis was in Algorithmic Mechanism Design and dealt with the design of approximation algorithms and truthful mechanisms for optimization problems in network mechanism design. In 2017, he taught postgraduate modules in the Computer Science and Informatics Division at London South Bank University, UK. From 2018 to 2021, he was a postdoctoral researcher at the Transport Systems and Logistics Laboratory in the Department of Civil and Environmental Engineering at Imperial College London, UK. Since 2023, he has been a postdoctoral researcher designing control algorithms for Wireless Sensor Networks in the microSENSES Laboratory of the Department of Electrical and Electronics Engineering at the University of West Attica, Greece. His research interests include approximation algorithms and mechanism design for network optimization problems, wireless sensor networks, and agent-based simulation for autonomous vehicle fleets.

electronics

Editorial

Advanced Wireless Sensor Networks: Applications, Challenges and Research Trends

Dionisis Kandris * and Eleftherios Anastasiadis

Department of Electrical and Electronics Engineering, Faculty of Engineering, University of West Attica, Thivon Av. 250, GR-12241 Athens, Greece; e.anastasiadis@uniwa.gr
* Correspondence: dkandris@uniwa.gr

1. Introduction to the Applications, Challenges, and Research Trends in Wireless Sensor Networks

A typical wireless sensor network (WSN) contains wirelessly interconnected devices, called sensor nodes, which have sensing, processing, and communication abilities and are disseminated within an area of interest. A WSN also includes at least one sink node, called the base station, which has enhanced energy, computational, and communication resources. Within a WSN, while sensor nodes monitor ambient conditions, process the relative data, and transmit them to other sensor nodes and the base station, the latter controls the operation of the specific WSN and its communication with other WSNs and/or the final user [1].

Taking advantage of the combined capabilities of its constituting elements, WSNs can monitor the conditions existing in areas of interest of almost any kind and size. This is the reason why, although WSNs were initially invented to be used exclusively in military sector, currently they are not only considered to be the basis of the Internet of Things (IoT), but also support a continuously growing range of applications that are associated with almost any sector of human activity, ranging from the environment and flora and fauna, to industry, urban activities, and healthcare [2].

On the other hand, the operation of WSNs is obstructed because of various reasons. First of all, WSNs have certain restrictions. Specifically, the energy sufficiency of typical sensors nodes is extremely limited. This is because their energy is typically supplied by batteries which, in most cases, are impractical to either recharge or replace, since the positions of the sensor nodes are usually difficult or even impossible to reach. Therefore, the attainment of energy conservation is a vital issue for WSNs. That is why, while energy saving is necessitated, energy sustainability is pursued through many different methodologies and means [3–6].

Additionally, sensor nodes have limited resources in terms of the storage and processing of data. Thus, data management in WSNs is by itself a very challenging issue of scientific research [7–10].

Moreover, wireless communications have inborn limitations regarding transmission power, transmission speed, the capacity of communication channels, and their vulnerability to interferences and intrusion that impede WSNs. Consequently, numerous challenges regarding WSNs arise [11–13].

Furthermore, in most cases, the incorporation of a large number of sensor nodes in WSNs makes it particularly challenging to achieve specific goals associated with tasks such as connectivity preservation with coverage maximization [14–16], congestion avoidance [17–19], quality of service attainment [20,21], security provision [22,23], data aggregation [24,25], fault tolerance [26,27], and node localization [28,29]. In many cases, the performance optimization of WSNs concerns more than one of the aforementioned metrics, thus necessitating the usage of multi-objective optimization algorithms [30,31].

Citation: Kandris, D.; Anastasiadis, E. Advanced Wireless Sensor Networks: Applications, Challenges and Research Trends. *Electronics* **2024**, *13*, 2268. https://doi.org/10.3390/electronics13122268

Received: 3 June 2024
Accepted: 6 June 2024
Published: 9 June 2024

At the same time, emerging developments in several sectors of science and technology, such as the Internet of Things [32,33], machine learning [34,35], deep learning [36,37], big data [38–40], 5G [41–43], edge computing [44], energy harvesting [3,45,46], and wireless power transfer [3,46,47] seem to be promising to support and enhance the operation of WSNs, thus triggering corresponding research trends.

2. Overview of this Special Issue

The Special Issue, entitled "Advanced Wireless Sensor Networks: Applications, Challenges and Research Trends", attracted the interest of many researchers associated with the topics mentioned in the previous section, and finally, after a double-blind review process, ten high-quality papers were selected for publication. In this section, a brief overview of these ten contributions is provided in order to encourage the reader to explore them in more detail.

The research article by Singh et al., the first contribution of this Special Issue, investigates the integration of WSNs with machine learning and deep learning techniques to enhance soil moisture estimations for agricultural and environmental management purposes. Specifically, this study evaluates five machine learning/deep learning methods and demonstrates the effectiveness of the long short-term memory (LSTM) model in accurately estimating soil moisture levels across different regions. By leveraging WSN-driven data alongside satellite observations and climate models, the proposed methodology offers a practical approach for high-resolution soil moisture estimation, with implications for precision agriculture and environmental monitoring. The paper concludes by identifying future research directions to further improve soil moisture estimation models and their applicability in real-world scenarios.

The second contribution, by Pinto Neves et al., introduces a novel firmware update method for microcontrollers, aiming to minimize downtime and optimize data transmission during updates. Unlike traditional methods that replace the entire program, this approach enables updating specific code segments without interrupting ongoing operations. Implemented and validated on a PIC18F27K42 microcontroller, the method showcases reduced downtime, less than 10 ms, and good recoverability in failure scenarios. However, it has limitations, such as updating only up to eight rows at a time and requiring full control over functionalities, excluding compatibility with operating systems or hardware abstraction layers. Despite these limitations, the method demonstrates easy replication across various microcontrollers, indicating broad applicability. Future research directions include exploring radio transmission options and automating memory partitioning for improved efficiency, suggesting a promising avenue for advancing firmware update practices in microcontroller-based systems.

The third contribution of this Special Issue is a paper by Wang et al. that addresses the challenge of channel collisions in dense long-range wide-area networks (LoRaWANs) by proposing a novel time-allocation adaptive data rate (TA-ADR) algorithm. By introducing the concept of time intervals for node transmissions, the TA-ADR algorithm aims to allocate independent time slots to each node, mitigating data collision issues in densely populated scenarios and optimizing network performance. Practically, the specific algorithm dynamically adjusts the spreading factor (SF) and transmission power (TP) for LoRa nodes, intelligently scheduling transmission times to reduce the risk of data collisions and enhance transmission efficiency. Simulations conducted in a dense LoRaWAN environment demonstrate significant improvements over existing algorithms, achieving an approximate 30.35% enhancement in data transmission rate, 24.57% reduction in energy consumption, and 31.25% increase in average network throughput compared to the ADR+ algorithm.

The paper by Tu et al., referred to as the fourth contribution of this Special Issue, addresses the challenge of complexity in multi-user detection (MUD) schemes for uplink massive multiple-input multiple-output (M-MIMO) systems by proposing a novel mixed over-relaxation (MOR) algorithm, combining the advantages of successive over-relaxation (SOR) and accelerated over-relaxation (AOR) methods. The MOR algorithm aims to reduce the bit error rate (BER), computational complexity, and adapt to both 4G and beyond fifth-generation (B5G) environments. By dividing MOR into initial and collaboration stages, the algorithm achieves rapid convergence and refinement performance through alternating iterations. Simulations demonstrate significant improvements in BER performance compared to traditional SOR and AOR methods, achieving approximately 99.999% and 99.998% improvement, respectively, while keeping the complexity at O(N^2). The collaborative architecture of MOR effectively balances BER performance and computational complexity, making it suitable for M-MIMO orthogonal frequency division multiplexing (OFDM) and universal filtered multi-carrier (UFMC) systems in both 4G and B5G environments, presenting a promising solution for future wireless communication systems.

The fifth contribution of the Special Issue is a paper by Xu et al. that introduces a Q-learning and efficient low-quantity charge (QL-ELQC) method tailored for the smoke alarm unit within a power system, aiming to enhance the lifetime of wireless sensor network nodes. Actually, traditional medium-access control protocols often overlook the alarm state, prompting the need for an optimized approach. The QL-ELQC method, considering the relationship between sensor data conditions and RF module activation, indeed optimizes the standby and active periods of nodes based on quantity charge models. By effectively managing the duty cycle, the proposed method mitigates the continuous state–action space limitations of Q-learning. Simulation results demonstrate significant improvements in latency and energy efficiency compared to existing schemes, with experimental validation aligning with theoretical expectations. The extension of the lifetime of nodes provided by the proposed method is particularly beneficial in scenarios where battery replacement or recharging is impractical, thus offering a promising solution for enhancing WSN longevity in alarm systems under harsh environmental conditions.

The paper by Kovtun et al. is the sixth contribution of this Special Issue. It investigates the process of information transfer between sensor network end IoT devices and hubs at the transport protocol level, focusing on leveraging the 5G platform. Viewing this process as a semi-Markov model with nested Markov chains, the study derives a stationary distribution of the sliding window size, crucial for determining information flow intensity. A recursive method with linear computational complexity is formalized to calculate this distribution. Using this, a distribution function characterizing communication channel bandwidth is formulated. The study showcases the potential of TCP protocol in handling massive IoT traffic but highlights security concerns. Future research aims to optimize TCP parameters for precise Quality of Service (QoS) policies in 5G clusters supporting sensor networks, contributing to advancements in ultra-reliable low-latency communications (URLLCs), massive machine-type communications (mMTCs), and enhanced mobile broadband (eMBB) technologies.

The seventh contribution of this Special Issue is a research article by El Boudani et al. It introduces a novel approach for enhancing indoor positioning accuracy in 5G IoT networks, crucial for identity and context-aware applications such as simultaneous localization and mapping (SLAM). Utilizing a K-nearest neighbors and deep neural network (K-DNN) algorithm, the study proposes a method that incorporates a fusion of Bluetooth low-energy (BLE) and wireless local area network (WLAN) signals, along with a unique data augmentation concept for received signal strength (RSS)-based fingerprinting, resulting in a 3D fused hybrid radiomap. This hybrid approach aims to improve 3D localization accuracy by addressing challenges such as outlier detection and reducing labor costs during data collection. The implementation demonstrates promising results, achieving a 91% classification accuracy in 1D and submeter accuracy in 2D positioning. The study underscores the potential of cooperative machine learning localization and suggests future directions for expanding the model's capabilities, including integrating data from different azimuth angles and incorporating floor-level detection for multi-story buildings.

The eighth contribution is an article by Kou et al. that addresses authentication and key negotiation challenges in unmanned aerial vehicle mobile ad hoc networks (UAVMANETs) for secure communication among multiple UAVs. By introducing a Latin square approach, the authentication process is simplified, enhancing the efficiency of signature aggregation within the Boneh–Lynn–Shacham (BLS) signature scheme and aggregating keys negotiated via the elliptic curve Diffie–Hellman (ECDH) protocol into new keys. This innovative protocol ensures secure communication over insecure channels, crucial for UAVMANETs operating in open wireless environments. Through security analysis and simulations, the proposed scheme demonstrates improved efficiency in authentication and key negotiation while meeting stringent security requirements. However, future research is suggested to address scenarios involving dynamic changes in group membership, aiming to design a more flexible protocol tailored to the dynamic nature of UAV networks.

Christakis et al., in their research article, which is the ninth contribution of this Special Issue, investigate the use of low-cost electrochemical sensors in WSNs for air quality monitoring in urban environments, addressing the challenge of sensor aging that affects measurement accuracy. Through a long-term experimental study, the researchers compared sensor data with official air monitoring instruments, revealing that aging due to factors such as gas exposure and temperature fluctuations degrades sensor performance. To mitigate this, they developed novel corrective formulae using specific coefficients, which adjust for aging and temperature variations respectively. Their methodology demonstrated high reliability and accuracy for nitrogen dioxide (NO_2) and ozone (O_3) sensors without the need for frequent recalibration, making it feasible to deploy cost-effective and dense air quality monitoring networks in smart cities.

Finally, the tenth contribution of this Special Issue is a paper by Pu et al. that addresses the challenge of ensuring data freshness in industrial wireless sensor networks (IWSNs). Specifically, this research article proposes a scheduling algorithm that maintains the age of information (AoI) of each data packet within a bounded interval. Recognizing that optimizing the average AoI alone is insufficient for industrial applications, the authors developed a low-complexity AoI-bounded scheduling algorithm that guarantees timely data delivery, critical for the stability of industrial control systems. The algorithm adjusts the transmission intervals and superframe lengths based on the nodes' sampling periods, ensuring schedulability and reducing peak AoI by allocating additional time slots to nodes with higher requirements. Numerical examples demonstrate the effectiveness of this approach in maintaining bounded AoI, thus enhancing the reliability and real-time performance of IWSNs in industrial settings.

3. Conclusions

The Guest Editors of this Special Issue believe that WSNs will continue being at the epicenter of scientific interest, and hope that this collection of articles will be helpful to scientists who focus their research efforts on this challenging domain.

Author Contributions: Conceptualization, D.K. and E.A.; writing—original draft preparation, D.K. and E.A.; writing—review and editing, D.K. and E.A. All authors have read and agreed to the published version of the manuscript.

Funding: This article received no external funding.

Acknowledgments: The Guest Editors of this Special Issue sincerely thank all the scientists who submitted their research articles, the reviewers who assisted in evaluating these manuscripts, and both the Editorial Board Members and the Editors of *Electronics* for their overall support.

Conflicts of Interest: The authors declare no conflicts of interest.

List of Contributions:

1. Singh, T.; Kundroo, M.; Kim, T. WSN-Driven Advances in Soil Moisture Estimation: A Machine Learning Approach. *Electronics* **2024**, *13*, 1590. https://doi.org/10.3390/electronics13081590.
2. Neves, B. P.; Santos, V. D. N.; Valente, A. Innovative Firmware Update Method to Microcontrollers during Runtime. *Electronics* **2024**, *13*, 1328. https://doi.org/10.3390/electronics13071328.

3. Wang, K.; Wang, K.; Ren, Y. Time-Allocation Adaptive Data Rate: An Innovative Time-Managed Algorithm for Enhanced Long-Range Wide-Area Network Performance. *Electronics* **2024**, *13*, 434. https://doi.org/10.3390/electronics13020434.
4. Tu, Y.-P.; Jian, P.-S.; Huang, Y.-F. Novel Hybrid SOR- and AOR-Based Multi-User Detection for Uplink M-MIMO B5G Systems. *Electronics* **2024**, *13*, 187. https://doi.org/10.3390/electronics13010187.
5. Xu, K.; Li, Z.; Cui, A.; Geng, S.; Xiao, D.; Wang, X.; Wan, P. Q-Learning and Efficient Low-Quantity Charge Method for Nodes to Extend the Lifetime of Wireless Sensor Networks. *Electronics* **2023**, *12*, 4676–4676. https://doi.org/10.3390/electronics12224676.
6. Kovtun, V.; Grochla, K.; Połys, K. Investigation of the Information Interaction of the Sensor Network End IoT Device and the Hub at the Transport Protocol Level. *Electronics* **2023**, *12*, 4662. https://doi.org/10.3390/electronics12224662.
7. Brahim El Boudani; Tasos Dagiuklas; Kanaris, L.; Iqbal, M.; Christos Chrysoulas. Information Fusion for 5G IoT: An Improved 3D Localisation Approach Using K-DNN and Multi-Layered Hybrid Radiomap. *Electronics* **2023**, *12*, 4150–4150. https://doi.org/10.3390/electronics12194150.
8. Kou, G.; Wei, G.; Yuan, Z.; Li, S. Latin-Square-Based Key Negotiation Protocol for a Group of UAVs. *Electronics* **2023**, *12*, 3131. https://doi.org/10.3390/electronics12143131.
9. Christakis, I.; Odysseas Tsakiridis; Dionisis Kandris; Ilias Stavrakas. Air Pollution Monitoring via Wireless Sensor Networks: The Investigation and Correction of the Aging Behavior of Electrochemical Gaseous Pollutant Sensors. *Electronics* **2023**, *12*, 1842–1842. https://doi.org/10.3390/electronics12081842.
10. Pu, C.; Yang, H.; Wang, P.; Dong, C. AoI-Bounded Scheduling for Industrial Wireless Sensor Networks. *Electronics* **2023**, *12*, 1499. https://doi.org/10.3390/electronics12061499.

References

1. Yick, J.; Mukherjee, B.; Ghosal, D. Wireless Sensor Network Survey. *Comput. Netw.* **2008**, *52*, 2292–2330. [CrossRef]
2. Kandris, D.; Nakas, C.; Vomvas, D.; Koulouras, G. Applications of Wireless Sensor Networks: An Up-To-Date Survey. *Appl. Syst. Innov.* **2020**, *3*, 14. [CrossRef]
3. Evangelakos, E.A.; Kandris, D.; Rountos, D.; Tselikis, G.; Anastasiadis, E. Energy Sustainability in Wireless Sensor Networks: An Analytical Survey. *J. Low Power Electron. Appl.* **2022**, *12*, 65. [CrossRef]
4. Rault, T.; Bouabdallah, A.; Challal, Y. Energy Efficiency in Wireless Sensor Networks: A Top-down Survey. *Comput. Netw.* **2014**, *67*, 104–122. [CrossRef]
5. Nakas, C.; Kandris, D.; Visvardis, G. Energy Efficient Routing in Wireless Sensor Networks: A Comprehensive Survey. *Algorithms* **2020**, *13*, 72. [CrossRef]
6. Lin, D.; Wang, Q.; Min, W.; Xu, J.; Zhang, Z. A Survey on Energy-Efficient Strategies in Static Wireless Sensor Networks. *ACM Trans. Sens. Netw.* **2021**, *17*, 1. [CrossRef]
7. Wang, F.; Liu, J. Networked Wireless Sensor Data Collection: Issues, Challenges, and Approaches. *IEEE Commun. Surv. Tutor.* **2011**, *13*, 673–687. [CrossRef]
8. Zhu, C.; Shu, L.; Hara, T.; Wang, L.; Nishio, S.; Yang, L.T. A Survey on Communication and Data Management Issues in Mobile Sensor Networks. *Wirel. Commun. Mob. Comput.* **2011**, *14*, 19–36. [CrossRef]
9. Ang, K.L.-M.; Seng, J.K.P.; Zungeru, A.M. Optimizing Energy Consumption for Big Data Collection in Large-Scale Wireless Sensor Networks with Mobile Collectors. *IEEE Syst. J.* **2018**, *12*, 616–626. [CrossRef]
10. Sasirekha, S.P.; Priya, A.; Anita, T.; Sherubha, P. Data Processing and Management in IoT and Wireless Sensor Network. *J. Phys. Conf. Ser.* **2020**, *1712*, 012002. [CrossRef]
11. Rappaport, T.S. *Wireless Communications: Principles and Practice*; Cambridge University Press: Cambridge, UK, 2024.
12. Bi, S.; Ho, C.K.; Zhang, R. Wireless Powered Communication: Opportunities and Challenges. *IEEE Commun. Mag.* **2015**, *53*, 117–125. [CrossRef]
13. Jin, L.; Hu, X.; Lou, Y.; Zhong, Z.; Sun, X.; Wang, H.; Wu, J. Introduction to Wireless Endogenous Security and Safety: Problems, Attributes, Structures and Functions. *China Commun.* **2021**, *18*, 88–99. [CrossRef]
14. Al-Karaki, J.N.; Gawanmeh, A. The Optimal Deployment, Coverage, and Connectivity Problems in Wireless Sensor Networks: Revisited. *IEEE Access* **2017**, *5*, 18051–18065. [CrossRef]
15. Tripathi, A.; Gupta, H.P.; Dutta, T.; Mishra, R.; Shukla, K.K.; Jit, S. Coverage and Connectivity in WSNs: A Survey, Research Issues and Challenges. *IEEE Access* **2018**, *6*, 26971–26992. [CrossRef]
16. Farsi, M.; Elhosseini, M.A.; Badawy, M.; Arafat Ali, H.; Zain Eldin, H. Deployment Techniques in Wireless Sensor Networks, Coverage and Connectivity: A Survey. *IEEE Access* **2019**, *7*, 28940–28954. [CrossRef]
17. Ghaffari, A. Congestion Control Mechanisms in Wireless Sensor Networks: A Survey. *J. Netw. Comput. Appl.* **2015**, *52*, 101–115. [CrossRef]
18. Bohloulzadeh, A.; Rajaei, M. A Survey on Congestion Control Protocols in Wireless Sensor Networks. *Int. J. Wirel. Inf. Netw.* **2020**, *27*, 365–384. [CrossRef]

19. Ploumis, S.E.; Sgora, A.; Kandris, D.; Vergados, D.D. *Congestion Avoidance in Wireless Sensor Networks: A Survey*; IEEE Xplore: New York, NY, USA, 2012. [CrossRef]
20. Kaur, T.; Kumar, D. A Survey on QoS Mechanisms in WSN for Computational Intelligence Based Routing Protocols. *Wirel. Netw.* **2019**, *26*, 2465–2486. [CrossRef]
21. Chiwariro, R.; Thangadurai, N. Quality of Service Aware Routing Protocols in Wireless Multimedia Sensor Networks: Survey. *Int. J. Inf. Technol.* **2020**, *14*, 789–800. [CrossRef]
22. Yu, J.-Y.; Lee, E.; Oh, S.-R.; Seo, Y.-D.; Kim, Y.-G. A Survey on Security Requirements for WSNs: Focusing on the Characteristics Related to Security. *IEEE Access* **2020**, *8*, 45304–45324. [CrossRef]
23. Tomic, I.; McCann, J.A. A Survey of Potential Security Issues in Existing Wireless Sensor Network Protocols. *IEEE Internet Things J.* **2017**, *4*, 1910–1923. [CrossRef]
24. Sirsikar, S.; Anavatti, S. Issues of Data Aggregation Methods in Wireless Sensor Network: A Survey. *Procedia Comput. Sci.* **2015**, *49*, 194–201. [CrossRef]
25. Randhawa, S.; Jain, S. Data Aggregation in Wireless Sensor Networks: Previous Research, Current Status and Future Directions. *Wirel. Pers. Commun.* **2017**, *97*, 3355–3425. [CrossRef]
26. Chouikhi, S.; El Korbi, I.; Ghamri-Doudane, Y.; Azouz Saidane, L. A Survey on Fault Tolerance in Small and Large Scale Wireless Sensor Networks. *Comput. Commun.* **2015**, *69*, 22–37. [CrossRef]
27. Adday, G.H.; Subramaniam, S.K.; Zukarnain, Z.A.; Samian, N. Fault Tolerance Structures in Wireless Sensor Networks (WSNs): Survey, Classification, and Future Directions. *Sensors* **2022**, *22*, 6041. [CrossRef]
28. Paul, A.; Sato, T. Localization in Wireless Sensor Networks: A Survey on Algorithms, Measurement Techniques, Applications and Challenges. *J. Sens. Actuator Netw.* **2017**, *6*, 24. [CrossRef]
29. Sneha, V.; Nagarajan, M. Localization in Wireless Sensor Networks: A Review. *Cybern. Inf. Technol.* **2020**, *20*, 3–26. [CrossRef]
30. Kandris, D.; Alexandridis, A.; Dagiuklas, T.; Panaousis, E.; Vergados, D.D. Multiobjective Optimization Algorithms for Wireless Sensor Networks. *Wirel. Commun. Mob. Comput.* **2020**, *2020*, 4652801. [CrossRef]
31. Singh, O.; Rishiwal, V.; Chaudhry, R.; Yadav, M. Multi-Objective Optimization in WSN: Opportunities and Challenges. *Wirel. Pers. Commun.* **2021**, *121*, 127–152. [CrossRef]
32. Shahraki, A.; Taherkordi, A.; Haugen, O.; Eliassen, F. A Survey and Future Directions on Clustering: From WSNs to IoT and Modern Networking Paradigms. *IEEE Trans. Netw. Serv. Manag.* **2021**, *18*, 2242–2274. [CrossRef]
33. Gulati, K.; Kumar Boddu, R.S.; Kapila, D.; Bangare, S.L.; Chandnani, N.; Saravanan, G. A Review Paper on Wireless Sensor Network Techniques in Internet of Things (IoT). *Mater. Today Proc.* **2021**, *51*, 161–165. [CrossRef]
34. Eldar, Y.C.; Goldsmith, A.; Gündüz, D.; Poor, H.V. *Machine Learning and Wireless Communications*; Cambridge University Press: Cambridge, UK, 2022.
35. Jiang, C.; Zhang, H.; Ren, Y.; Han, Z.; Chen, K.-C.; Hanzo, L. Machine Learning Paradigms for Next-Generation Wireless Networks. *IEEE Wirel. Commun.* **2017**, *24*, 98–105. [CrossRef]
36. Yu, W.; Sohrabi, F.; Jiang, T. Role of Deep Learning in Wireless Communications. *IEEE BITS Inf. Theory Mag.* **2022**, *2*, 56–72. [CrossRef]
37. Qiu, Y.; Ma, L.; Priyadarshi, R. Deep Learning Challenges and Prospects in Wireless Sensor Network Deployment. *Arch. Comput. Methods Eng.* **2024**, 1–24. [CrossRef]
38. Bi, S.; Zhang, R.; Ding, Z.; Cui, S. Wireless Communications in the Era of Big Data. *IEEE Commun. Mag.* **2015**, *53*, 190–199. [CrossRef]
39. Dai, H.-N.; Wong, R.C.-W.; Wang, H.; Zheng, Z.; Vasilakos, A.V. Big Data Analytics for Large-Scale Wireless Networks. *ACM Comput. Surv.* **2019**, *52*, 99. [CrossRef]
40. Djedouboum, A.; Abba Ari, A.; Gueroui, A.; Mohamadou, A.; Aliouat, Z. Big Data Collection in Large-Scale Wireless Sensor Networks. *Sensors* **2018**, *18*, 4474. [CrossRef]
41. Sufyan, A.; Khan, K.B.; Khashan, O.A.; Mir, T.; Mir, U. From 5G to beyond 5G: A Comprehensive Survey of Wireless Network Evolution, Challenges, and Promising Technologies. *Electronics* **2023**, *12*, 2200. [CrossRef]
42. Le, N.T.; Hossain, M.A.; Islam, A.; Kim, D.; Choi, Y.-J.; Jang, Y.M. Survey of Promising Technologies for 5G Networks. *Mob. Inf. Syst.* **2016**, *2016*, 2676589. [CrossRef]
43. Dash, L.; Khuntia, M. *Energy Efficient Techniques for 5G Mobile Networks in WSN: A Survey*; IEEE Xplore: New York, NY, USA, 2020. [CrossRef]
44. Wang, T.; Liang, Y.; Shen, X.; Zheng, X.; Mahmood, A.; Sheng, Q.Z. Edge Computing and Sensor-Cloud: Overview, Solutions, and Directions. *ACM Comput. Surv.* **2023**, *55*, 1–37. [CrossRef]
45. Williams, A.J.; Torquato, M.F.; Cameron, I.M.; Fahmy, A.A.; Sienz, J. Survey of Energy Harvesting Technologies for Wireless Sensor Networks. *IEEE Access* **2021**, *9*, 77493–77510. [CrossRef]
46. Ijemaru, G.K.; Ang, K.L.-M.; Seng, J.K. Wireless Power Transfer and Energy Harvesting in Distributed Sensor Networks: Survey, Opportunities, and Challenges. *Int. J. Distrib. Sens. Netw.* **2022**, *18*, 15501477211067740. [CrossRef]
47. Huda, S.M.A.; Arafat, M.Y.; Moh, S. Wireless Power Transfer in Wirelessly Powered Sensor Networks: A Review of Recent Progress. *Sensors* **2022**, *22*, 2952. [CrossRef]

electronics

Article

WSN-Driven Advances in Soil Moisture Estimation: A Machine Learning Approach

Tinku Singh , Majid Kundroo and Taehong Kim *

School of Information and Communication Engineering, Chungbuk National University, Cheongju 28644, Republic of Korea; tinku.singh@cbnu.ac.kr (T.S.); kundroomajid@cbnu.ac.kr (M.K.)
* Correspondence: taehongkim@cbnu.ac.kr

Abstract: Soil moisture estimation is crucial for agricultural productivity and environmental management. This study explores the integration of Wireless Sensor Networks (WSNs) with machine learning (ML) and deep learning (DL) techniques to optimize soil moisture estimation. By combining data from WSN nodes with satellite and climate data, this research aims to enhance the accuracy and resolution of soil moisture estimation, enabling more effective agricultural planning, irrigation management, and environmental monitoring. Five ML models, including linear regression, support vector machines, decision trees, random forests, and long short-term memory networks (LSTM), are evaluated and compared using real-world data from multiple geographical regions, which includes a dataset from NASA's SMAP project, supplemented by climate data, which employs both active and passive sensors for data collection. The outcomes demonstrate that the LSTM model consistently outperforms other ML algorithms across various evaluation metrics, highlighting the effectiveness of WSN-driven approaches to soil moisture estimation. The study contributes to the advancement of soil moisture monitoring technologies, offering insights into the potential of WSNs combined with ML and DL for sustainable agriculture and environmental management practices.

Keywords: soil moisture estimation; wireless sensor networks (WSNs); precision agriculture; remote sensing; LSTM (long short-term memory)

Citation: Singh, T.; Kundroo, M.; Kim, T. WSN-Driven Advances in Soil Moisture Estimation: A Machine Learning Approach. *Electronics* **2024**, *13*, 1590. https://doi.org/10.3390/electronics13081590

Academic Editors: Dionisis Kandris and Eleftherios Anastasiadis

Received: 25 March 2024
Revised: 19 April 2024
Accepted: 19 April 2024
Published: 22 April 2024

1. Introduction

Soil moisture estimation, which involves quantifying the water content present in the soil, is a critical parameter in agriculture and crop management, as it directly influences various aspects of plant growth, health, and overall agricultural productivity. Soil moisture content is an essential indicator of water availability for plants. Optimal soil moisture levels are essential for seed germination, crop emergence, and supporting vigorous field activity during critical growth stages [1]. Soil moisture levels not only affect the physical and chemical processes of soil but also influence the global ecological environment, hydrological patterns, and climate change [2].

Monitoring soil moisture levels can provide valuable insights to support precision agriculture techniques. Studies have shown that soil moisture estimation can aid in precision irrigation management [3]. By accurately quantifying soil moisture content, farmers can optimize water application, preventing over-watering or under-watering of crops. This leads to improved water use efficiency, reduced agricultural water consumption, and enhanced crop yields. Additionally, soil moisture data can help detect the onset of agricultural droughts, enabling timely interventions and mitigation strategies to support crop resilience during water-stressed conditions [4].

Soil moisture estimation also plays a critical role in crop planning and management. Knowing the available soil moisture content can guide decisions on crop selection, planting schedules, and the application of fertilizers and pesticides [1]. This information helps farmers maximize the use of limited water resources and ensures optimal growing conditions for their crops. The importance of soil moisture estimation extends beyond agriculture, encompassing domains such as water resource management, environmental monitoring, and environmental management. Soil moisture plays a crucial role in evaluating erosion risks and assessing the potential for geological hazards like landslides and sinkholes [3]. Monitoring soil moisture is essential for managing water resources, detecting environmental stresses, and supporting sustainable ecosystem management [3]. For instance, soil moisture information can aid in flood prediction and control efforts. By monitoring soil saturation levels, authorities can better anticipate the risk of flooding and take appropriate mitigation measures [5]. Additionally, soil moisture data can help evaluate erosion risks, as excessive moisture can lead to soil instability and increased erosion rates. Moreover, soil moisture estimation contributes to improved weather forecasting and climate modeling. Incorporating soil moisture data into these models enhances their accuracy, enabling more reliable predictions of precipitation patterns, temperature fluctuations, and other climate-related phenomena [3]. This information is crucial for managing water resources, planning climate adaptation strategies, and supporting sustainable ecosystem management.

Soil moisture encompasses capillary, gravitational, and hygroscopic water, with its dynamics shaped by variables like temperature, vegetation, soil composition, land use, topography, and precipitation patterns [6]. The level of soil moisture emerges as a pivotal factor impacting plant development, nutrient uptake, and the physical and chemical properties of the soil. Surface soil moisture levels ranging from 20 to 25 mm are conducive to germination and emergence of crops but may hinder fieldwork and damage newly-seeded crops if prolonged. Optimal vigorous field activity is associated with 15–20 mm of surface soil moisture, while levels below 10 mm may not support seed germination or early growth. Subsurface soil moisture values above 100 mm indicate favorable moisture conditions, while levels below 25 mm may lead to crop stress and reduced yields, especially during critical growth stages [7]. Moreover, it serves as a valuable indicator for detecting agricultural droughts, early signs of water scarcity, and aids in strategic crop planning and management strategies. Furthermore, soil moisture plays a crucial role in evaluating erosion risks, assessing the potential for geological hazards such as landslides and sinkholes, contributing to improved weather prediction accuracy, and facilitating flood control initiatives [3].

The traditional methods for estimating soil moisture have several limitations. Even though gravimetric methods are accurate, they are time-consuming and labor-intensive, providing moisture content only at specific depths. Similarly, hand-feel techniques and moisture blocks are qualitative methods and may produce varying results based on users and conditions [6]. Remote sensing methods like satellite imaging offer a cost-effective and non-invasive approach for monitoring soil moisture over large areas [8,9]. However, remote measurements typically have lower resolution compared to point measurements and estimate moisture content only for the top few centimeters of soil. Complex data processing is also required to filter out the effects of vegetation, terrain, soil type, and other factors influencing remote sensing data.

To address these shortcomings, this study explores the integration of ML and DL techniques with WSN and remote sensing data for soil moisture estimation. ML and DL models are data-driven and can integrate relevant input features like brightness temperature, synthetic aperture radar (SAR) backscatter, sensor properties, geographical information, and meteorological variables from WSNs and remote sensing sources to map the output [10]. These advanced models have shown promising results in accurately predicting surface soil moisture with high spatial and temporal resolution. By leveraging their ability to extract complex patterns and relationships from diverse data sources, including WSNs and remote sensing data, ML and DL techniques can overcome the limitations of traditional methods in terms of spatial and temporal resolution, coverage, and adaptability to non-linear and dynamic relationships [11].

This study aims to develop an optimal methodology for accurately determining soil moisture content over large agricultural areas using active and passive microwave satellite data sensitive to topsoil moisture integrated with data from WSNs. Five ML/DL methods are evaluated as part of the proposed methodology, including linear regression (LR), support vector machines (SVM), decision trees (DT), random forests (RF), and LSTM. The models are trained using data from NASA's SMAP satellite mission, which offers global measurements of moisture in the top 5 cm of soil. Additionally, data from WSNs are incorporated. Models are assessed using various metrics such as Mean Square Error (MSE), Root Mean Square Error (RMSE), Mean Absolute Error (MAE), and Mean Absolute Percentage Error (MAPE). These evaluations help identify the most effective algorithm for continuously estimating and predicting real-time soil moisture content by leveraging both microwave data and WSN data. The main contributions of this study are as follows:

- This paper presents an efficient method for estimating soil moisture content using ML and DL techniques for active-passive microwave remote sensing data.
- By consolidating data from ground-based sensors embedded in wireless sensor network platforms, specifically targeting soil moisture levels, with inputs from remote sensing sources like satellite observations and climate models, this approach elevates the precision of soil moisture estimation through comprehensive data integration.
- It provides a practical approach to soil moisture estimation that combines cutting-edge machine-learning techniques with real-world data sources, making it applicable in agricultural and environmental management contexts.
- Developing a framework leveraging the best-performing ML model to provide accurate, high-resolution soil moisture quantification, which can aid farmers, water management authorities, and stakeholders in irrigation planning, drought monitoring, and crop yield forecasting.

The structure of this paper is as follows. Section 2 provides an overview of recent advancements in soil moisture estimation using WSNs, remote sensing, and machine learning/deep learning techniques. Section 3 introduces the basic tools and techniques used in this study. Section 4 describes the methodology, including data consolidation, pre-processing, feature selection, model building, and the evaluation metrics employed. Subsequently, Section 5 presents the results and provides a comprehensive discussion of the obtained findings. Finally, the paper is concluded in Section 6, and potential future research directions are identified.

2. Literature Review

Soil moisture is pivotal in agriculture and environmental studies, influencing plant growth and water balance. Precise assessment is crucial for tasks like precision agriculture, water resource management, and flood prediction. Traditional methods like gravimetric and TDR are laborious and spatially limited. WSN and IoT infrastructure also offers versatile tools for assessing various environmental parameters like water quality and soil moisture with precision and efficiency [12,13]. Remote sensing has become popular, with recent advancements leveraging WSN for real-time monitoring. Additionally, ML models have gained traction due to their ability to handle complex relationships, enhancing soil moisture prediction accuracy. SVM has significantly advanced soil moisture prediction, exhibiting superior accuracy and effectiveness in various studies. Utilizing a kernel function to map input data into a higher-dimensional space, SVM has outperformed traditional regression models in predicting soil moisture levels. Studies by [14–16] have demonstrated SVM's efficacy in this domain. However, SVM's computational complexity, particularly with large datasets, and sensitivity to parameter tuning remain as significant limitations. W. Wu et al. [17] employed the Random Forest Regression (RFR) model for soil moisture prediction. While effective in surpassing traditional regression models, RFR faces challenges in handling complex relationships or noisy datasets, necessitating careful consideration in soil moisture estimation tasks.

N. Zhu et al. [18] introduced a multilayer neural network model, utilizing data from the Heihe River Basin in China. This model outperformed multiple linear regression and SVM in terms of accuracy. The data used here are obtained from a small area that lacks diversity, and spatial representations and hence may have serious challenges in scaling up. Similarly, Song et al. [19] proposed a DL-based cellular automata model for spatiotemporal soil moisture distribution, achieving high accuracy across various spatial and temporal scales in China; however, the study area is relatively small, only 22 km^2, and the data used in the study are of only two months, August and September. Furthermore, LSTM, a type of recurrent neural network (RNN) known for handling long-term dependencies, has gained attention in soil moisture prediction. Q. Yuan et al. [20] applied LSTM in the Yellow River Basin of China, demonstrating superior performance compared to other ML models. However, despite their effectiveness, these methods may encounter challenges related to noise in the data and computational complexity, which should be taken into account when utilizing them for soil moisture estimation.

Remote sensing techniques, coupled with ML models, have also emerged as promising alternatives for accurate soil moisture estimation. For instance, H. Adab et al. [21] utilized an RF model for estimating surface soil moisture using remote sensing data from the Soil Moisture Active Passive (SMAP) satellite mission. Similarly, P. Leng et al. [22] presented a framework utilizing a combination of a land surface model and an RF model for all-weather fine-resolution soil moisture estimation. Additionally, M. A. Rajib et al. [5] proposed a drought evaluation and forecast model based on soil moisture simulation, employing a hybrid approach combining a hydrological model with remote sensing. William et al. [23] also used a multilayer perceptron model trained on sensor data from local meteorological departments to predict droughts in the coastal regions of Ecuador. Despite the advancements in soil moisture estimation, there are still several limitations to consider in remote sensing methods, such as limited spatial resolution, which is specifically relevant when it comes to estimating soil moisture on a finer scale.

Despite the significant contributions to soil moisture estimation, several research gaps remain to be addressed. The literature lacks a focus on refining the algorithms and models used to analyze WSN data and incorporating them effectively into remote sensing data for more precise soil moisture estimation. There is a need to develop methodologies for the seamless integration and interoperability of heterogeneous data sources. Additionally, research should address the challenges associated with data quality, consistency, and compatibility to ensure reliable and accurate soil moisture estimation across diverse geographical regions and environmental conditions. The integration of ML models with the fusion of remote sensing and WSN sources may enhance the reliability of soil moisture estimation. Overall, addressing these research gaps will contribute to advancing the state-of-the-art in soil moisture estimation and its applications in agricultural sustainability, water resource management, and environmental conservation.

3. Preliminaries

This section introduces key concepts and technologies driving soil moisture estimation advancements. It begins with an overview of NASA's SMAP project, employing active and passive sensors for global soil moisture data collection. The discussion follows on the Google Earth Engine, facilitating satellite data access and geospatial analysis. Additionally, the section highlights the Power Access Climate Data platform, integrating ground-based sensors and satellite imagery for comprehensive climate analysis. Further exploration covers Wireless Sensor Networks, notably IMD's deployment for real-time monitoring of essential meteorological parameters, including soil moisture content.

3.1. Soil Moisture Active Passive (SMAP)

The SMAP project [24–26], led by NASA, employs a combination of active and passive sensors to gather soil moisture data. An active synthetic aperture radar (SAR) transmits microwave pulses and measures their return signal strength, while a passive radiometer records natural microwave emissions from the Earth's surface. Equipped with a 6-m reflector antenna rotating every three seconds, the observatory scans a 1000 km-wide swath of the Earth's surface. Collaboratively, the radar and radiometer produce high-resolution soil moisture maps with a spatial accuracy of approximately 10 km. Orbiting at an altitude of about 685 km, the SMAP observatory completes global coverage every 2–3 days over a 3-year period, ensuring frequent and comprehensive soil moisture measurements. Data collected undergo processing and analysis at ground-based stations utilizing specialized algorithms to estimate soil moisture content within the top 5 cm of the soil. Validation occurs through ground-based measurements and other data sources, such as climate models.

3.2. Google Earth Engine (GEE)

GEE is a cloud computing platform that assesses, stores, and analyzes data from a variety of satellites, including Sentinel, Landsat, and MODIS. The collection includes climate, atmosphere, surface temperature, land cover, terrain, cropland, and other geophysical data that are openly and freely available. The web-based Interactive Development Environment and internet-based Application Programming Interface are available in Python and JavaScript. It helps the researchers to reduce the burden of storing a large number of big data files locally. It saves the data pre-processing and formatting time with the advantage of accessing earth observation data. Earth Engine Explorer lets users manage and visualize data from several satellites, while Earth Engine Time-lapse lets them see the Earth's evolution over 40 years. GEE can process large geospatial datasets with global coverage.

3.3. Power Access Climate Data

Power Access Climate Data [27] is a sophisticated web platform designed to offer comprehensive access to a wide range of climate data and analysis tools. Built on state-of-the-art technologies, it integrates data from ground-based sensors, satellite imagery, and climate models. Notably, the platform provides an extensive array of climate models with various resolutions, outputs, and time scales, available in formats like NetCDF, Comma Separated Values, and JSON. Users can access historical climate data spanning decades. It offers a wealth of climate information, including temperature, precipitation, humidity, wind speed and direction, and atmospheric pressure. Additionally, it provides real-time weather data sourced from an extensive network of weather stations located across the globe. This combination of historical and real-time data empowers users to perform comprehensive analyses and make accurate predictions regarding climate patterns and trends. The platform's visualization tools include interactive maps, time-series charts, and scatter plots for spatial and temporal data exploration. Moreover, its analysis and forecasting tools employ advanced statistical and ML algorithms to identify patterns and forecast future climate scenarios, catering to researchers, policymakers, and businesses in need of informed decision-making based on climate data.

3.4. WSN Based Data

The monitoring stations deployed by the Indian Meteorological Department (IMD) are strategically distributed to capture essential meteorological parameters using WSN. WSN technology enables the seamless collection of data related to temperature, humidity, rainfall, and wind speed from various geographical locations. Each monitoring station within the network is equipped with sensors capable of measuring these parameters in real-time [28]. IMD employs a variety of sensors within the WSNs to capture different meteorological parameters. Thermometers and hygrometers are utilized for measuring temperature and humidity. Rainfall is measured using rain gauges, while anemometers are employed for wind speed measurement. Soil moisture content is assessed using soil moisture sensors embedded in the ground at appropriate depths. To ensure the accuracy and reliability of the collected data, IMD follows stringent calibration processes for all deployed sensors. Calibration involves comparing sensor readings with known reference values under controlled conditions to adjust for any systematic errors or discrepancies. This calibration process is regularly conducted to maintain the accuracy of sensor measurements over time. Additionally, IMD implements various quality control measures to assess and maintain the integrity of the collected data. These measures include outlier detection, data validation checks, and sensor health monitoring. Outlier detection algorithms identify anomalous data points that may indicate sensor malfunctions or environmental disturbances. Data validation checks verify the consistency and plausibility of sensor readings based on predefined thresholds and ranges. WSNs are programmed to collect data at regular intervals depending on the specific requirements of the monitoring stations and the parameters being measured. The experiments in this study are based on daily frequency data. WSNs play a crucial role in facilitating the transmission of sensor data from remote locations to centralized data acquisition systems. These networks utilize wireless communication protocols to relay information over long distances, enabling IMD to gather comprehensive meteorological data across diverse terrains and regions in India. By leveraging WSN technology, IMD ensures the efficient operation of its monitoring stations and the timely acquisition of critical weather and soil moisture information for agricultural, environmental, and disaster management applications.

4. Methodology

Figure 1 provides an overview of the proposed methodology of this study, which begins with data consolidation, involving the integration of active and passive microwave satellite data sensitive to topsoil moisture with data obtained from WSNs. The data from NASA's SMAP satellite mission, which provides global measurements of moisture in the top 5 cm of soil, are acquired. WSN data are collected concurrently with satellite data to supplement the training dataset. The raw SMAP data are processed into usable estimates utilizing the GEE platform. Furthermore, the WSN data augment the SMAP-derived moisture levels with additional climate parameters such as temperature and precipitation. The integrated dataset consolidates satellite soil moisture observations from SMAP, and ground-based climate measurements, offering comprehensive insights into soil moisture dynamics and their relationships with weather, water availability, vegetation health, and climate patterns.

In the next step, the consolidated data undergo preprocessing, which includes handling missing values in the dataset using techniques like mean before-after and multivariate imputation. Subsequently, the dataset is further preprocessed to clean noise and ensure consistency, followed by feature selection, where the most relevant features from the integrated dataset are identified for the soil moisture estimation task. The data used in this study are recorded at regular intervals, indicating its time-series nature. To effectively utilize ML and DL techniques, the time-series data are transformed into a supervised learning problem. This transformation involves shifting the time series data and selecting appropriate lag values to create a dataset suitable for forecasting using supervised learning algorithms. This process ensures that the temporal relationships within the data are preserved and utilized for accurate estimation. Before concluding the data preprocessing step, the preprocessed data are split into training and testing sets in an 80:20 ratio.

Next, in the model building and training step, five different models (Logistic Regression, Support Vector Machines, Decision Trees, Random Forests, and LSTM) are selected for training. Each selected model undergoes training for a predefined number of epochs using the training subset to optimize parameters and weights for accurate predictions. Hyperparameter tuning is also performed simultaneously to achieve better results. The performance of each model is evaluated using standard evaluation metrics such as MSE, RMSE, MAE, and MAPE. These metrics enable the identification of the most reliable algorithm for continuously estimating and predicting real-time soil moisture content from microwave and WSN data. In the following sections, the steps involved in the proposed methodology are discussed in detail.

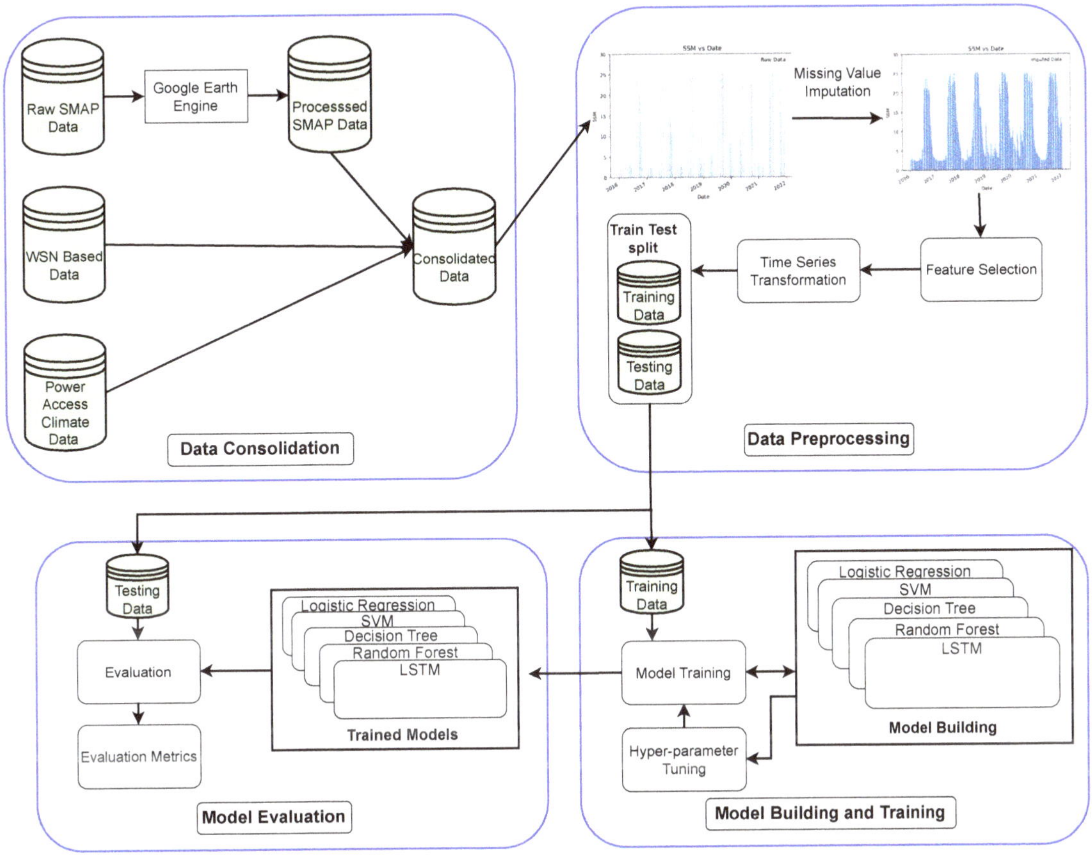

Figure 1. Methodology overview for soil moisture estimation via WSN-driven ML approaches.

4.1. Study Area

India is selected as the study area due to its geographical diversity and status as an agriculture-based country. It is geographically diverse, encompassing a wide range of environmental conditions, soil types, and land uses with a variety of agricultural practices prevailing in different regions. The study aimed to capture this diversity by selecting study areas from different regions across the country with a variety of agricultural practices prevailing in different regions. By including areas from various states and geographical regions, such as coastal areas, plains, and hilly terrains, the study ensures representation of the country's geographical diversity. The selected study areas represent a diverse range of agricultural practices found across India. For instance, some regions may specialize in rice cultivation, while others may focus on wheat, sugarcane, or pulses. Additionally, it provides real-time weather data sourced from an extensive network of weather stations located across the globe. This combination of historical and real-time data empowers users to perform comprehensive analyses and make accurate predictions regarding climate patterns and trends. By selecting areas with different dominant crops and agricultural techniques, the study aims to capture the variability in agricultural practices across the country. Furthermore, India exhibits diverse climatic zones, ranging from tropical in the south to temperate in the north, along with arid and semi-arid regions. These climatic variations influence soil moisture dynamics and agricultural productivity. The selected study spans different climatic zones, ensuring the representation of the full spectrum of climatic conditions in India. For instance, areas like Talcher and Angul may experience tropical climates, while regions like Kandhamal may have temperate climates. Additionally, consideration was given to areas with distinct industrial and agricultural activities. Talcher and Angul are known for their industrial and mining activities, which can have implications for soil moisture dynamics and land use patterns. On the other hand, regions like Cuttack, Dhenkanal, and Kandhamal are predominantly agricultural areas with diverse soil types and land uses.

4.2. Dataset Preparation

The process of consolidating the dataset involves bringing together a diverse array of data sources to create a unified soil moisture dataset. This includes incorporating raw data from various sources such as NASA's SMAP project, Power Access Climate Data, and data from WSNs. The soil moisture dataset was compiled by amalgamating data from multiple sources and processing tools. Initially, soil moisture data were sourced from NASA's SMAP project [29], utilizing GEE. This dataset encompasses features such as surface soil moisture levels, land surface temperature, soil texture, land cover/land use, and other pertinent parameters collected by the satellite mission. To enhance the dataset's richness, additional parameters from Power Access Climate Data were integrated. These parameters encompass temperature, wind direction (degrees), wind speed (m/s), surface pressure (kPa), dew point (°C), temperature at 2 meters height (°C), earth temperature (°C), and precipitation (mm per day). Incorporating these parameters offers a more comprehensive understanding of the factors influencing soil moisture levels. Moreover, parameters from WSNs-based data collected through the Indian Meteorological Department (IMD) were also included to augment the global dataset. These parameters comprise humidity, precipitation (rainfall/snowfall), sunshine duration, and cloud cover, further enriching the dataset with localized environmental data. The methodology includes preprocessing to standardize units and resolve inconsistencies, alignment, and integration to match data points based on spatial and temporal dimensions, and spatial interpolation to harmonize resolutions. Additional features are engineered, such as soil moisture anomalies, to enrich the dataset. Quality control checks ensure accuracy, with discrepancies addressed. The output is a consolidated dataset ready for analysis and applications in agriculture, environment, and disaster management.

4.2.1. Data Preprocessing

For time series data preprocessing, visual techniques are used to explore the dataset, with line plots specifically useful for identifying seasonality and trends. In cases where non-stationarity is observed, differencing methods are applied to stabilize the mean and variance of the data. The stationarity of the time series is further validated using the Augmented Dickey-Fuller (ADF) test, ensuring the suitability of the data for subsequent modeling steps. Feature engineering plays a vital role in extracting informative features from the time series, with lag values computed using autocorrelation function (ACF) plots to capture temporal dependencies. Missing values within the dataset are addressed using forward or backward-filling techniques, ensuring the continuity of the temporal sequence. Following preprocessing, the dataset is partitioned into training and testing sets while preserving the temporal order of the observations. Through these comprehensive preprocessing steps, the time series data are prepared for training and evaluation, enabling accurate forecasting or classification tasks.

4.2.2. Missing Value Imputation

For the efficient handling of the missing values in the dataset, two widely used imputation methods mean-before-after and multivariate imputation have been employed. The mean-before-after technique replaces null values at time i with the mean of adjacent values at times $i-1$ and $i+1$.

$$\bar{x}_i = \frac{x_{i-1} + x_{i+1}}{2}$$

This approach may not perform well when there is a continuous sequence of null values. On the other hand, multivariate imputation can fill each null value with multiple potential values. Compared to a single imputation, this method accounts for uncertainties associated with missing value imputation [30].

4.2.3. Lag Values

Lag values in soil moisture estimation denote the time intervals between current and historical soil moisture measurements. Optimal lag values are crucial for capturing temporal dependencies and improving the accuracy of soil moisture forecasting models. The correlation indicates significant temporal dependencies, crucial for capturing soil moisture dynamics accurately. It can be utilized to identify optimal lag values for maximizing correlation. This can be expressed as:

$$\hat{\rho}_\tau = \frac{\sum_{t=1}^{n}(r_t - \bar{r})(r_{t-\tau} - \bar{r})}{\sqrt{\sum_{t=1}^{n}(r_t - \bar{r})^2 \sum_{t=1}^{n}(r_{t-\tau} - \bar{r})^2}}$$

where $\hat{\rho}_\tau$ represents the sample autocorrelation coefficient at lag τ, measuring the linear relationship between soil moisture instances at time t and $t - \tau$. r_t represents the soil moisture instance at time t, capturing its value in the time series. $\bar{r}$ is the mean of all soil moisture instances. n is the total number of instances in the time series.

By utilizing the sample autocorrelation coefficient ($\hat{\rho}_\tau$), which quantifies the linear relationship between soil moisture instances at time t and $t - \tau$, we can identify optimal lag values. It helps in understanding the past temporal dependencies. The autocorrelation analysis allows us to assess how closely related current soil moisture values are to their historical counterparts at different time lags. By calculating $\hat{\rho}_\tau$ for various lag values (τ), we identified the lag that maximizes correlation, indicating the most influential historical time points for predicting future soil moisture levels. Once optimal lag values are determined, they can be incorporated into forecasting models. These models utilize historical soil moisture data at specific lag intervals to make accurate predictions about future soil moisture dynamics. Incorporating such lag values enhances the predictive performance of machine learning models trained on soil moisture time series data.

4.3. Machine Learning Models

To incorporate ML and DL techniques into the soil moisture estimation approach, a variety of models were utilized to capture the intricate data relationships. The models included linear regression, support vector machine (SVM), decision tree, random forest, and LSTM networks. Linear regression served as a fundamental model, capturing linear data relationships. SVM was chosen for its capacity to handle nonlinear relationships via kernel functions, while decision trees provided interpretability and the ability to model complex interactions. Random forest, an ensemble method, enhanced accuracy by aggregating predictions from multiple decision trees. LSTM networks, a form of recurrent neural network (RNN), were employed to capture temporal dependencies crucial for time series analysis. The details of the models employed in the study are provided in the following subsections.

4.3.1. Linear Regression

Linear regression is employed as one of the ML models. It can be represented by the equation:

$$y = \beta_0 + \beta_1 x_1 + \beta_2 x_2 + \ldots + \beta_p x_p + \epsilon$$

where:

- y represents the soil moisture content, our dependent variable.
- $x_1, x_2, \ldots, x_p$ are independent variables, such as wind speed, wind direction, pressure, and temperature.

- β_0 is the y-intercept, indicating the soil moisture content when all independent variables are zero.
- $\beta_1, \beta_2, \ldots, \beta_p$ are coefficients representing the change in soil moisture content for a one-unit change in each independent variable.
- ϵ is the error term, accounting for the difference between the predicted and actual soil moisture values.

To train the linear regression model, we utilize the least squares method to estimate the coefficients β:

$$\beta = (X_{\text{train}}^T X_{\text{train}})^{-1} X_{\text{train}}^T y_{\text{train}}$$

where X_{train}^T represents the transpose of the matrix of input features X_{train} for training, and y_{train} represents the corresponding observed soil moisture values.

Once trained, the model can make predictions on new data by multiplying the matrix of input features for testing X_{test} by the vector of coefficients β:

$$\hat{y} = X_{\text{test}} \beta$$

Here, $\hat{y}$ represents the vector of predicted soil moisture values for the test dataset. This approach allows us to estimate soil moisture levels based on various environmental factors.

The theoretical basis for using linear regression is its ability to capture the linear relationship between the input features (e.g., weather data, soil properties) and the target variable (soil moisture content). Linear regression assumes a linear model, which is often a reasonable approximation for many soil moisture estimation problems, where the factors influencing soil moisture exhibit relatively straightforward, linear dependencies.

4.3.2. Support Vector Machine

To train an SVM model for soil moisture regression, a kernel function is selected to map input features into a higher-dimensional space. For this purpose, the radial basis function (RBF) kernel is chosen, defined as:

$$K(x_i, x_j) = \exp(-\gamma ||x_i - x_j||^2)$$

where x_i and x_j represent the input feature vectors for two data points, and γ is a hyperparameter controlling the kernel function's width.

With the kernel function defined, the SVM model is trained to determine the hyperplane maximizing the margin between support vectors and the decision boundary. The decision function for the SVM regression model is expressed as:

$$f(x) = \sum_{i=1}^{n} (\alpha_i - \alpha_i^*) K(x, x_i) + b$$

Here, x denotes the input feature vector for a new data point, n is the number of training examples, α_i and α_i^* are the Lagrange multipliers for the ith training example and its corresponding slack variable, $K(x, x_i)$ is the kernel function evaluated at x and x_i, and b is a bias term.

To train the SVM model, the optimization problem is solved:

$$\min_{\alpha, \alpha^*, b} \frac{1}{2} (\alpha - \alpha^*)^T K (\alpha - \alpha^*) + C \sum_{i=1}^{n} (\xi_i + \xi_i^*)$$

subject to the constraints:

$$\sum_{i=1}^{n}(\alpha_i - \alpha_i^*) = 0$$
$$0 \leq \alpha_i, \alpha_i^* \leq C$$
$$y_i - f(x_i) \leq \epsilon + \xi_i$$
$$f(x_i) - y_i \leq \epsilon + \xi_i^*$$

where C is a hyperparameter controlling the trade-off between maximizing the margin and minimizing the training error, ξ_i and ξ_i^* are slack variables allowing points to fall on the wrong side of the decision boundary, and ϵ controls the margin width.

To predict soil moisture levels for new data, the decision function $f(x)$ is evaluated for each data point. The predicted values are obtained as:

$$\hat{y} = f(X_{\text{test}})$$

where X_{test} represents the matrix of input features for the testing data.

SVMs are well-suited for soil moisture estimation due to their capacity to handle non-linear relationships between the input features and the target variable. By employing kernel functions, SVMs can map the input data into a higher-dimensional feature space where linear models can be used to capture the complex non-linear patterns in the data. This makes SVMs particularly effective in modeling the intricate relationships between various environmental factors and soil moisture dynamics.

4.3.3. Decision Tree

To apply decision trees for soil moisture estimation, a training dataset comprising feature-target pairs $(\mathbf{x}_1, y_1), \ldots, (\mathbf{x}_N, y_N)$ is used where $\mathbf{x}_i$ represents a D-dimensional feature vector and y_i denotes the corresponding soil moisture content. The decision tree algorithm recursively partitions the feature space into regions R_j where soil moisture levels exhibit similar characteristics. This process involves defining a recursive function T that takes a dataset D and a set of candidate splitting functions $\mathcal{F}$ as inputs. The function T constructs a decision tree minimizing impurity or maximizing information gain. At each step, T selects the optimal splitting function f^* from $\mathcal{F}$ to partition the data into subsets $D_1, \ldots, D_K$ based on $f^*(\mathbf{x})$. These subsets are recursively used to generate child nodes. Termination occurs when a maximum depth is reached, impurity or information gain drops below a threshold, or the number of samples in a node falls below a threshold. For predicting soil moisture for new inputs $\mathbf{x}$, the decision tree traverses from the root to a leaf node, guided by the splitting functions. The prediction is obtained by averaging the soil moisture levels of training examples falling within the leaf node's region.

The hierarchical, tree-like structure of decision trees aligns well with the task of soil moisture estimation. Decision trees can effectively capture the complex interactions between multiple input features, as they recursively partition the feature space into regions with similar soil moisture characteristics. This ability to handle non-linear relationships and model higher-order feature interactions makes decision trees a suitable choice for soil moisture modeling.

4.3.4. Random Forest

For soil moisture estimation, a dataset $(\mathbf{x}_1, y_1), \ldots, (\mathbf{x}_N, y_N)$ incorporates various soil and environmental features. Here, each $\mathbf{x}_i$ encapsulates soil attributes, weather conditions, and environmental variables, while y_i signifies the corresponding soil moisture content. This dataset serves as the basis for employing the random forest regression algorithm to build an ensemble of decision trees for predictive modeling. Each decision tree within the random forest acts as a model capturing the intricate relationship between input features and soil moisture content. Techniques like bootstrapping and random feature subset selection during training foster diversity among the trees, enhancing the overall predictive capability of the model.

During prediction, the random forest aggregates the outputs of individual trees, providing an ensemble prediction that is more robust and less prone to overfitting compared to a single decision tree model. This ensemble approach enables more accurate estimation of soil moisture levels across various environmental contexts and geographical regions, bolstering agricultural planning, water resource management, and environmental surveillance efforts. Examining the algorithmic framework of random forests for soil moisture estimation, the procedural steps are outlined as follows:

- Define a hyperparameter B, representing the number of trees in the forest.
- For each tree $b = 1, \ldots, B$:
  - Draw a bootstrap sample of size N from the training set, denoted as D_b.
  - Randomly select a subset of features of size m, where $m << D$, for training the decision tree. This fosters diversity and mitigates overfitting.
  - Train a decision tree on the bootstrap sample D_b using the selected features. The decision tree is constructed by recursively partitioning the feature space into rectangles, similar to the decision tree algorithm. This tree is denoted as T_b.
- To predict soil moisture for a new input $\mathbf{x}$, the random forest regression algorithm aggregates predictions from all trees in the forest. Mathematically, the predicted soil moisture content $\hat{y}$ is computed as the average of predictions from each tree:

$$\hat{y} = \frac{1}{B} \sum_{b=1}^{B} T_b(\mathbf{x})$$

where $T_b(\mathbf{x})$ represents the prediction of the b-th decision tree for input $\mathbf{x}$.

Random forests, as an ensemble of decision trees, leverage the strengths of individual decision trees while mitigating their potential to overfit. By training multiple decision trees on random subsets of the data and features, random forests can capture a more comprehensive representation of the underlying relationships between the input features and soil moisture. This ensemble approach enhances the robustness and reliability of soil moisture estimation.

4.3.5. Long Short-Term Memory (LSTM)

In the domain of soil moisture estimation, the LSTM model emerges as a potent tool, adept at handling the sequential data inherent in meteorological conditions and environmental factors over time. In the context of moisture estimation, the input dataset $\boldsymbol{X}$ comprises features like wind speed, wind direction, pressure, temperature, and time, spanning multiple time steps. These features encapsulate critical environmental information affecting soil moisture dynamics. The LSTM model processes these sequential data to capture temporal dependencies and intricate patterns, thereby enhancing its predictive capability.

During training, the LSTM model learns to map the input features to the corresponding soil moisture levels through the computation of hidden states h_t. This hidden state encapsulates the historical context of the input features up to time step t, allowing the model to capture long-term dependencies crucial for accurate soil moisture estimation. The output layer of the LSTM model transforms the hidden state h_t into the predicted soil moisture content $\boldsymbol{y}_t$ through a linear transformation:

$$\boldsymbol{y}_t = \boldsymbol{W}\boldsymbol{h}_t + \boldsymbol{b}$$

where $\boldsymbol{W} \in \mathbb{R}^{1 \times H}$ is the weight matrix, and $\boldsymbol{b} \in \mathbb{R}^1$ is the bias vector.

Moreover, by incorporating techniques such as dropout regularization during training, the LSTM model mitigates overfitting concerns, ensuring robust performance across diverse environmental conditions. Once trained, the LSTM model can be deployed to make predictions on new input data, providing valuable insights into soil moisture dynamics over time.

LSTMs are particularly well-suited for soil moisture estimation due to their ability to model temporal dependencies and long-term patterns in time-series data. Soil moisture dynamics are often influenced by historical weather conditions, soil properties, and other time-dependent factors. LSTMs can effectively capture these long-term dependencies, enabling more accurate predictions of soil moisture levels compared to models that treat each time step independently.

4.4. *Evaluation Metrics*

The performance of soil moisture estimation models was evaluated using several key metrics, including the MAE, RMSE, MSE, and MAPE. These metrics provide insights into the accuracy and reliability of the forecasting models.

MAE, a common metric for regression models, quantifies the average magnitude of errors between the actual and predicted soil moisture values. It is calculated as:

$$MAE = \frac{1}{n}\sum_{j=1}^{n}|y_j - \hat{y}_j| \tag{1}$$

MSE is another important metric that quantifies the average squared difference between actual and predicted values:

$$MSE = \frac{1}{n}\sum_{j=1}^{n}(y_j - \hat{y}_j)^2 \tag{2}$$

RMSE is a quadratic measure that penalizes larger errors more heavily. It is computed as the square root of the average of squared differences between actual and predicted values:

$$RMSE = \sqrt{\frac{1}{n}\sum_{j=1}^{n}(y_j - \hat{y}_j)^2} \tag{3}$$

MAPE provides a normalized measure of prediction accuracy by considering the percentage difference between predicted and actual values relative to the actual values:

$$MAPE = \frac{100\%}{n}\sum_{i=1}^{n}\frac{|y_i - \hat{y}_i|}{y_i} \tag{4}$$

In Equations (1)–(4), y_i and $\hat{y}_i$ represent the actual and predicted soil moisture values, respectively, and n signifies the total number of predictions. MAPE offers insights into how closely the model's predictions align with the actual soil moisture values on average. These metrics collectively offer insights into the performance and predictive accuracy of soil moisture estimation models, aiding in model selection and refinement for effective environmental monitoring and agricultural planning.

The metrics discussed above are well-suited for evaluating the performance of ML models for a task like soil moisture estimation due to the following reasons:

- Practical Relevance: Soil moisture is a continuous variable, so regression-based metrics like RMSE, MAE, and MAPE are appropriate to quantify the model's ability to accurately predict the actual soil moisture values.
- Interpretability: These metrics are widely used and understood in tasks like soil moisture estimation, making it easier to compare the results to other studies and understand the practical implications of the model's performance.

- Error Characteristics: RMSE and MAE provide complementary information about the models; RMSE is sensitive to large errors, while MAE gives a sense of the average error magnitude. This helps assess both the overall accuracy and typical error levels
- Practical Applications: For the many real-world applications of soil moisture estimation, such as irrigation scheduling or drought monitoring, having a good understanding of the typical error magnitudes (via RMSE and MAE) and the overall model fit (via R^2) is crucial to ensure the practical usefulness of the predictions.

In summary, the choice of RMSE, MAE, and MAPE as evaluation metrics is well-justified for this soil moisture estimation study, as they provide a comprehensive assessment of the model's predictive performance in a way that is directly relevant to the practical applications of the technology.

4.5. *Hyper-Parameter Tuning*

The tuning of hyperparameters is crucial to achieving optimal model performance. Grid search [31] was employed to systematically evaluate models across predefined hyperparameter search spaces. Grid Search uses a different combination of all the specified hyperparameters and their values and calculates the performance for each combination and selects the best value for the hyperparameters. For each model, the hyper-parameters used, the range of values tested and the optimal selected hyperparameter are given in Table 1.

Table 1. Optimal hyper-parameters for models identified through grid search.

Model	Hyper-Parameter	Search Space	Optimal Value
LR	None	None	None
SVM	C	{0.1, 1, 10, 100, 1000}	10
	Gamma	{0.0001, 0.001, 0.01, 0.1, 1}	0.001
DT	Max depth	{2, 3, 5, 10, 20}	3
	Min samples leaf	{5, 10, 20, 50, 100}	50
RF	B	{25, 50, 100, 150}	25
LSTM	Learning rate	{0.001, 0.01, 0.1, 0.2}	0.001
	Batch Size	{8, 16, 32, 64}	32
	Dropout	{0.2, 0.3, 0.4}	0.3
	Optimizer	{SGD, RMSprop, Adam}	Adam

The thorough hyper-parameter tuning process ensured that each ML model was optimized for the specific characteristics of the soil moisture estimation problem, enhancing their performance and reliability in practical applications.

5. Results and Discussion

This section presents the results of experiments estimating soil moisture conducted using Google Colaboratory (GC) and GEE. The GC environment is powered by Python 3.7 and is equipped with a two-core Intel(R) Xeon(R) CPU running at 2.0 GHz, along with 13 GB of RAM and an NVIDIA Tesla T4 GPU.

Data from all the studied locations discussed in Section 4.1 are selected, and all the models are trained and evaluated using these data. In the following sub-sections, we analyze and discuss the results based on evaluation metrics listed in Table 2. Moreover, boxplots of the evaluation matrices are given in Figure 2.

Table 2. Location-wise performance of different models using varied matrices.

Location:	Cuttack			
Model	**MSE**	**RMSE**	**MAE**	**MAPE**
LR	0.59	0.77	0.51	7.52%
SVM	0.26	0.51	0.25	4.24%
DT	1.02	1.01	0.75	7.17%
RF	0.69	0.83	0.46	5.57%
LSTM	**0.06**	**0.24**	**0.16**	**2.80%**
Location:	**Kandhamal**			
LR	0.53	0.73	0.53	6.69%
SVM	0.27	0.52	0.25	2.98%
DT	0.94	0.97	0.48	5.05%
RF	0.47	0.68	0.36	3.75%
LSTM	**0.08**	**0.28**	**0.18**	**2.00%**
Location:	**Dhenkanal**			
LR	0.2	0.45	0.29	6.57%
SVM	0.2	0.44	0.18	3.93%
DT	0.56	0.75	0.31	5.00%
RF	0.26	0.51	0.23	3.88%
LSTM	**0.03**	**0.17**	**0.11**	**2.56%**
Location:	**Talcher**			
LR	0.37	0.61	0.44	6.27%
SVM	0.18	0.42	0.22	3.06%
DT	0.78	0.89	0.49	5.17%
RF	0.46	0.68	0.37	3.85%
LSTM	**0.06**	**0.24**	**0.17**	**2.28%**
Location:	**Angul**			
LR	0.76	0.87	0.59	9.31%
SVM	0.26	0.51	0.24	3.80%
DT	0.96	0.99	0.51	5.07%
RF	0.49	0.72	0.40	4.07%
LSTM	**0.05**	**0.21**	**0.18**	**2.75%**

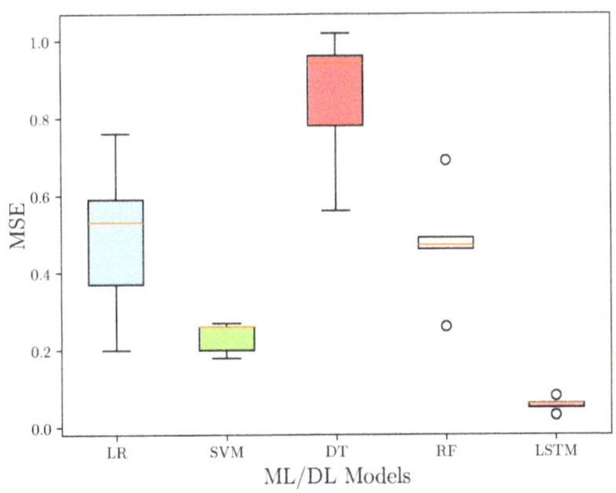

(a)

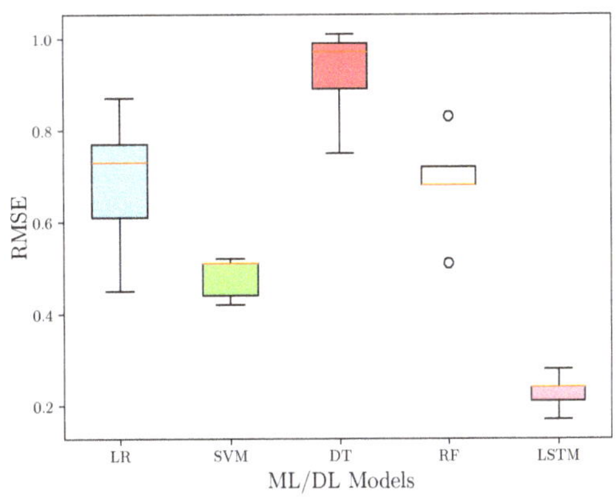

(b)

Figure 2. *Cont.*

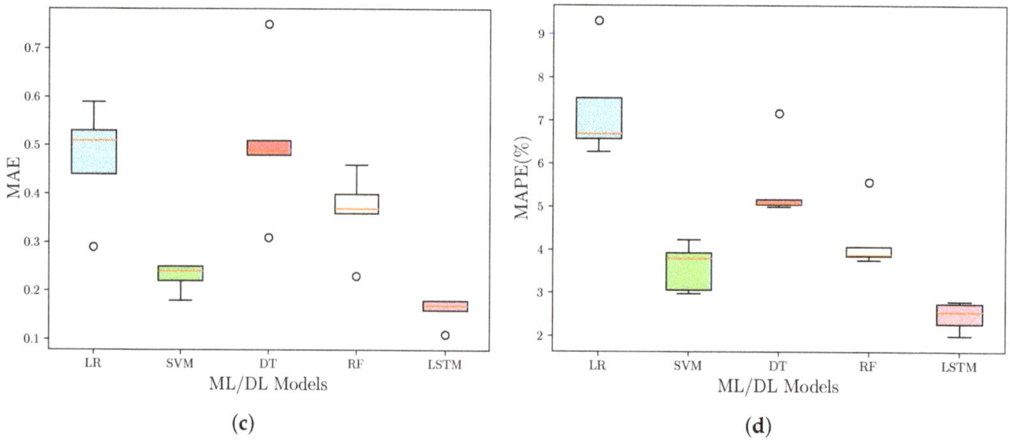

Figure 2. Comparison of model performance using (**a**) MSE, (**b**) RMSE, (**c**) MAE, and (**d**) MAPE for LR, SVM, DT, RF, and LSTM models.

5.1. Cuttack

The LSTM model exhibited superior performance compared to the other models in the Cuttack area, achieving the lowest MSE of 0.06, RMSE of 0.24, MAE of 0.16, and MAPE of 2.80%. The SVM model also performed reasonably well, with the second-lowest MAE (0.25) and MAPE (4.24%). In contrast, the decision tree model demonstrated the poorest performance, with the highest MSE (1.02), RMSE (1.01), and MAE (0.75), While its MAPE (7.17%) was lower compared to the linear regression model, which exhibited a MAPE of 7.52%.

5.2. Kandhamal

Consistent with the results in Cuttack, the LSTM model exhibited the best performance in the Kandhamal area, with the lowest MSE (0.08), RMSE (0.28), MAE (0.18), and MAPE (2.00%). The Random Forest (RF) model also performed well, securing the second-lowest MSE (0.47), RMSE (0.68), MAE (0.36), and MAPE (3.75%). The DT model showed the highest MSE (0.94) and RMSE (0.97), while the LR model had the highest MAE (0.53) and MAPE (6.69%).

5.3. Dhenkanal

In the Dhenkanal area, the LSTM model maintained its superior performance, achieving the lowest MSE (0.03), RMSE (0.17), MAE (0.11), and MAPE (2.56%). The RF and SVM models also exhibited good performance, with comparable MSE, RMSE, MAE, and MAPE values. The DT model had the highest MSE (0.56), RMSE (0.75), and MAE (0.31), while the LR model had the highest MAPE (6.57%).

5.4. Talcher

For the Talcher area, the LSTM model continued to outperform the other models, with the lowest MSE (0.06), RMSE (0.24), MAE (0.17), and MAPE (2.28%). The SVM model exhibited the second-best performance, with the lowest MSE (0.18), RMSE (0.42), and MAE (0.22), while the RF model had the second-lowest MAPE (3.85%). The DT model showed the highest MSE (0.78), RMSE (0.89), and MAE (0.49), and the LR model had the highest MAPE (6.27%).

5.5. Angul

In the Angul area, the LSTM model once again demonstrated superior performance, with the lowest MSE (0.05), RMSE (0.21), MAE (0.18), and MAPE (2.75%). The SVM model was the second-best performer, with the second-lowest MSE (0.26), RMSE (0.51), MAE (0.24), and MAPE (3.80%). The DT model showed the highest MSE (0.96), RMSE (0.99), and MAE (0.51), while the LR model had the highest MAPE (9.31%).

Figure 2 presents a comprehensive comparison of the performance of various ML and DL models, including LR, SVM, DT, RF, and LSTM, using different evaluation metrics for all the regions under consideration. Figure 2a shows the box plot of the MSE values for each model, and we can observe that the LSTM model exhibits the lowest MSE, indicating its superior performance in minimizing the squared differences between predicted and actual values. The DT model, on the other hand, shows the highest MSE, suggesting a poorer fit to the data. The SVM and RF models perform moderately well, with MSE values lower than the DT model but higher than the LSTM model. Figure 2b illustrates the RMSE, which is the square root of the MSE and provides a more interpretable measure of the model's average prediction error. Consistent with the MSE results, the LSTM model demonstrates the lowest RMSE, followed by the SVM, RF, LR, and DT models, respectively. In Figure 2c, MAE is presented, which measures the average absolute difference between predicted and actual values. The LSTM model again outperforms the other models with the lowest MAE, indicating its ability to minimize the magnitude of prediction errors. The DT model exhibits the highest MAE, while the SVM and RF models perform moderately well. Finally, Figure 2d compares the models based on the MAPE, which expresses the prediction error as a percentage of the actual value. The LSTM model continues to excel, achieving the lowest MAPE, suggesting its superior performance in capturing the relative magnitude of prediction errors. The DT model shows the highest MAPE, indicating a larger relative error compared to the other models.

Based on Figure 2 and Table 2, it is evident that the LSTM model consistently outperformed the other ML and DL models across all evaluation metrics and geographical areas. The LSTM model demonstrated its effectiveness in minimizing prediction errors, achieving higher accuracy, and capturing the long-term dependencies and temporal patterns within the data, which are crucial for accurate forecasting or prediction tasks. Soil moisture data exhibit complex temporal patterns and long-term dependencies, making LSTM's memory cells crucial for retaining relevant information over multiple time steps. Its sequential data processing capability ensures it captures subtle temporal trends often overlooked by traditional models. Furthermore, LSTMs dynamically adapt to changing patterns, making them robust in capturing non-linear relationships and abrupt changes in soil moisture dynamics. Additionally, they excel at modeling seasonal trends and cyclical patterns inherent in soil moisture data, rendering them superior for accurate time series forecasting of soil moisture estimation. Furthermore, as shown in Table 2, the LSTM model maintained its dominance across various geographic locations, consistently achieving the lowest MSE, RMSE, MAE, and MAPE values in areas such as Cuttack, Kandhamal, Dhenkanal, Talcher, and Angul. This consistency in performance highlights the robustness and adaptability of the LSTM model to different contexts and data patterns.

In contrast, the DT model generally exhibited the poorest performance across both Figure 2 and Table 2. The DT model had the highest MSE, RMSE, MAE, and MAPE values in most cases, potentially due to its tendency to overfit the data or its inability to capture complex patterns effectively. The SVM and RF models performed moderately well, often outperforming the LR model but falling short of the LSTM model's exceptional performance. This suggests that while SVM and RF models can capture intricate patterns and relationships within the data to some extent, the DL techniques employed by the LSTM model offer a distinct advantage in handling complex, sequential, or time-series data.

Overall, the consistent out-performance of the LSTM model across various evaluation metrics and geographical areas highlights its suitability for accurate forecasting and prediction tasks, particularly in scenarios involving sequential or time-dependent data. Although overfitting can occur if the model memorizes noise in the training data, generalization may be hindered by insufficiently diverse training data. Additionally, meticulous data preprocessing is essential, and training can be computationally demanding. Sensitivity to hyperparameters requires careful tuning, and interpreting model decisions can be challenging due to its black-box nature. Despite these limitations, LSTM models remain powerful tools for capturing temporal dependencies in soil moisture data and making accurate predictions, provided careful attention is given to model development and evaluation.

Additionally, soil moisture estimation models can be compared not only for their predictive accuracy but also for the computing resources they require as shown in Table 3, where n refers to the number of data samples, d refers to the number of input features, T is specific to Random Forests referring to the number of trees, and t represents the number of time steps for LSTM.

Table 3. Computational cost of different models used in this study.

Model	Time Complexity
Linear Regression	O(nd)
Support Vector Machine	O(nd)
Decision Tree	O(ndlog(n))
Random Forest	O(Tndlog(n))
LSTM	O(tnd)

Traditional ML algorithms like SVM, Random Forests, Decision Trees, and Linear Regression typically have a time complexity that scales linearly with the number of samples and features, as indicated by the O(nd) time complexity. This suggests that these models can require significant computational resources for feature engineering and training, especially when dealing with large datasets with numerous features. In contrast, the LSTM model has a time complexity of O(tnd), where the additional time step factor t results in higher computational requirements compared to the traditional ML algorithms. This is due to the complex architecture and sequential data processing capabilities of LSTM networks, which demand extensive computational power during the training phase. Despite the higher training time associated with LSTM, its advantages in accuracy and its capacity to capture temporal dependencies in soil moisture data make it a viable option, even if at a greater computational cost. The trade-off between model performance and computational efficiency is an important consideration when selecting the appropriate machine-learning technique for soil moisture estimation tasks.

6. Conclusions and Future Work

In this study, we have explored the utilization of WSNs in conjunction with ML techniques for advancing soil moisture estimation. By leveraging data from WSNs integrated with remote sensing sources such as satellite observations and ground-based sensors, we have demonstrated the effectiveness of ML models in accurately estimating soil moisture content across various geographical regions. Our analysis reveals that models trained on WSN data supplemented with satellite observations exhibit robust performance in estimating soil moisture levels. Specifically, the LSTM model consistently outperforms other ML algorithms across different regions, achieving lower error rates in terms of MSE, RMSE, MAE, and MAPE. The LSTM's MSE of 0.06 and MAPE of 2.8% indicate a remarkable performance, suggesting that the predicted outcomes closely align with the actual values. The MSE being close to 0 further underscores the accuracy of the predictions, highlighting the model's ability to minimize the squared differences between predicted and actual values. This underscores the importance of leveraging WSN-driven data to enhance soil moisture estimation accuracy. The findings of our study hold significant implications for agriculture and environmental management. Using the proposed approach, soil moisture can be estimated more accurately. This precision enables optimized irrigation practices, enhancing water use efficiency and crop yields. Moreover, the timely detection of low soil moisture levels facilitates proactive drought mitigation efforts, safeguarding agricultural productivity and rural livelihoods. Additionally, the insights gained support sustainable land management practices, aiding in ecosystem conservation and resilience-building against climate change impacts. It is important to note that our study does not claim to have provided the "best" model for soil moisture estimation [10]. Instead, we aim to offer a comprehensive analysis of the performance of different ML and DL models under similar conditions.

Future research can improve soil moisture estimation models by integrating additional data sources like remote sensing satellites, exploring alternative machine learning algorithms, incorporating climate and environmental factors, and evaluating model uncertainty and robustness. The researchers may focus on the deployment of real-time monitoring systems based on WSNs which can enable continuous monitoring of soil moisture levels, facilitating timely interventions and adaptive management practices. Furthermore, conducting comprehensive validation and calibration studies across diverse environmental conditions and geographical regions will be crucial for ensuring the reliability and applicability of WSN-driven soil moisture estimation models in real-world scenarios.

Author Contributions: T.S.: Writing—original draft preparation, Methodology, data curation. M.K.: Writing—original draft, Conceptualization, Software. T.K.: review and editing, Conceptualization, Supervision. All authors have read and agreed to the published version of the manuscript.

Funding: This research was partly supported by Basic Science Research Program through the National Research Foundation of Korea (NRF) funded by the Ministry of Education (No. NRF-2022R1I1A3072355, 50%) and Innovative Human Resource Development for Local Intellectualization program through the Institute of Information & Communications Technology Planning & Evaluation(IITP) grant funded by the Korea government(MSIT) (IITP-2024-2020-0-01462, 50%).

Data Availability Statement: The datasets generated and/or analyzed during the current study are available from the corresponding author on reasonable request.

Conflicts of Interest: The authors declare that they have no known competing financial interests or personal relationships that could have appeared to influence the work reported in this paper.

References

1. Acharya, U.; Daigh, A.L.M.; Oduor, P.G. Machine Learning for Predicting Field Soil Moisture Using Soil, Crop, and Nearby Weather Station Data in the Red River Valley of the North. *Soil Syst.* **2021**, *5*, 57. [CrossRef]
2. Kumar, S.V.; Dirmeyer, P.A.; Peters-Lidard, C.D.; Bindlish, R.; Bolten, J. Information theoretic evaluation of satellite soil moisture retrievals. *Remote Sens. Environ.* **2018**, *204*, 392–400. [CrossRef] [PubMed]
3. Mittelbach, H.; Casini, F.; Lehner, I.; Teuling, A.J.; Seneviratne, S.I. Soil moisture monitoring for climate research: Evaluation of a low-cost sensor in the framework of the Swiss Soil Moisture Experiment (SwissSMEX) campaign. *J. Geophys. Res. Atmos.* **2011**, *116*, D05111. [CrossRef]
4. Mladenova, I.E.; Bolten, J.D.; Crow, W.; Sazib, N.; Reynolds, C. Agricultural drought monitoring via the assimilation of SMAP soil moisture retrievals into a global soil water balance model. *Front. Big Data* **2020**, 3, 10. [CrossRef] [PubMed]
5. Rajib, M.A.; Merwade, V.; Yu, Z. Multi-objective calibration of a hydrologic model using spatially distributed remotely sensed/in-situ soil moisture. *J. Hydrol.* **2016**, *536*, 192–207. [CrossRef]
6. Saxton, K.E.; Rawls, W.J. Soil water characteristic estimates by texture and organic matter for hydrologic solutions. *Soil Sci. Soc. Am. J.* **2006**, *70*, 1569–1578. [CrossRef]
7. Brady, N.C.; Weil, R.R.; Weil, R.R. *The Nature and Properties of Soils*; Prentice Hall: Upper Saddle River, NJ, USA, 2008; Volume 13.
8. Entekhabi, D.; Njoku, E.G.; O'neill, P.E.; Kellogg, K.H.; Crow, W.T.; Edelstein, W.N.; Entin, J.K.; Goodman, S.D.; Jackson, T.J.; Johnson, J.; et al. The soil moisture active passive (SMAP) mission. *Proc. IEEE* **2010**, *98*, 704–716. [CrossRef]
9. Singh, T.; Sharma, N.; Satakshi; Kumar, M. Analysis and forecasting of air quality index based on satellite data. *Inhal. Toxicol.* **2023**, *35*, 24–39. [CrossRef]
10. Singh, A.; Gaurav, K. Deep learning and data fusion to estimate surface soil moisture from multi-sensor satellite images. *Sci. Rep.* **2023**, *13*, 2251. [CrossRef]
11. Orth, R. Global soil moisture data derived through machine learning trained with in-situ measurements. *Sci. Data* **2021**, *8*, 1–14.
12. Romano, E.; Bergonzoli, S.; Bisaglia, C.; Picchio, R.; Scarfone, A. The Correlation between Proximal and Remote Sensing Methods for Monitoring Soil Water Content in Agricultural Applications. *Electronics* **2023**, *12*, 127. [CrossRef]
13. Kumar, M.; Singh, T.; Maurya, M.K.; Shivhare, A.; Raut, A.; Singh, P.K. Quality Assessment and Monitoring of River Water Using IoT Infrastructure. *IEEE Internet Things J.* **2023**, *10*, 10280–10290. [CrossRef]
14. Cai, Y.; Zheng, W.; Zhang, X.; Zhangzhong, L.; Xue, X. Research on soil moisture prediction model based on deep learning. *PLoS ONE* **2019**, *14*, e0214508. [CrossRef]

15. Ge, X.; Wang, J.; Ding, J.; Cao, X.; Zhang, Z.; Liu, J.; Li, X. Combining UAV-based hyperspectral imagery and machine learning algorithms for soil moisture content monitoring. *PeerJ* **2019**, *7*, e6926. [CrossRef] [PubMed]
16. Feng, Y.; Hao, W.; Li, H.; Cui, N.; Gong, D.; Gao, L. Machine learning models to quantify and map daily global solar radiation and photovoltaic power. *Renew. Sustain. Energy Rev.* **2020**, *118*, 109393. [CrossRef]
17. Wu, W.; Zucca, C.; Muhaimeed, A.S.; Al-Shafie, W.M.; Fadhil Al-Quraishi, A.M.; Nangia, V.; Zhu, M.; Liu, G. Soil salinity prediction and mapping by machine learning regression in C entral M esopotamia, I raq. *Land Degrad. Dev.* **2018**, *29*, 4005–4014. [CrossRef]
18. Lu, Z.; Chai, L.; Liu, S.; Cui, H.; Zhang, Y.; Jiang, L.; Jin, R.; Xu, Z. Estimating Time Series Soil Moisture by Applying Recurrent Nonlinear Autoregressive Neural Networks to Passive Microwave Data over the Heihe River Basin, China. *Remote Sens.* **2017**, *9*, 574. [CrossRef]
19. Song, X.; Zhang, G.; Liu, F.; Li, D.; Zhao, Y.; Yang, J. Modeling spatio-temporal distribution of soil moisture by deep learning-based cellular automata model. *J. Arid. Land* **2016**, *8*, 734–748. [CrossRef]
20. Yuan, Q.; Shen, H.; Li, T.; Li, Z.; Li, S.; Jiang, Y.; Xu, H.; Tan, W.; Yang, Q.; Wang, J.; et al. Deep learning in environmental remote sensing: Achievements and challenges. *Remote Sens. Environ.* **2020**, *241*, 111716. [CrossRef]
21. Adab, H.; Morbidelli, R.; Saltalippi, C.; Moradian, M.; Ghalhari, G.A.F. Machine learning to estimate surface soil moisture from remote sensing data. *Water* **2020**, *12*, 3223. [CrossRef]
22. Leng, P.; Yang, Z.; Yan, Q.Y.; Shang, G.F.; Zhang, X.; Han, X.J.; Li, Z.L. A framework for estimating all-weather fine resolution soil moisture from the integration of physics-based and machine learning-based algorithms. *Comput. Electron. Agric.* **2023**, *206*, 107673. [CrossRef]
23. Villegas-Ch, W.; García-Ortiz, J. A Long Short-Term Memory-Based Prototype Model for Drought Prediction. *Electronics* **2023**, *12*, 3956. [CrossRef]
24. Sazib, N.; Bolten, J.D.; Mladenova, I.E. Leveraging NASA Soil Moisture Active Passive for Assessing Fire Susceptibility and Potential Impacts Over Australia and California. *IEEE J. Sel. Top. Appl. Earth Obs. Remote Sens.* **2022**, *15*, 779–787. [CrossRef]
25. Mladenova, I.E.; Bolten, J.D.; Crow, W.T.; Sazib, N.; Cosh, M.H.; Tucker, C.J.; Reynolds, C. Evaluating the Operational Application of SMAP for Global Agricultural Drought Monitoring. *IEEE J. Sel. Top. Appl. Earth Obs. Remote Sens.* **2019**, *12*, 3387–3397. [CrossRef]
26. Sazib, N.; Mladenova, I.E.; Bolten, J.D. Assessing the Impact of ENSO on Agriculture Over Africa Using Earth Observation Data. *Front. Sustain. Food Syst.* **2020**, *4*, 509914. [CrossRef]
27. NASA Power Data Access Viewer. Available online: https://power.larc.nasa.gov/data-access-viewer/ (accessed on 18 February 2024).
28. IMD Pune. Available online: https://dsp.imdpune.gov.in/ (accessed on 18 February 2024).
29. NASA. *SMAP: Soil Moisture Active Passive Mission*; National Aeronautics and Space Administration. Available online: https://smap.jpl.nasa.gov/ (accessed on 18 February 2024).
30. Zhang, Z. Multiple imputation with multivariate imputation by chained equation (MICE) package. *Ann. Transl. Med.* **2016**, *4*, 1–5.
31. LaValle, S.M.; Branicky, M.S.; Lindemann, S.R. On the relationship between classical grid search and probabilistic roadmaps. *Int. J. Robot. Res.* **2004**, *23*, 673–692. [CrossRef]

electronics

Article

Innovative Firmware Update Method to Microcontrollers during Runtime

Bernardino Pinto Neves [1], Victor D. N. Santos [2,3] and António Valente [1,4,*]

1 Engineering Department, School of Sciences and Technology, University of Trás-os-Montes and Alto Douro (UTAD), Quinta de Prados, 5000-801 Vila Real, Portugal; bernardino.p.neves@gmail.com
2 Polytechnic Institute of Coimbra, Coimbra Institute of Engineering, Rua Pedro Nunes-Quinta da Nora, 3030-199 Coimbra, Portugal; vsantos@isec.pt
3 INESC Coimbra, DEEC, Polo II, 3030-290 Coimbra, Portugal
4 INESC Technology and Science, 4200-465 Porto, Portugal
* Correspondence: avalente@utad.pt

Abstract: This article presents a new firmware update paradigm for optimising the procedure in microcontrollers. The aim is to allow updating during program execution, without interruptions or restarts, replacing only specific code segments. The proposed method uses static and absolute addresses to locate and isolate the code segment to be updated. The work focuses on Microchip's `PIC18F27K42` microcontroller and includes an example of updating functionality without affecting ongoing applications. This approach is ideal for band limited channels, reducing the amount of data transmitted during the update process. It also allows incremental changes to the program code, preserving network capacity, and reduces the costs associated with data transfer, especially in firmware update scenarios using cellular networks. This ability to update the normal operation of the device, avoiding service interruption and minimising downtime, is of remarkable value.

Keywords: firmware update; partial update; runtime; internet of things; microcontrollers

1. Introduction

In electronic devices that incorporate microcontrollers, it is common to implement firmware update mechanisms to correct errors and make new services available after the product has been launched. Firmware updates often involve risks related with downtime, failure of the update itself, and costs associated with communications to support those updates. The article aims to address these limitations by presenting an innovative firmware update method that minimises or eliminates downtime and optimises the data to be updated. Despite the importance of this topic, there is little research into efficient firmware update methods that minimise or eliminate downtime. There are devices for which interruption of operation is critical, for example, the digital control of the power supply of a data centre (or other critical system) in a non-redundant configuration. In this scenario, firmware updates on the power supply unit can lead to temporary service interruptions [1]. Kilpeläinen [2] presents an innovative method for dynamic firmware updates, addressing updates without the need to reboot the device and modify the program code during execution. With regard to the efficient use of the communications channel, the literature refers to methods for optimising the data transmission to be updated. Bogdan [3] focuses on optimising data transmission in firmware update processes, detailing the concept of delta transmission and its combination with data compression. That work is based on the use of opcodes instead of addresses, offering an innovative perspective to efficiently transmit the updates. The system inactivity time present in the aforementioned methods, which assume a reboot after the update, led to the proposal of an innovative firmware update method based on block updates, with the aim of replacing specific code segments the program's memory, which is done during runtime and without the need for a reboot.

Citation: Neves, B.P.; Santos, V.D.N.; Valente, A. Innovative Firmware Update Method to Microcontrollers during Runtime. *Electronics* **2024**, *13*, 1328. https://doi.org/10.3390/electronics13071328

Academic Editors: Eleftherios Anastasiadis and Dionisis Kandris

Received: 23 February 2024
Revised: 25 March 2024
Accepted: 28 March 2024
Published: 1 April 2024

The originality of this study lies in the innovative approach of updating firmware by blocks, enabling an efficient and secure implementation while minimising negative impacts on system operations. This new method is expected to significantly reduce downtime and the use of communication channels. A circuit with a PIC18F27K42 microcontroller [4] was developed to validate the method. The firmware that comprises the applications and the update process was initially uploaded to that circuit using a RS232 serial channel and a serial terminal. The article is organised in sections. Section 2 describes several similar related studies. Section 3 gives a detailed description of the implemented block oriented firmware update method and the assumptions that allow the method to be successfully replicated. The communications protocol used to perform the update file transfer is also described as well the update process. In Section 4, the results obtained are described. Section 5 presents the main conclusions derived from the findings of this study.

2. Related Work

Several notable studies were analysed related to the firmware updates management, optimisation of the update files transmission, and improving the process of writing to the microcontroller's program memory. In the field of firmware update management, Mahfoudhi [5] describes an over-the-air firmware update management model for NB-IoT networks as the number of end devices increases significantly, seeking improvements in flexibility, installation time, efficiency, and cost reduction. In a similar context, Frisch [6] proposes a set of models and rules for the firmware update process based on secure distribution and automatic installation mechanisms. Kachman [7] addresses energy efficiency and its impact on firmware update processes as well as explores the evolution of this method based on delta transmission. In the area of optimising the transmission of update files, several significant studies stand out. Wee [8] presents a methodology for transmitting update files that is based on the differences between the new and old firmware, with the aim of optimising the firmware update process. Moreover, a high speed compression and decompression algorithm to significantly speed up the update time is described. Ji [9] refers to a study that focuses on the incremental firmware update method by modules. This method is based on assigning memory zones to each module and introducing the concept of static allocation of functions and relevant security considerations. This innovative approach improves the efficiency and security of firmware updates. Regarding the optimising of the writing process of to the microcontroller's program memory, several studies have made significant contributions. Jisu Kwon [10] presents a method of updating the microcontroller's program memory based on updating by functional blocks. This makes possible a partial update of the program memory instead of completely rewriting it, avoiding downtime during the update process. Xia [11] presents the concept of function addressing by means of a module orientated programming model. In this model, the code is organised around modes and modules for a generic dispatching procedure. Xia also introduces the concept of multimode application management, grouping together applications with similar behaviour and analysing performance evaluation techniques and metrics. Dhakal [12] presents an architecture based on delta updates and incremental mode for large scale IoT systems and refers to the ability to verify firmware integrity, highlighting the advantages of delta updates and identifying scenarios in which this method may not be efficient. Sun [13] reveals the limits of conventional firmware update methods and proposes a method that uses partial updates, optimising the lifetime of program memory. This method is based on partitioning the program memory into several sections, updating only the relevant section, and classifying each partition as a component. The study addresses security mechanisms, such as encryption, signing, and validation before and after the update, as well as solutions for the static allocation of functions in scenarios where the function addresses are different between the two firmware versions; in addition, the update method is based on packets that include the functions or modules to be updated, and the study presents a statistical analysis of update times as a function of the transmission channel. Kwon [14] proposes partitioning the firmware into functional blocks, introducing the concept of a function map. The method aims to update only the functional blocks with differences, reducing the use of program memory, energy consumption, and update time. This involves sending a functional block, where the updating application checks for differences and updates only what is necessary, then updating the function map to reflect the new state. Baldassari [15] explores delta firmware updates in scenarios with bandwidth constraints by updating only small memory files of the firmware. The study details the delta update process, which requires one application to build the delta file and another to rebuild the new firmware from the received deltas. Although this approach offers the advantage of updating the firmware with small memory files, it also has disadvantages, such as greater complexity compared to traditional methods, a higher probability of failure, and the need to keep a copy of the original version of the firmware in the microcontroller. In addition, it requires substantial resources on the microcontroller side, including memory and processing to handle delta updates and corrections.

3. Method Development

The underlying idea of the proposed new method consists of the `Non Volatile Memory (NVM)` controller usage to directly update parts of the existent program code. The `NVM` controller is a hardware resource present in the majority of microcontrollers that is responsible for the management of non-volatile memory—also known as flash memory—the type of memory that retains data even when the microcontroller is turned off. The above mentioned `NVM` controller acts over the available flash memory blocks allowing one to read, write, and erase the existing data in memory. The use of this `NVM` controller allow us to update the existent firmware during runtime in the same way we can read and write `NVM` user data without compromising the operation of the applications. Consequently, an update task application is added that aims to receive the data blocks associated with the code of a particular application and update them in the flash program memory, as illustrated in Figure 1.

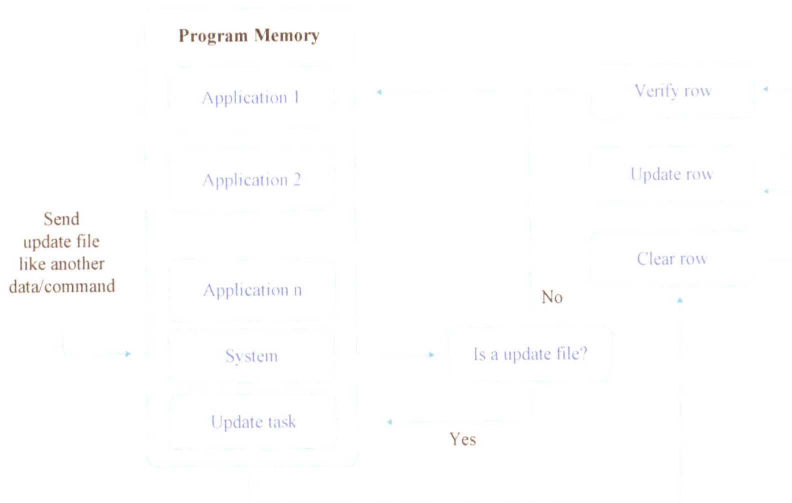

Figure 1. Runtime firmware updates method.

The non-volatile memory of a microcontroller is usually segmented or organised into several sectors, most of them devoted to the program memory. The program memory can be configured with different partitions, sizes, and write protection attributes. These partitions can be configured to implement the boot area, the application area, and the user memory data. In this paper, a `PIC18F27K42` microcontroller is used as a testbed platform to validate the proposed techniques. This microcontroller has a non-volatile memory control mechanism that uses an internal timer and voltage generator to perform writing operations. Reading program memory is executed byte by byte. The writing process is, however, more complex, as it requires the operation to be performed on a row of bytes. The content of this row must be previously erased or available for writing if it is its first use. The writing operation also requires that a write unlock sequence be activated [4]. Writing or erasing program memory will halt the microcontroller central processing unit CPU, making it impossible to execute instructions from the memory row that is being erased, as the microcontroller CPU is blocked until the process is completed [4]. For the above mentioned PIC18F27K42, the measured erasing and writing procedures take 10 ms per row. Table 1 illustrates the size and number of rows [4].

Table 1. Size and number of rows, PIC18F27K42.

Description	Value	Units
Erase Row Size	64	Word
Length Row	128	Byte
User Rows	1024	Byte

The program memory read operation does not modify data; therefore, it is very simple to carry out, simply defining the memory area to be accessed. To complete this operation, we need to previously select the program flash memory and set the address to be read using the TBLPTR register, then read the contents of that position. Note that the reading is performed byte to byte, but each program memory position has a size of two bytes; therefore, it is necessary to increment the pointer of the reading table TBLRD for each byte read. The result is in the register TABLAT: the first byte corresponds to the less significant byte and the second to the most significant byte of the specified memory position content [4]. To read the contents of a particular program memory address, the following sequence of operations must be completed, as illustrated in the flowchart of Figure 2.

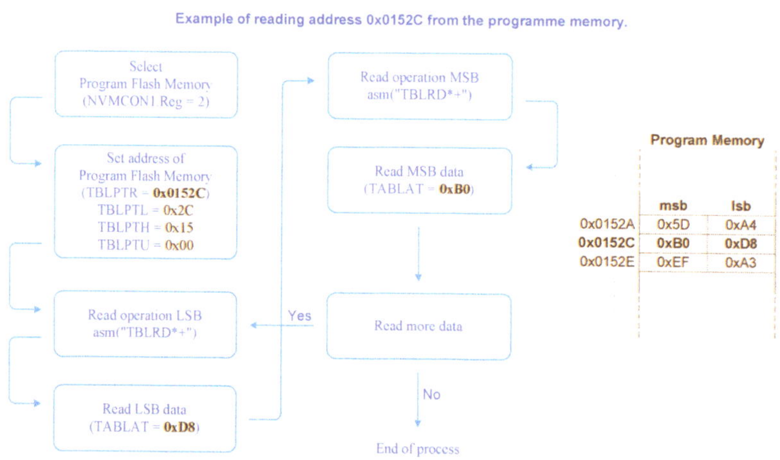

Figure 2. Reading the contents of program memory PIC18F27K42.

The write operation follows the same principle as the read operation, but operates over rows instead of bytes. The write operation is performed on an entire row, but it is implemented byte by byte [4]. As a recommended practice, in a write operation in which only part of the row is changed, it is suggested that the row be read and stored in volatile memory RAM before being erased. The copied row is then updated with the portion of the data that differs from the original version. Finally, the NVM row should be deleted and rewritten with the updated version. For the writing process to be successful, we must first make sure that the row is available for writing; in other words, the row is formatted. Thereafter, it is necessary to define the NVM area to be used for writing, where through the TBLPTR register we define the address we want to write; as with reading, the writing is also done byte by byte, and, in the writing process, the least significant byte is copied to the register TABLAT followed by the increment of the writing table TBLWR. That process is repeated for the most significant byte. After copying the row, the next step involves activating the NVMCON1bits.WREN write permission bit as well as selecting the NVMCON1bits.FREE write bit command, followed by sending the write unblock sequence to the NVM. The actual write is initiated by activating the NVMCON1bits.WR bit [4]; see the flowchart in Figure 3.

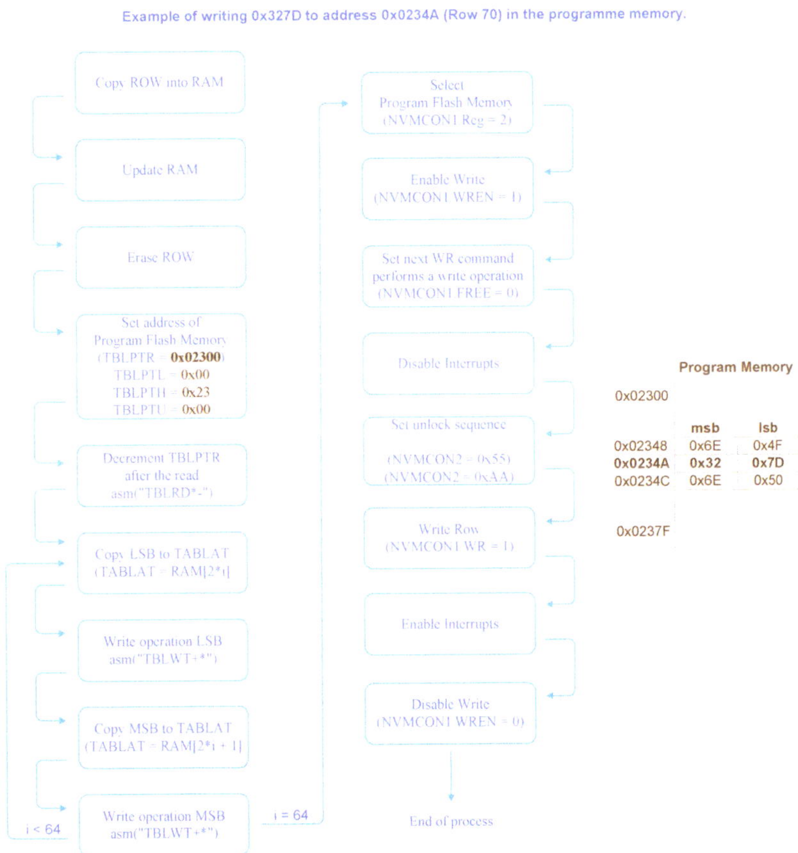

Figure 3. Writing process to program memory PIC18F27K42.

To erase a row of non-volatile memory, a specific NVM controller command is used devoted for that purpose. The FREE bit of the NVMCON1 register, if enabled, indicates that on the next enable the WR bit of the same register will erase the row specified by the address contained in the TBLPTR register. Moreover, it is necessary to previously unlock a specific range of rows to accommodate the program code and thereafter complete the erase procedure [4], as depicted in the flowchart in Figure 4.

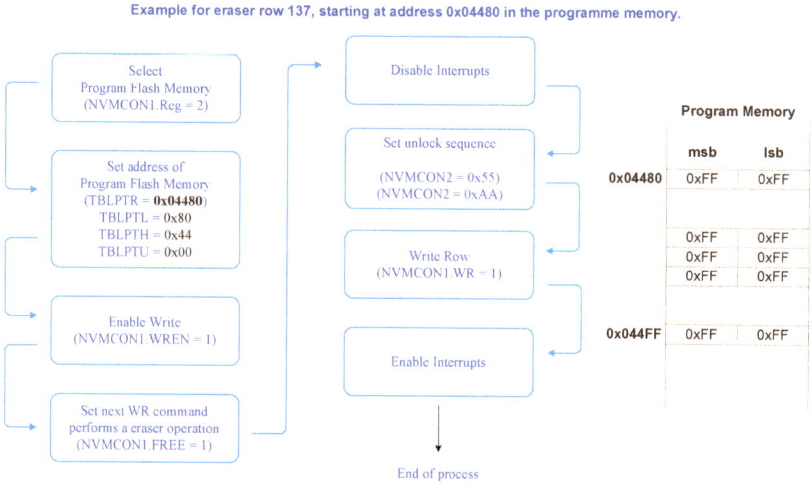

Figure 4. Erasing a row of program memory `PIC18F27K42`.

The `NVM` memory locking mechanism prevents unintended self-write programming or erasing. Thus, to promote memory integrity, any write and erase operation performed by the `NVM` controller must be preceded by an unlocking process. This process must be executed sequentially and without interruptions. If the sequence, for some reason, is interrupted, the writing or erasing process is cancelled [4]. To implement this method successfully, two non-mandatory but highly recommended requirements must be met to facilitate its implementation. The first one concerns the static and absolute allocation of the functions. Typically, a compiler, in order to optimise the space of the memory of the program, leans all the code to minimise the used memory space, making it more difficult to identify the location of the block of code that will need to be updated. By allocating the function's code in a static and absolute way, an absolute reference of the location of each function of program is set, facilitating the identification of the code block in an `Intel Hex` file (see Figure 5).

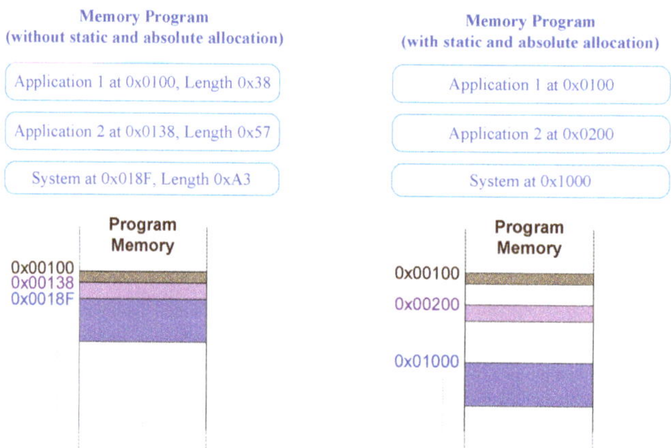

Figure 5. Example of memory allocation with and without static and absolute allocation.

The usage of static and absolute function allocation also improves the code organisation. Without static and absolute allocation, even small changes in source code can result in a hex file completely reformulated by the compiler. The usage of static and absolute allocation avoids major changes. Now, small code changes in specific functions will only affect the associated allocated memory areas, as illustrated in Figure 6.

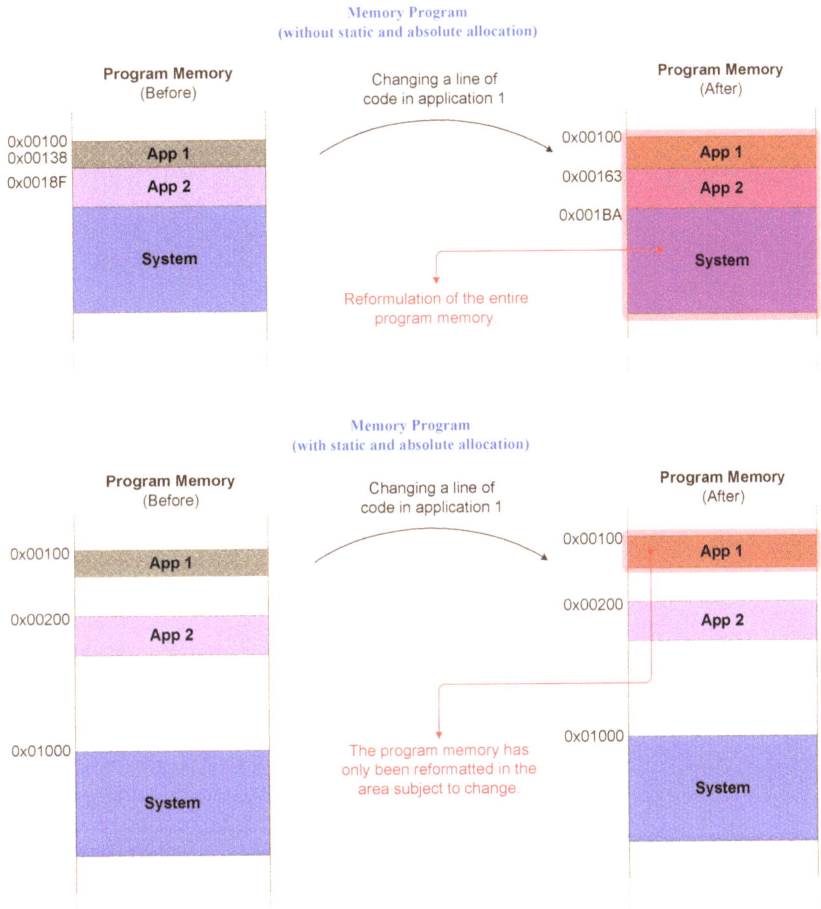

Figure 6. Result of the hexadecimal file with the change of only one line of code.

Static and absolute allocation of functions requires well designed system architecture and a complete knowledge of the program's memory map in order to avoid overlaps between the functions or applications code blocks. In order to prevent an accidental overlap of two or more functions, the compiler warns us by displaying a message with the functions that are at stake, promoting the necessary changes in the memory map. The following error message was generated by the compiler under the above mentioned conditions [16].

error: (596)
segment "_Reset_CNT_TMR_text" (19574-195A3)
overlaps segment "_TMR0_Interrupt_Handling_text" (194F6-1958F)

The second requirement concerns the size allocated to each function, which must be an integer multiple of row size; in the considered microcontroller, that size is equal to 128 bytes [4]. An example of a program memory map is depicted in Figure 7.

```
Flah_Memory_Map.h

#define Row                                   128

#define APP_1_Start_Address                   0x01000
#define APP_1_Init_Address                    (APP_1_Start_Address + 2*Row)
#define APP_1_Wait_Address                    (APP_1_Init_Address + Row)
#define APP_1_Process_Address                 (APP_1_Wait_Address + Row)

#define APP_2_Start_Address                   0x05000
#define APP_2_Init_Address                    (APP_2_Start_Address + 2*Row)
#define APP_2_Wait_Address                    (APP_2_Init_Address + Row)
#define APP_2_Process_Address                 (APP_2_Wait_Address + Row)

#define APP_3_Start_Address                   0x09000

#define APP_4_Start_Address                   0x0D000

#define Update_Process_Start_Address          0x15000
#define Update_Process_Init_Address           (Update_Process_Start_Address + 5*Row)

//Addressing system functions
#define System_Start_Address                  0x19000
#define Main_Address                          (System_Start_Address)

/Hardware_Startup
#define Hardware_StartUp_Address              (Main_Address + Row)
#define Oscilator_Initialize_Address          (Hardware_StartUp_Address + Row)
```

Figure 7. Example of program memory map.

The following step, after the program memory map definition, comprises setting the function's indexes addresses in the above mentioned range. To allocate a function in a static and absolute way, one simply needs to add before the function name the method `__at(address)`; from here, the compiler will place that function in that specific address, as illustrated in the following function prototype.

void __at(APP_1_Start_Address) App_1(void)

To validate this method, a testbed was developed comprising a circuit board with the microcontroller, two push buttons, and an `ICSP` header (depicted in Figure 8).

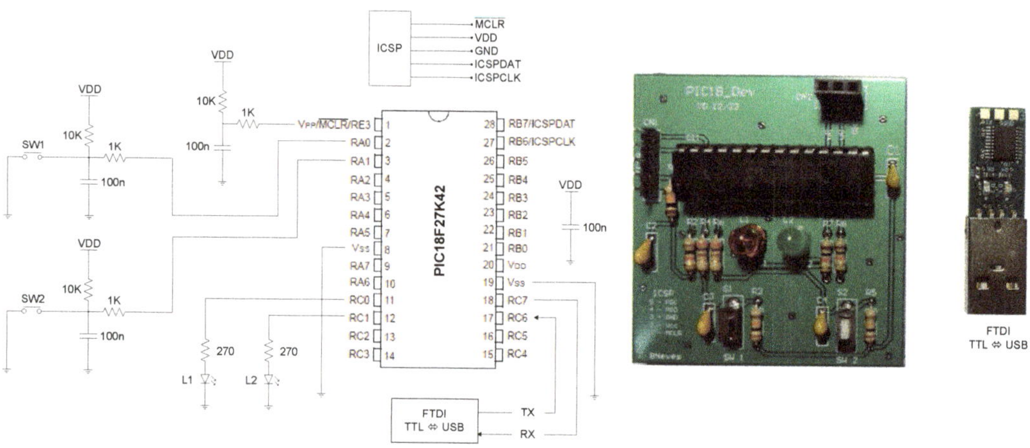

Figure 8. Layout of the circuit implemented to validate the method.

The firmware project comprises three applications: two similar applications associated to different hardware resources, in this case push buttons, and In Application Programming

(IAP) that performs an update of the firmware by means of a runtime self programming process. The first application prints in the serial port the message "`Button 1 has been pressed`" when button 1 is pressed. Similarly, the second application prints the message "`Button 2 has been pressed`" in the serial port when button 2 is pressed. These messages are defined and saved in the microcontroller flash memory. Figure 9 illustrates the flowchart of the implemented program.

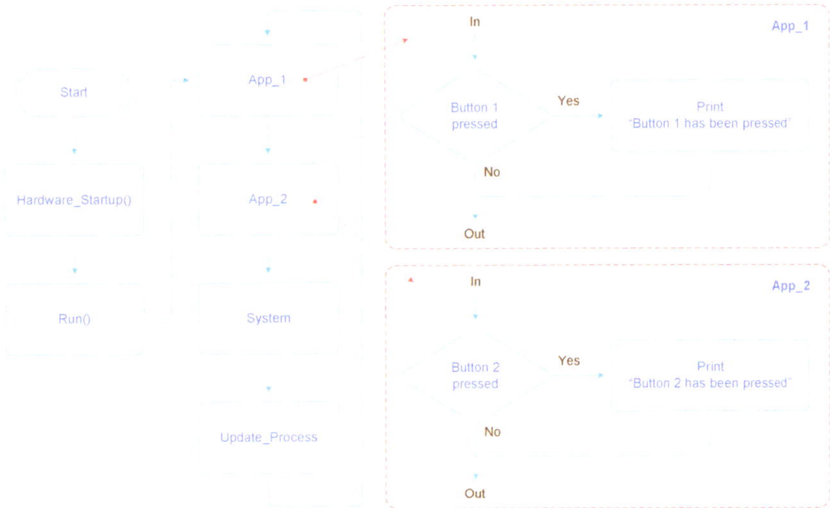

Figure 9. Implemented system and applications used to validate the proposed method.

After executing a firmware upgrade operation, it is intended to update the message printed by the first application from "`Button 1 has been pressed`" to "`This string has been changed by update at run time`", whenever the hardware button 1 is pressed (see Figure 10).

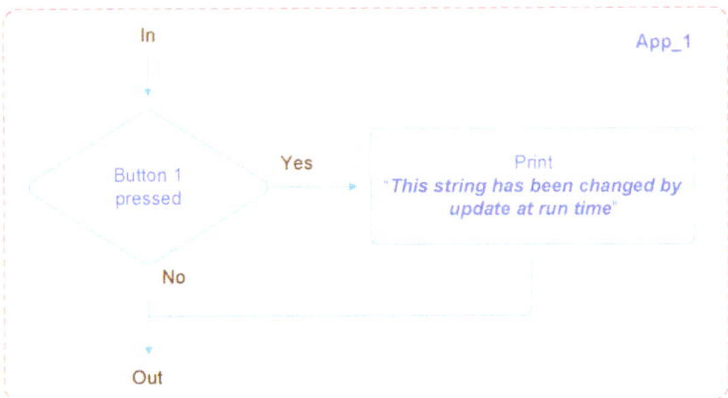

Figure 10. Proposed amendment for App_1.

From the analysis of the compiled program code, it can be seen where each function of the first application is allocated in the program memory (see Figure 11 and example of program memory map in Figure 7).

App_1.c

```
//Include project files
#include "..\main.h"

//Declaration of variables and structures
unsigned char app_1_sm = app_1_init;
const char string_1[] __at( APP_1_Process_Address + Row) = "Button 1 has been pressed\r\n";

void __at(APP_1_Start_Address) App_1(void)
{
    switch(app_1_sm)
    {
        case app_1_init: App_1_Init(); break;

        case app_1_wait: App_1_Wait(); break;

        case app_1_process: App_1_Process(); break;

        default: app_1_sm = app_1_init; break;
    }

    unsigned long *ptr;
    ptr = (unsigned long *)App_2;
    if(*ptr != 0xFFFFFFFF)
        Run = App_2;
    else
        Run = App_1;
}

void __at(APP_1_Init_Address) App_1_Init(void)
{
    Input_1_SetDigitalInput();
    Input_1_SetDigitalMode();

    Output_1_SetLow();
    Output_1_SetDigitalOutput();
    Output_1_SetDigitalMode();

    app_1_sm = app_1_wait;
}

void __at(APP_1_Wait_Address) App_1_Wait(void)
{
    if(Input_1_GetValue() == LOW)
    {
        Clear_CNT_TMR(TMR_APP_1);
        app_1_sm = app_1_process;
    }
}

void __at(APP_1_Process_Address) App_1_Process(void)
{
    if(Read_CNT_TMR(TMR_APP_1) > 1000)
    {

        Send_String_By_Uart( UART_1, (char *)string_1);
        Clear_CNT_TMR(TMR_APP_1);
        app_1_sm = app_1_wait;
    }
}
```

```
#define APP_1_Start_Address
0x01000
```

```
#define APP_1_Init_Address
(APP_1_Start_Address + 2*Row)
```

```
#define APP_1_Wait_Address
(APP_1_Init_Address + Row)
```

```
#define APP_1_Process_Address
(APP_1_Wait_Address + Row)
```

Address	00	02	04	06	08	0A	0C	0E
00FF0	FFFF	FFFF	FFFF	FFFF	FFFF	FFFF	FFFF	FFFF
01000	EF12	F008	EC80	F008	EF2B	F008	ECC0	F008
01010	EF2B	F008	EC00	F009	EF2B	F008	0E00	6E33
01020	EF2B	F008	5033	6E4B	6A4C	504C	0A00	B4D8
01030	EF1C	F008	EF0E	F008	504B	0A00	B4D8	EF02
01040	F008	0A01	B4D8	EF06	F008	0A03	B4D8	EF0A
01050	F008	EF0E	F008	0E00	6E4F	0E50	6E50	0060
01060	F13F	FFD9	0060	F143	FFDA	28DE	E10D	28DE
01070	E10B	28DE	E109	0E0F	18DE	B4D8	EF42	F008
01080	EF44	F008	EF4D	F008	0E00	0100	6FAF	0E50
01090	6FB0	0E00	6FB1	EF54	F008	0E00	0100	6FAF
010A0	0E10	6FB0	0E00	6FB1	0012	FFFF	FFFF	FFFF
010B0	FFFF	FFFF	FFFF	FFFF	FFFF	FFFF	FFFF	FFFF
010C0	FFFF	FFFF	FFFF	FFFF	FFFF	FFFF	FFFF	FFFF
010D0	FFFF	FFFF	FFFF	FFFF	FFFF	FFFF	FFFF	FFFF
010E0	FFFF	FFFF	FFFF	FFFF	FFFF	FFFF	FFFF	FFFF
010F0	FFFF	FFFF	FFFF	FFFF	FFFF	FFFF	FFFF	FFFF
01100	86C2	013A	9740	9EBA	9EC2	9F40	0E01	6E33
01110	0012	FFFF	FFFF	FFFF	FFFF	FFFF	FFFF	FFFF
01120	FFFF	FFFF	FFFF	FFFF	FFFF	FFFF	FFFF	FFFF
01130	FFFF	FFFF	FFFF	FFFF	FFFF	FFFF	FFFF	FFFF
01140	FFFF	FFFF	FFFF	FFFF	FFFF	FFFF	FFFF	FFFF
01150	FFFF	FFFF	FFFF	FFFF	FFFF	FFFF	FFFF	FFFF
01160	FFFF	FFFF	FFFF	FFFF	FFFF	FFFF	FFFF	FFFF
01170	FFFF	FFFF	FFFF	FFFF	FFFF	FFFF	FFFF	FFFF
01180	B6CA	EFC5	F008	EFC7	F008	EFCC	F008	0E01
01190	EC40	F0CC	0E02	6E33	0012	FFFF	FFFF	FFFF
011A0	FFFF	FFFF	FFFF	FFFF	FFFF	FFFF	FFFF	FFFF
011B0	FFFF	FFFF	FFFF	FFFF	FFFF	FFFF	FFFF	FFFF
011C0	FFFF	FFFF	FFFF	FFFF	FFFF	FFFF	FFFF	FFFF
011D0	FFFF	FFFF	FFFF	FFFF	FFFF	FFFF	FFFF	FFFF
011E0	FFFF	FFFF	FFFF	FFFF	FFFF	FFFF	FFFF	FFFF
011F0	FFFF	FFFF	FFFF	FFFF	FFFF	FFFF	FFFF	FFFF
01200	0E01	EC00	F0CC	5043	1042	E10B	0EE9	5C40
01210	0E03	5841	A0D8	EF0F	F009	EF11	F009	EF1F
01220	F009	0E80	6E45	0E12	6E46	0E00	6E47	0E01
01230	EC00	F0CF	0E01	EC40	F0CC	0E01	6E33	0012
01240	FFFF	FFFF	FFFF	FFFF	FFFF	FFFF	FFFF	FFFF
01250	FFFF	FFFF	FFFF	FFFF	FFFF	FFFF	FFFF	FFFF
01260	FFFF	FFFF	FFFF	FFFF	FFFF	FFFF	FFFF	FFFF
01270	FFFF	FFFF	FFFF	FFFF	FFFF	FFFF	FFFF	FFFF
01280	7542	7474	6E6F	3120	6820	7361	6220	6565
01290	206E	7270	7365	6573	0D64	000A	FFFF	FFFF
012A0	FFFF	FFFF	FFFF	FFFF	FFFF	FFFF	FFFF	FFFF

Figure 11. Sample code and location of functions in program flash memory.

Additionally, it is also possible to identify and locate application 1 in the hexadecimal file generated from the compiler (see Figure 12).

From the analysis of the modified program hexadecimal file, it can be concluded that only a well defined area of the program memory was changed; all the remaining program memory stays intact. Figure 13 presents the original and upgraded code versions of the aforementioned application 1, demonstrating the code blocks that have been removed on the original version and the ones that have been inserted on the modified one.

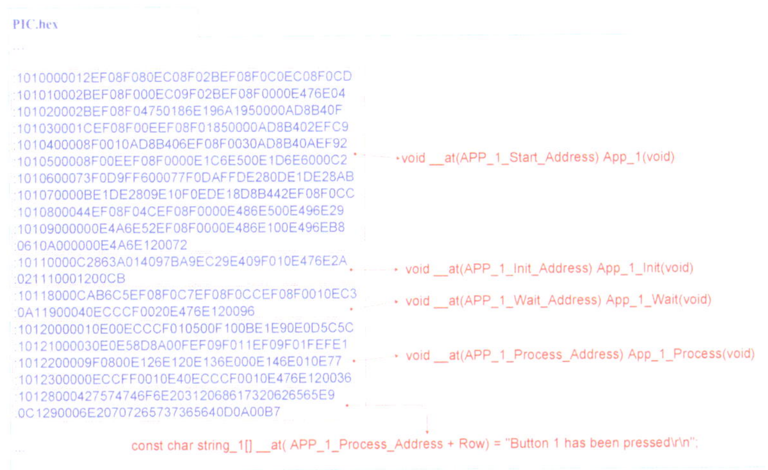

Figure 12. Location of functions in program memory in the hexadecimal file.

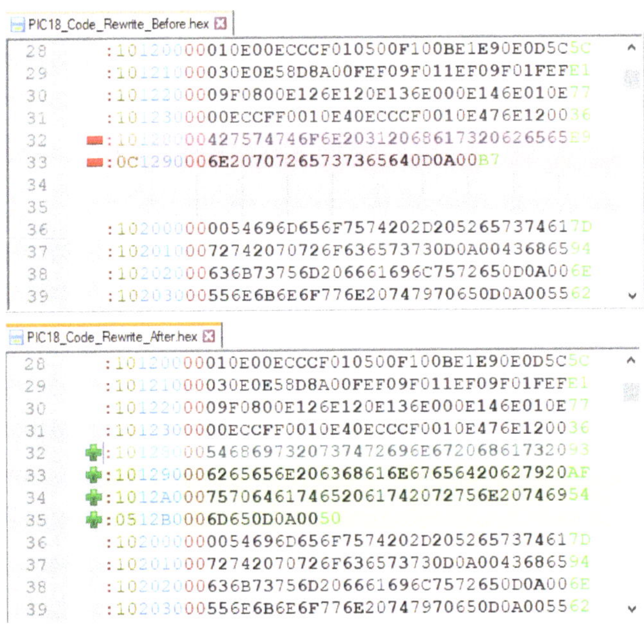

Figure 13. Changed program memory area viewed from hexadecimal file.

Using the static and absolute function allocation allows one to control and manipulate the entire program memory, making the updating task easier and keeping the firmware update circumscribed to a well defined block of program memory between addresses 0x00001280 and 0x000012B0. The update process consists of receiving a hexadecimal file in the `Intel Hex File` format [17] over the serial channel. However, as only one block of the program's memory is to be updated and the hexadecimal file is not formatted to send a single block but the entire file, some changes need to be made so that the update process application can interpret the file correctly. Those changes include the addition of a start file, followed by the most significant word of the address and the end of file, as illustrated in Figure 14.

Figure 14. Update file in `Intel Hex File` format, adapted to the application.

The `Intel Hex File` format is one of the formats used to update microcontrollers' firmware, but there are also other possible formats, such as the binary `.bin` file. The `Intel Hex File` format is characterized by the lines being in hexadecimal format; all the lines start with the character ':' followed by the data field length, start address, data type, the associated data (for each specific data type), and, finally, the error control checksum mechanism [17].

Figure 13 depicts the hexadecimal file obtained from the compiled modified code, which is sent to the microcontroller according to the aforementioned Intel Hex File format described in Figure 15 and Table 2. As explained previously, with the inclusion of all the fields, the file sent to the microcontroller is the one presented in Figure 14. The update file results from the extraction of a block of program memory of the hexadecimal file with the updated code; the file is started with a start file named `Update`, followed by the most significant word of the address, the data to update, and, finally, the end of file indicator. The update process application is responsible for the file receiving and processing. The initial state of the update process app waits for the reception of a start file, "Update" string, to proceed to the data acquisition state. In this state, the process waits until it receives a complete record and verifies its integrity using the checksum mechanism. If the line is valid, the process thereafter extracts the address, the type, and the data contained in the line. Depending on the type of data, the process reacts accordingly. For `Extended Linear Address` type, the MSW of the address is defined; if the type is `"Data_Record"`, it updates the LSW of the address and copies the data to a process buffer; finally, if the type is `End of file`, the process proceeds to the next stage, updating the program memory block. First, it copies the area of the program memory block to be updated to volatile memory `RAM` for final verification purposes of the update integrity; in the next operation, it erases the memory block to be updated, followed by updating with the data received by the update file; finally, a verification is performed between the data in the update file received and the data stored in the updated memory block. The update process application can be seen in the flowchart in Figure 16.

Figure 15. `Intel Hex File` record format [17].

Table 2. Line fields structure, `Intel Hex File` format [17].

Field	Designation
Start frame	Record start character
Length	Two ASCII digits to specify the record data field size
Address	Four ASCII digits to define the starting address of this data record.
Type	Data type: 0—Data record; 1—End of file record; 2—Extended segment address record; 4—Extended linear address record.
Data	Data bytes.
Checksum	Two ASCII digits representing the checksum calculated as 2s complement of all preceding bytes in data record except the colon.

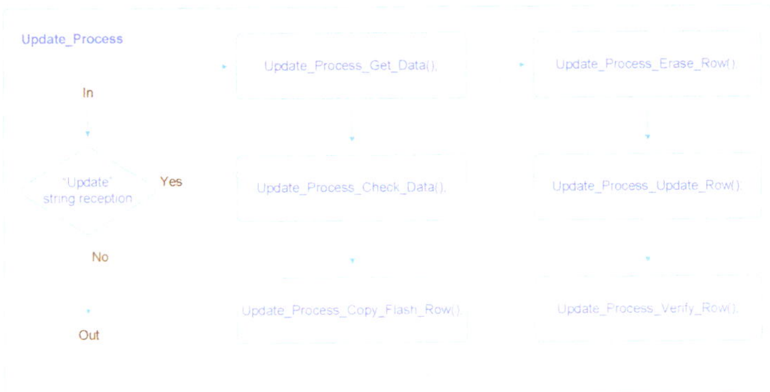

Figure 16. Updating process application state diagram.

If the process was completed successfully, it reports '`Update success`'; otherwise, it reports '`Update failure`' through the serial channel. Figure 17 illustrates the update file transfer protocol implemented between the host and the device microcontroller.

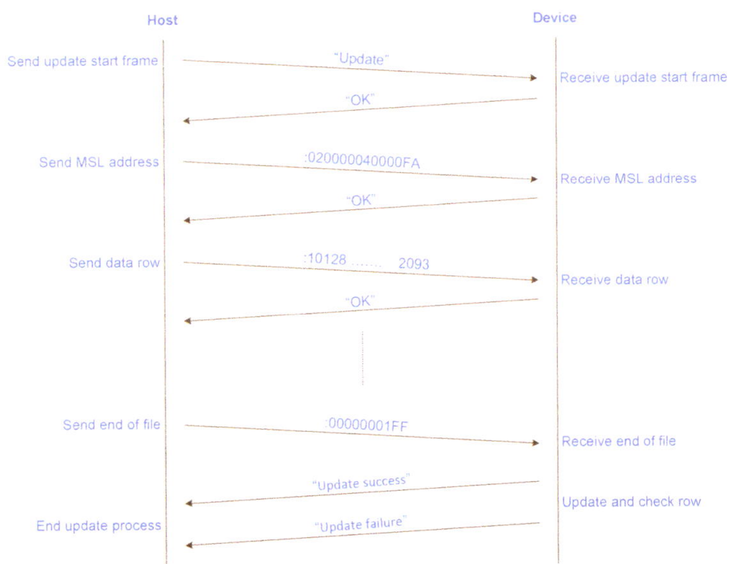

Figure 17. Implemented update file transfer protocol.

4. Results

This section presents the results obtained by the proposed novel firmware update method during runtime. This method was validated using the previously presented example program and testbed. First, an Intel hex file was sent to the `PIC18F27K42` microcontroller using an RS232 terminal to change program memory, updating the message that occurs when the push button is activated. The update process application receives and verifies the integrity of the Intel hex file, producing the desired modification in one particular block of the flash memory (see Figure 18). The updated applications now operate accordingly with the performed changes. The application assigned to button, 1 instead of printing the message "`Button 1 has been pressed`" on the serial port, starts to print a different message: "`This string has been changed by update at run time`". It is also possible to verify through the terminal log time that the update completion time took around 63 ms, which is the expected value for an update with a size of 128 bytes. The 63 ms corresponds to about 52 ms spent in the transmission (about 200 bytes at a rate of 38.4 kbps), 10 ms in the block update [4], and about 1 ms in the update process. One of the main contributions of this study is the significant reduction in downtime during the update process as well as the elimination of the need for rebooting the end device after the update. Moreover, this method aims to overcome some limitations associated with the delta firmware update method described in [3,12,15], namely, the requirement to reconstruct the firmware from the deltas, leading to resource savings and process simplification.

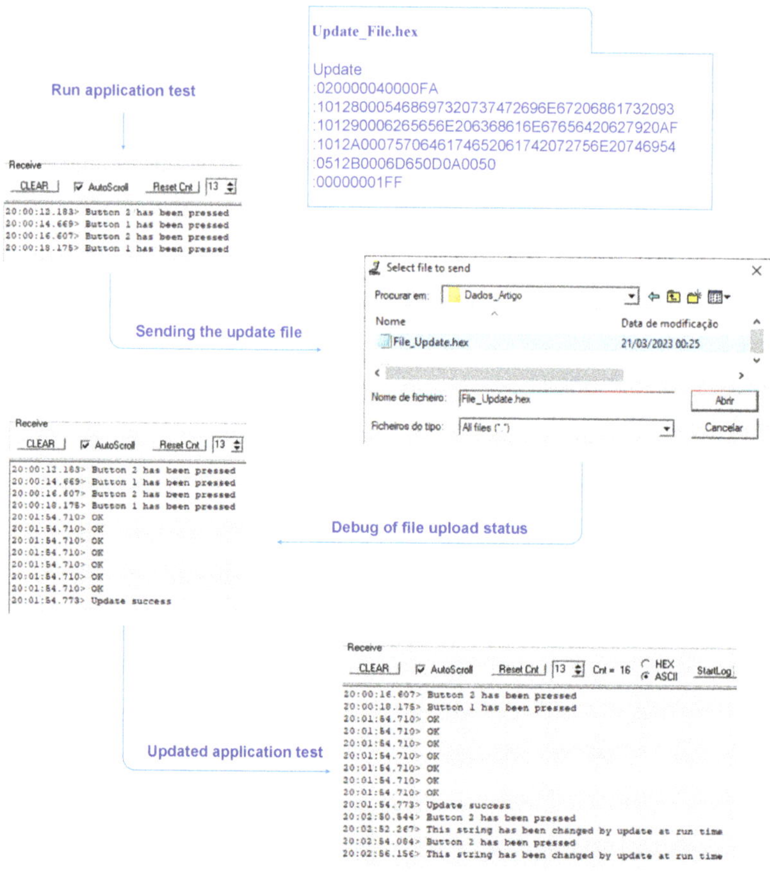

Figure 18. Microcontroller update during runtime.

5. Conclusions

In this paper, a new firmware update method for microcontrollers is presented, implemented, and validated. This new method differs from existing ones because it allows for updating only specific code lines, blocks, or functions instead of replacing the entire program during runtime. This method is suited to band limited channels that take into account the attained reduction on the amount of data transmitted. The proposed update procedure offers additional advantages, such as a reduced downtime, less than 10 ms, and good recoverability in a failure scenario.

The planned method also presents some limitations; the update process was designed to update only up to eight rows (1024 bytes' maximum), so it is therefore impossible to update the entire program memory at once.

This firmware update method is also incompatible with operating systems and/or intermediate hardware abstraction layers; it requires full control over all functionalities. Moreover, under a power failure event, the success of the update process is not guaranteed. Thus, it is advisable to include a supercapacitor-based backup power circuit to maintain module power and the upgrade process integrity.

This method was successfully and easily replicated on several microcontrollers, such as the MSP430, STM8, STM32, ATtiny, ATmega, SAMD21, and PIC32. This observation emphasises the feasibility and applicability of the method on a broad set of microcontrollers, thus increasing the scope of its potential usefulness. Future advances on the proposed method must consider the inclusion of radio transmission, using LoRaWAN or available cellular networks, to send the update file to remote sensor end-devices. An automated process to manage the partitioning of program memory and assign to each specific function an area of appropriated size based on its likelihood of being updated will also be investigated in the future. In conclusion, this article leaves an open door to a new generation of firmware updates for microcontrollers.

Author Contributions: Conceptualization, A.V.; Formal analysis, V.D.N.S.; Investigation, B.P.N.; Methodology, B.P.N., V.D.N.S. and A.V.; Resources, V.D.N.S.; Writing—original draft, B.P.N.; Writing—review and editing, V.D.N.S. and A.V. All authors have read and agreed to the published version of the manuscript.

Funding: This research received no external funding.

Data Availability Statement: Data are contained within the article.

Acknowledgments: This work was financed by National Funds through the Portuguese funding agency, FCT—Fundação para a Ciência e a Tecnologia, within project LA/P/0063/2020. DOI 10.54499/LA/P/0063/2020 | https://doi.org/10.54499/LA/P/0063/2020. This work was supported by the Portuguese Foundation for Science and Technology under the project grant UIDB/00308/2020 with the DOI 10.54499/UIDB/00308/2020 | https://doi.org/10.54499/UIDB/00308/2020.

Conflicts of Interest: The authors declare no conflict of interest.

References

1. Dumais, A.; Schlunder, H. *AN2601—Online Firmware Updates in Timing-Critical Applications*; Microchip Technology Inc.: Chandler, AZ, USA, 2018. Available online: https://ww1.microchip.com/downloads/en/Appnotes/Live%20Update%20Application%20Note.pdf (accessed on 27 March 2024).
2. Kilpeläinen, H. Dynamic Firmware Updating of an Embedded System. Bachelor's Thesis, Information and Communications Technology, Metropolia University of Applied Sciences, Metropolia, Finland, 2023.
3. Bogdan, D.; Bogdan, R.; Popa, M. Delta flashing of an ECU in the automotive industry. In Proceedings of the IEEE 11th International Symposium on Applied Computational Intelligence and Informatics (SACI), Timisoara, Romania, 12–14 May 2016. [CrossRef]
4. *Low-Power High-Performance Microcontrollers with XLP Technology*; PIC18(L)F26/27/45/46/47/55/56/57K42, Datasheet; Microchip Technology Inc.: Chandler, AZ, USA, 2021. Available online: https://ww1.microchip.com/downloads/aemDocuments/documents/MCU08/ProductDocuments/DataSheets/PIC18%28L%29F26-27-45-46-47-55-56-57K42-Data-Sheet-40001919G.pdf (accessed on 27 March 2024).
5. Mahfoudhi, F.; Sultania, A.K.; Famaey, J. Over-the-air firmware updates for constrained Nb-IOT devices. *Sensors* **2022**, *22*, 7572. [CrossRef] [PubMed]

6. Frisch, D.; Reißmann, S.; Pape, C. An Over the Air Update Mechanism for ESP8266 Microcontrollers. 2017. Available online: https://www.researchgate.net/publication/320335879 (accessed on 27 March 2024).
7. Kachman, O.; Balaz, M. Optimized differencing algorithm for firmware updates of low-power devices. In Proceedings of the IEEE 19th International Symposium on Design and Diagnostics of Electronic Circuits & Systems (DDECS), Kosice, Slovakia, 20–22 April 2016. [CrossRef]
8. Wee, Y.; Kim, T. A new code compression method for FOTA. *IEEE Trans. Consum. Electron.* **2010**, *56*, 2350–2354. [CrossRef]
9. Ji, Z.; Xiangyu, Z.; Yong, P. Implementation and research of bootloader for automobile ECU remote incremental update. In Proceedings of the AASRI International Conference on Industrial Electronics and Applications, London, UK, 27–28 June 2015. . [CrossRef]
10. Kwon, J. Available online: https://ai-soc.github.io/m_jskwon.html (accessed on 10 January 2024).
11. Xia, M.; Chi, K.; Wang, X.; Cheng, Z. Mode-oriented hybrid programming of sensor network nodes for supporting rapid and Flexible Utility Assembly. *Comput. Netw.* **2019**, *158*, 77–97. [CrossRef]
12. Dhakal, S.; Jaafar, F.; Zavarsky, P. Private Blockchain Network for IOT device firmware integrity verification and update. In Proceedings of the IEEE 19th International Symposium on High Assurance Systems Engineering (HASE), Hangzhou, China, 3–5 January 2019. [CrossRef]
13. Sun, S. Design and Implementation of Partial Firmware Upgrade. Master's Thesis, School of Electrical Engineering and Computer Science, KTH Royal Institute of Technology, Stockholm, Sweden, 2019.
14. Kwon, J.; Cho, J.; Park, D. Function block-based robust firmware update technique for additional flash-area/energy-consumption overhead reduction. In Proceedings of the International Symposium on Intelligent Signal Processing and Communication Systems (ISPACS), Taipei, Taiwan, 3–6 December 2019. [CrossRef]
15. Baldassari, F. Saving Bandwidth with Delta Firmware Updates. Interrupt. 2022. Available online: https://interrupt.memfault.com/blog/ota-delta-updates (accessed on 10 January 2024).
16. MPLAB XC8 C Compiler. *MPLAB® XC8 C Compiler User's Guide for PIC® MCU*; Microchip Technology Inc.: Chandler, AZ, USA, 2020. Available online: https://ww1.microchip.com/downloads/en/devicedoc/50002053g.pdf (accessed on 27 March 2024).
17. *Bootloader Generator User's Guide*; DS400001779B; Microchip Technology Inc.: Chandler, AZ, USA, 2020. Available online: https://ww1.microchip.com/downloads/en/DeviceDoc/40001779B.pdf (accessed on 27 March 2024).

electronics

Article

Time-Allocation Adaptive Data Rate: An Innovative Time-Managed Algorithm for Enhanced Long-Range Wide-Area Network Performance

Kunzhu Wang [1], Kun Wang [2] and Yongfeng Ren [1,*]

1 Science and Technology on Electronic Test and Measurement Laboratory, North University of China, Taiyuan 030051, China; wkz1070512015@163.com
2 College of Software, Shanxi Agricultural University, Jinzhong 030801, China; wangkun@sxau.edu.cn
* Correspondence: renyongfeng@nuc.edu.cn

Abstract: Currently, a variety of Low-Power Wide-Area Network (LPWAN) technologies offer diverse solutions for long-distance communication. Among these, Long-Range Wide-Area Network (LoRaWAN) has garnered considerable attention for its widespread applications in the Internet of Things (IoT). Nevertheless, LoRaWAN still faces the challenge of channel collisions when managing dense node communications, a significant bottleneck to its performance. Addressing this issue, this study has developed a novel "time allocation adaptive Data Rate" (TA-ADR) algorithm for network servers. This algorithm dynamically adjusts the spreading factor (SF) and transmission power (TP) of LoRa (Long Range) nodes and intelligently schedules transmission times, effectively reducing the risk of data collisions on the same frequency channel and significantly enhancing data transmission efficiency. Simulations in a dense LoRaWAN network environment, encompassing 1000 nodes within a 480 m × 480 m range, demonstrate that compared to the ADR+ algorithm, our proposed algorithm achieves substantial improvements of approximately 30.35% in data transmission rate, 24.57% in energy consumption, and 31.25% in average network throughput.

Keywords: LPWAN; LoRaWAN; ADR algorithm; time allocation

Citation: Wang, K.; Wang, K.; Ren, Y. Time-Allocation Adaptive Data Rate: An Innovative Time-Managed Algorithm for Enhanced Long-Range Wide-Area Network Performance. *Electronics* **2024**, *13*, 434. https://doi.org/10.3390/electronics13020434

Academic Editors: Dionisis Kandris and Eleftherios Anastasiadis

Received: 15 December 2023
Revised: 17 January 2024
Accepted: 18 January 2024
Published: 20 January 2024

1. Introduction

In today's digital landscape, the Internet of Things (IoT) has become a pivotal force in bridging the gap between the virtual and physical worlds, driving connectivity across a myriad of devices [1]. With the exponential growth of IoT endpoints, conventional wireless networks are hitting their limits in scalability and energy efficiency [2]. This has prompted a shift toward more sustainable, low-energy, and expansive communication frameworks. Within this paradigm, Low-Power Wide-Area Networks (LPWANs) stand out as a beacon of IoT connectivity, offering a trifecta of benefits: minimal power requirements, extensive coverage, and economical operation. As a member of the numerous Low-Power Wide-Area Network (LPWAN) technologies, LoRaWAN is widely used in modern smart cities due to its advantages such as simple and flexible network formation, operation in unlicensed frequency bands, and the ability to communicate remotely with extremely low power consumption. It is considered one of the most promising LPWAN technologies today [3–5] and is applied in areas like monitoring water consumption in urban buildings and checking [6] the structural health of buildings [7].

LoRaWAN is a Media Access Control (MAC) protocol built on top of the LoRa (Long-Range) physical layer. LoRa itself is a wireless modulation method at the physical layer, based on Chirp Spread Spectrum (CSS) [8] technology, and is focused on providing long-distance, low-power wireless transmission. The core advantage of LoRa technology lies in its ability to achieve low data rate communication over long distances while maintaining low energy consumption. In contrast, LoRaWAN defines how to establish a complete network on the foundation of LoRa, encompassing features like device addressing, encrypted communication, data rate management, and multiple access. To facilitate interoperability among LoRa devices from different manufacturers and to provide better support for the diversity and scalability of LoRa technology in IoT applications, the LoRa Alliance introduced the standardized LoRaWAN communication protocol in 2015. This protocol not only promotes compatibility among LoRa devices from various suppliers but also enhances the functionality and reliability of the entire LoRa network.

The classical LoRaWAN follows a star network topology, primarily consisting of end devices, gateways, and a network server, as shown in Figure 1. The end devices are IoT terminal nodes responsible for data collection and communication with the gateways. Gateways serve as intermediate devices, connecting the end devices to the network server and handling the reception and forwarding of data from the end devices. The network server is the core of the LoRaWAN network, responsible for managing and coordinating communication between end devices and gateways, as well as handling functions such as data transmission, device authentication, and security [9].

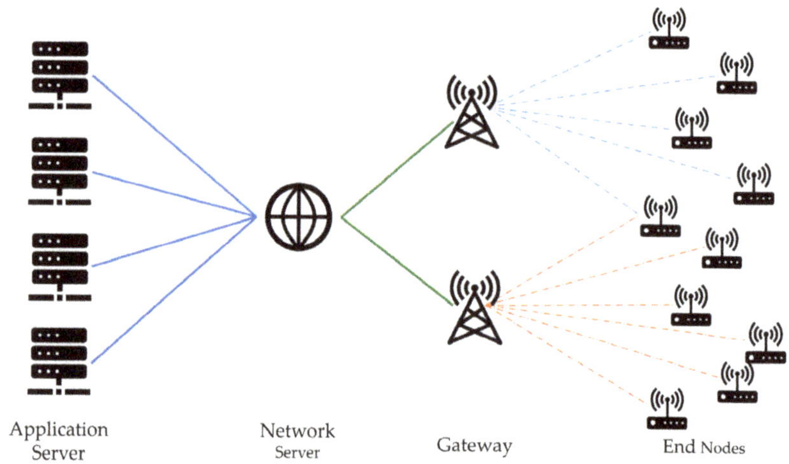

Figure 1. Classic LoRaWAN architecture.

The LoRaWAN standard protocol defines three classes of end devices, Class A, Class B, and Class C, to meet the power and latency requirements of different scenarios [10,11]. Among them, Class A is the most common and basic device class that all LoRa devices must support. It has the lowest power consumption and the simplest communication mode. Its operation is illustrated in Figure 2. In Class A mode, the end devices actively send data packets based on their own needs. After a certain airtime delay, these packets are received by one or multiple gateways. Following the completion of an uplink transmission, the device pauses for a period (Delay 1). During this time, the gateway sends downlink data to the end device, based on the frequency and data rate of the previous uplink transmission. After each data transmission, the end device sequentially opens two short receive windows, with their opening times determined by the end of the transmission. If the end device successfully receives data during the first receive window, it does not open the second receive window. If not, after the first window closes and following another period (Delay 2), the device opens the second receive window to continue receiving potential data. Since the gateway cannot ascertain whether the end device has actually opened the receive window, it proceeds to send data during the pre-scheduled periods of both receive windows.

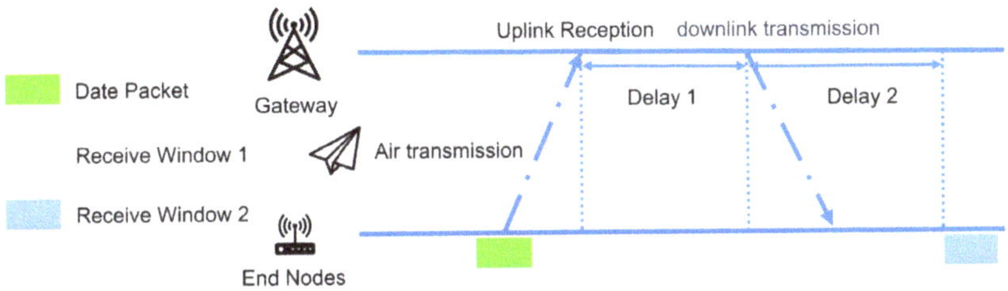

Figure 2. Information sending and receiving process of Class A node.

The Adaptive Data Rate (ADR) feature in the LoRaWAN protocol plays a crucial role in optimizing communication efficiency by dynamically adjusting two key parameters: spreading factor and transmit power. These parameters help regulate network capacity, coverage, power consumption, and device lifespan. Increasing SF levels enhances interference resistance and sensitivity, widening coverage but at the cost of lower data rates and increased energy requirements. Conversely, TP directly impacts communication range and energy consumption. To balance these factors, LoRaWAN employs a standard ADR algorithm to harmonize SF and TP, thereby improving overall network throughput [10]. Additionally, the quasi-orthogonal nature of different SFs in LoRaWAN allows for nearly interference-free transmissions on the same bandwidth when the SFs are different [12]. However, significant signal interference occurs between LoRa nodes using the same SF. However, as the density of nodes increases, more nodes with the same SF transmitting at the same time lead to escalated data collisions, significantly reducing the data transmission success rate.

To address this challenge, our proposed Time-Allocation Adaptive Data Rate (TA-ADR) algorithm innovates by calibrating the adjustment steps for SF and TP. It introduces the concept of transmission time intervals for LoRa nodes, allocating appropriate time slots for nodes with the same SF to transmit data, thereby reducing the risk of conflicts and enhancing the transmission success rate.

The rest of this article is structured as follows. In Section 2, we present an overview of the related work and introduce the main contributions of our study. Section 3 describes the flow of the ADR+ algorithm and the optimized TA-ADR algorithm. Section 4 introduces the simulation configuration and gives the analysis of the results after simulation. The conclusion of this paper is given in Section 5.

2. Related Work

The ADR algorithm, as a fundamental feature in the LoRaWAN protocol, is a key advantage and has received extensive attention from many research teams. Several research teams have focused on optimizing the ADR algorithm to adapt to different application scenarios and network requirements, proposing a series of corresponding improvement methods.

Slabicki et al. pointed out in the literature [13] that the basic ADR algorithm mentioned in the LoRaWAN standard protocol is to select the maximum signal-to-noise ratio in the last 20 packets as the calculation basis, but this method is too optimistic in a noisy channel. Therefore, they simply modified the ADR algorithm. In the proposed ADR+ algorithm, the average value of the SNR of the latest packet is used as the basis for the subsequent calculation, which improves the performance of ADR algorithm in noisy channels.

In reference [14], Babaki et al. introduced the Ordered Weighted Averaging (OWA) operator to optimize the ADR algorithm for accurately demodulating the suitable spreading factor based on the current channel conditions and external environment. This algorithm improves the transmission success rate and achieves almost the same energy consumption as other ADR algorithms, even in dense LoRa networks and high channel noise.

Reference [15] proposes a new and more efficient dynamic ADR algorithm called ND-ADR (New-Dynamic Adaptive Data Rate Algorithm). This algorithm introduces RSSI in addition to the basic ADR algorithm's selection of SF based solely on the maximum SNR value. By combining the average values of RSSI and SNR, ND-ADR dynamically adjusts the number of SNR values considered (denoted as "n"). Initially set to three frames, the value of "n" is dynamically increased based on certain conditions. This allows the server to quickly adapt to changes in the external environment, effectively addressing communication quality issues and high packet loss rates in mobile terminal devices operating in harsh environments.

In reference [16], the authors build upon the ADR+ algorithm by introducing an energy efficiency controller α, which is related to the total energy consumption of all nodes. The algorithm multiplies the average SNR value from the most recent 20 packets by α and then gradually decreases α from 1 in steps of 0.1. This approach aims to find the optimal α value that minimizes network energy consumption without compromising data delivery rates. The simulation results presented in the reference demonstrate that this algorithm outperforms the ADR+ algorithm in terms of energy consumption and data delivery rates.

In reference [17], Marini et al. propose a new ADR algorithm for LoRaWAN networks called CA-ADR (Collision-Aware ADR). This algorithm takes into account the collision probability at the MAC layer of the network. When allocating data rates, it minimizes the collision probability while maintaining controllable link performance by considering the set of nodes in the entire network. The feasibility of both cloud computing and fog computing architectures is also validated. The results demonstrate that the fog computing-based architecture is feasible and reduces end-to-end transmission latency.

In reference [18], Jeon et al. present a simple and energy-efficient uplink transmission rate control scheme for LoRaWAN. The aim is to support efficient communication for a large number of IoT devices over a wide area. This scheme enables devices to increase or decrease the transmission rate based on the changing link quality. It introduces a ping-pong mechanism to avoid frequent rate changes. Through modeling and simulation comparisons, the results show that this scheme outperforms other approaches in terms of transmission success rate, effective transmission rate, frame transmission delay, and energy consumption.

In reference [19], Anwar et al. discovered that fixed SF allocations are no longer efficient in LoRaWAN when the end devices (EDs) are in motion. The link conditions between the EDs and gateways change abruptly, resulting in significant packet loss and increased retransmission attempts. To address this issue, they propose a resource management ADR (RM-ADR) scheme that considers both packet transmission information and received power. The research findings indicate that in a mobile LoRaWAN network environment, RM-ADR achieves faster convergence time by reducing packet loss and retransmission attempts.

Reference [20] mentioned an EE-LoRa for spread spectrum factor selection and power control in multi-gateway LoRaWAN networks. The author first optimized the energy efficiency of the network, and then applied power control to minimize the transmit power of nodes while maintaining the reliability of communication.

In reference [21], Cuomo et al. proposed two LoRa spread spectrum channel allocation algorithms to solve the problems existing in the ADR algorithm allocation mechanism of LoRaWAN. Scenario 1 (EXPLoRa-SF) uses a heuristic algorithm to evenly allocate SFs to these nodes, with the same number of LoRa nodes for each spread spectrum factor. Scenario 2 (EXPLoRa-AT) is used to fairly distribute broadcast time among network nodes so that the various SFs transmit data at the same time.

We summarize the features of the ADR algorithms mentioned in Table 1.

Compared to the ADR algorithms proposed in the existing literature, our TA-ADR algorithm incorporates a time scheduling strategy, allowing for the allocation of communication windows within the LoRaWAN network to reduce channel conflicts among nodes with the same SF. This approach not only enhances the success rate of data transmission but also optimizes the network's energy consumption efficiency. Here are the primary contributions of our study:

(1) The TA-ADR algorithm diminishes channel conflicts by distributing non-overlapping communication time windows among nodes, thereby improving the data delivery rate and network throughput.
(2) The TA-ADR algorithm adapts the timetable to changes in node density, particularly as the number of nodes increases, enabling better management of communication loads.

Table 1. Features of the algorithms.

Algorithm	Features
Reference [13]	SNR is calculated from the average of the most recent frames.
Reference [14]	The SNR is calculated by the OWA operator.
Reference [15]	RSSI was introduced to the SNR and modified during the adjustment step.
Reference [16]	Introduction of the α of energy efficiency controllers.
Reference [17]	Consider the collision probability of the MAC layer to reduce collisions when allocating data rates.
Reference [18]	The transmission rate can be dynamically adjusted according to the link quality change, and the ping-pong mechanism is introduced to avoid frequent rate changes.
Reference [19]	In a mobile LoRaWAN environment, resource management is performed by combining packet transmission information and received power.
Reference [20]	Start by optimizing the energy efficiency of the network, and then apply power control.
Reference [21]	EXPLoRa-SF features: The heuristic algorithm is used to evenly distribute SF to nodes to avoid SF aggregation. EXPLoRa-AT features: Fairly allocates the broadcast time to ensure the simultaneous transmission of data from different SFs.

3. Introduction and Optimization of ADR+ Algorithm

3.1. Standard ADR Algorithm and ADR+ Algorithm

In this section, the speed regulation mechanism of the ADR algorithm is introduced, and then we describe the standard ADR algorithm and ADR+ algorithm. Finally, the algorithm of the SF and TP adjustment stage is optimized on the basis of the ADR+ algorithm, and the optimization process is described in detail.

The ADR mechanism can be divided into two parts: the network server (NS)-side algorithm is responsible for increasing the data transmission rate of end nodes, while the end node (ED)-side algorithm is responsible for decreasing the data transmission rate of end nodes. The ED-side algorithm for ADR is defined by the LoRa Alliance, while developers can choose basic ADR algorithms or configure their own algorithms for the NS side [15]. Additionally, the NS is located at the core of the network, allowing it to access global information. This enables the NS-side algorithm to dynamically adjust SF and TP of end nodes based on global information, leading to better optimization of network performance compared to node-side algorithms. The ADR algorithm in the LoRaWAN standard protocol includes both the ED-side algorithm and the NS-side algorithm. The ED-side algorithm relies on acknowledge character (ACK) feedback to determine if the data transmission is successful. If the node does not receive an acknowledgment from the gateway within two receiving windows, it considers the data transmission as failed and activates the retransmission mechanism. It automatically reduces the data transmission rate before retransmitting the data. The NS-side algorithm determines the link quality based on the signal-to-noise ratio (SNR) of the recently received data and adjusts the SF and TP accordingly. The ADR+ algorithm, compared to the ADR algorithm, only modifies the NS-side algorithm while keeping the node-side algorithm unchanged. Algorithm 1 describes the implementation steps of the NS-side ADR+ algorithm [14]. In Algorithm 1, the input SF value ranges from 7 to 12 in steps of 1, and the input TP value ranges from 2 to 14 in steps of 3. It involves calculating the required SNR ($SNR_{required}$) based on the current SF, averaging the SNR of the latest received 20 data packets to obtain SNR_{avg}, determining the appropriate SF based on SNR_{avg}, and then computing the SNR margin (SNR_{margin}) and the adjustment steps ($nsteps$) using Equations (1) and (2):

$$SNR_{margin} = SNR_{avg} - SNR_{required} - device_{margin} \quad (1)$$

$$nsteps = int(SNR_{margin}/3) \quad (2)$$

Ultimately, through iteration, the values of SF and TP are gradually adjusted until certain conditions are met, and then the adjusted values of SF and TP are output. Table 2 lists the minimum SNR required for different SFs. If the SNR margin (SNR_{margin}) is positive, it indicates that the current channel quality is good, and the node can increase the data rate or decrease the transmission power to reduce power consumption and extend the node's battery life. If the SNR margin is negative, it indicates that the node is currently using a transmission power that is too low, resulting in a low SNR for the uplink signal. Therefore, it is necessary to increase the transmission power or decrease the data rate.

Table 2. SNR required for different data rates (BW 125 KHz) [22].

Data Rate	Spreading Factor	SNR (dB)
DR5	SF7	−7.5
DR4	SF8	−10.0
DR3	SF9	−12.5
DR2	SF10	−15
DR1	SF11	−17.5
DR0	SF12	−20.0

Algorithm 1 NS ADR+ Algorithm

Input: $SF \in [7,12], TP \in [2,14]$,
Output: SF and TP

1. $SNR_{required} = demodulation\ floor\ (current\ data\ rate)$
2. $SNR_{avg} = avg\ (SNRs\ of\ last\ 20\ frames)$
3. $SF = demodulation\ floor\ (SNR_{avg})$
4. $SNR_{margin} = SNR_{avg} - SNR_{required} - device\_margin$
5. $nsteps = int\left(SNR_{margin}/3\right)$
6. $while\ nsteps > 0\ and\ SF > SF_{min}$

$SF = SF - 1$
$nsteps = nstep - 1$
$end\ while$

7. $while\ nsteps > 0\ and\ TP > TP_{min}$

$T = TP - 3$
$nsteps = nsteps - 1$
$end\ while$

8. $while\ nsteps < 0\ and\ TP < TP_{max}$

$TP = TP + 3$
$nsteps = nsteps + 1$
$end\ while$

The flowchart of the ADR+ algorithm is presented in Figure 3.

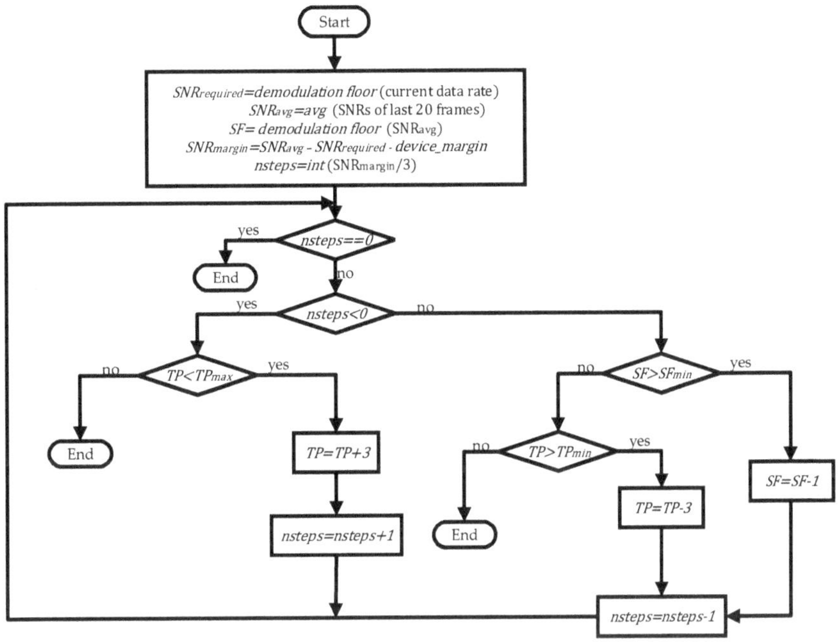

Figure 3. Flowchart of the standard ADR algorithm.

3.2. Algorithm Optimization

In this section, we will introduce the idea of algorithm optimization and the optimized algorithm.

The ADR+ algorithm has powerful capabilities in controlling data rates, which can improve the communication success rate and reduce the overall network energy consumption to some extent. However, from the ADR+ algorithm flow, it is evident that the ADR+ algorithm always starts by changing the spreading factor and tends to decrease it. In a network with a large number of deployed LoRa nodes, this can lead to numerous nodes operating on the same spreading channel, resulting in severe data collisions, increased triggering of the node's retransmission mechanism, and inevitably increasing network energy consumption while reducing the communication success rate.

To address these issues, we propose the communication time algorithm to allocate signal transmission times for nodes with the same spreading factor. The message types in LoRaWAN are divided into uplink messages and downlink messages. The data packet structure of uplink messages mainly consists of five parts as shown in Figure 4: Preamble, PHDR (Physical Header), PHDR_CRC (Physical Header Cyclic Redundancy Check), PHYPayload (Physical Payload), and CRC (Cyclic Redundancy Check). The PHDR_CRC is the Cyclic Redundancy Check for the PHDR, which is used to detect errors in the header information. The CRC is for the PHYPayload, ensuring the integrity of the data payload [10].

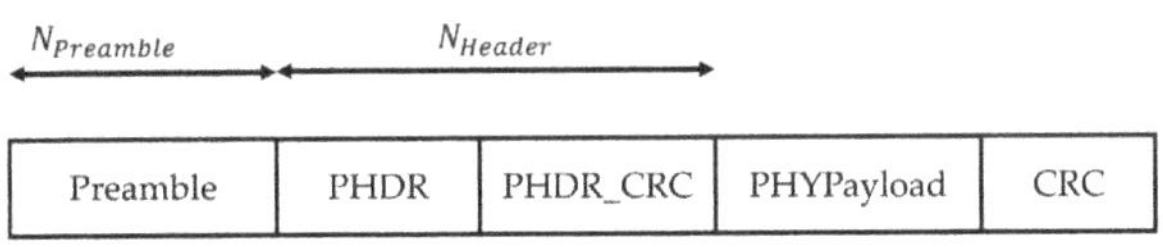

Figure 4. The data packet structure for uplink messages in LoRaWAN.

The transmission time of a LoRaWAN data packet is composed of the transmission time of the preamble and the transmission time of the payload. The transmission time of the preamble is determined by the symbol effective length $N_{Preamble}$ and the time to send a single symbol T_s, where T_s is related to the symbol rate R_s of LoRa. The specific calculation formula is as follows:

$$R_s = \frac{BW}{2^{SF}} \tag{3}$$

$$T_s = \frac{1}{R_s} \tag{4}$$

$$T_{Preamble} = (N_{Preamble} + 4.25) \times T_s \tag{5}$$

Here, BW represents bandwidth, and SF represents spreading factor.

The transmission time of the payload is related to the selected header type. In explicit header mode, the header contains information such as payload length, forward error correction rate, and whether CRC is used. In implicit header mode, the payload bytes, forward error correction rate, and CRC need to be manually set. The number of payload symbols ($N_{Payload}$) is calculated as follows:

$$N_{Payload} = 8 + \max\left(ceil\left(\frac{8PL - 4SF + 28 + 16 - 20H}{4(SF - 2DE)}(CR + 4)\right), 0\right) \tag{6}$$

where PL is the number of bytes in the payload. H represents the selected header type, where $H = 0$ indicates explicit header mode and $H = 1$ indicates implicit header mode. DE represents whether low data rate optimization is used during data transmission, where $DE = 1$ indicates it is used and $DE = 0$ indicates it is not used. max() denotes the maximum value function, and $ceil()$ represents the ceiling function for rounding up.

After obtaining the number of payload symbols, the formula to calculate the transmission time of the payload ($T_{Payload}$) is defined as:

$$T_{Payload} = N_{Payload} \times T_s \tag{7}$$

Finally, by adding the transmission time of the preamble and the transmission time of the payload, we can determine the transmission time of the LoRa data packet (T_{Packet}):

$$T_{Packet} = T_{Peamble} + T_{Payload} \tag{8}$$

From the derivation of the above data packet transmission time formulas, we can see that the factors affecting the data packet transmission time include BW, SF, CR, $N_{Preamble}$, PL, header type, and whether low data rate optimization is used. In this article, our algorithm only dynamically adjusts the SF and TP of the LoRa node. Therefore, we preset the variables W, CR, $N_{Preamble}$, header type, and whether low data rate optimization is used. We set BW to 125 kHz, CR to 1, $N_{Preamble}$ to 8, PL to 23 bytes, $H = 0$ for explicit header, and $DE = 0$ for not using the low data rate optimization configuration. By using the formulas, we can calculate the number of payload symbols $N_{Preamble}$ and the transmission time T_{Packet} of the LoRa data packet for different spreading factors, as shown in Table 3.

Table 3. The number of payload symbols and transmission time of LoRa node for different SFs.

SF	7	8	9	10	11	12
$N_{Payload}$ (symbol)	48	43	38	33	33	28
T_{Packet} (ms)	61.696	113.152	205.824	370.688	741.376	1318.912

From Table 3, it can be observed that if a LoRa node uses SF12 to transmit a data packet, then within the corresponding time, 21 LoRa nodes using SF7 can complete the transmission of one data packet. Next, we established a communication time algorithm for all nodes based on their channel, spreading factor, and node number. We calculated the communication time interval t_i^{SF} which characterizes the time interval from the beginning to the end of data transmission by a LoRa node i at the SF, where the time occupied by a spreading factor channel is denoted as T_{SF}, such as $T_7 = 61.696$ ms. The time interval between node communications is denoted as ΔT_{SF}, such as $\Delta T_7 = 123.392$ ms. Assuming the starting time of the first node is t_0, the formula to determine the interval of t_i^{SF} is defined as:

$$\Delta T_{SF} = 2 * T_{SF} \quad (9)$$

$$t_i^{SF} \in [t_0 + (T_{SF} + \Delta T_{SF})(i-1),\ t_0 + T_{SF}i + \Delta T_{SF}(i-1)] \quad (10)$$

Based on the time interval t_i^{SF}, a new algorithm called TA-ADR (Time Slot Adaptive Data Rate) is proposed. The algorithm flow of TA-ADR is detailed in Algorithm 2.

Algorithm 2 NS TA-ADR Algorithm

Input: $SF \in [7, 12]$; $TP \in [2, 14]$; Time range T of the current node; Timetable T_i^{SF} of communication of all nodes in LoRa gateway.
Output: SF, TP and update timetable T_i^{SF} of all node communications for all spread spectrum channels of LoRa Gateway.

1. $SNR_{required} = demodulation\ floor\ (current\ data\ rate)$
2. $SNR_{avg} = avg\ (SNRs\ of\ last\ 20\ frames)$
3. $SF = demodulation\ floor\ (SNR_{avg})$
4. $SNR_{margin} = SNR_{avg} - SNR_{required} - device\_margin$
5. $nsteps = int(SNR_{margin}/3)$
6. $while\ nsteps > 0\ \&\&\ TP > TP_{min}\ Do$

$TP = TP - 3$
$nsteps = nsteps - 1$
$end\ while$

7. $if\ nsteps > 0\ and\ SF - nsteps \geq SF_{min}$

$if\ T \cap t_i^{SF-nsteps} == \varnothing$
$SF = SF - nsteps, nsteps = 0;$
$else\ SF = SF - nsteps - 1, k = 1;$
$while\ T \cap t_i^{SF}! = \varnothing\ and\ SF > SF_{min}$
$SF = SF - 1, k = k + 1;$
$if\ T \cap t_i^{SF} == \varnothing\ and\ TP + k * 3 \leq TP_{max}$
$TP = TP + k * 3, nsteps = 0;$
$break;$
$end\ while$
$end\ if$

8. $while\ nsteps < 0\ and\ TP < TP_{max} Do$

$TP = TP + 3$
$nsteps = nsteps + 1$
$end\ while$

9. $if\ nsteps < 0\ and\ SF - nsteps \leq SF_{max}$

$if\ T \cap t_i^{SF-nsteps} == \varnothing$
$SF = SF - nsteps, nsteps = 0;$
$else\ SF = SF - nsteps - 1, k = 1;$
$while\ T \cap t_i^{SF}! = \varnothing\ and\ SF < SF_{max}$
$SF = SF + 1, k = k + 1;$
$if\ T \cap t_i^{SF} == \varnothing\ and\ TP + k * 3 \geq TP_{min}$
$TP = TP - k * 3, nsteps = 0;$
$break;$
$end\ while$
$end\ if$

3.3. Algorithm Implementation

In Section 3.2 of our paper, we delved into a novel ADR management algorithm, dubbed the Time-Allocation Adaptive Data Rate algorithm. This algorithm is designed to optimize data transmission and significantly reduce conflicts between nodes on the same frequency channel using the same SF. This section will elaborate on the details of implementing the TA-ADR algorithm.

The input parameters for the TA-ADR algorithm include the SF range of the current node, the TP range, the time range T of nodes that need to be optimized, and the communication schedule T_i^{SF} of all nodes within the LoRa gateway. The output of the algorithm is the adjusted SF and TP values, along with an updated communication schedule for all nodes across all spread spectrum channels. To illustrate the relationship between T, t_i^{SF}, and T_i^{SF}, let us consider an example in a LoRaWAN network with six LoRa nodes, all having a TP of 2 dBm. Among these, three nodes use a SF of 7, while the other three use SF8, and the initial send time is 0 s. Therefore, the communication time intervals for these six nodes are $t_1^{SF7} \in [0\ \text{s}, 0.063\ \text{s}]$, $t_2^{SF7} \in [0.189\ \text{s}, 0.252\ \text{s}]$, $t_3^{SF7} \in [0.378\ \text{s}, 0.441\ \text{s}]$, $t_1^{SF8} \in [0\ \text{s}, 0.114\ \text{s}]$, $t_2^{SF8} \in [0.342\ \text{s}, 0.456\ \text{s}]$, and $t_3^{SF8} \in [0.684\ \text{s}, 0.798\ \text{s}]$, respectively. From this, we can derive the theoretical communication timetable T_i^{SF} for these nodes, as presented in Table 4.

Table 4. The time communication table T_i^{SF} before updating.

T_i^{SF}	$i = 1$	$i = 2$	$i = 3$
SF7	t_1^{SF7}	t_2^{SF7}	t_3^{SF7}
SF8	t_1^{SF8}	t_2^{SF8}	t_3^{SF8}

After the nodes 2 and 3 using SF8 transmit their data, the NS calculates that these nodes have SNR_{margin} with both having an *nsteps* of 1. Therefore, NS optimizes the parameters for these nodes under SF8 using Algorithm 2. For node 2 under SF8 (with T being t_2^{SF8}), after evaluating intersections with t_i^{SF7} $(i = 1, 2, 3)$, it is found that T intersects with t_3^{SF7}, indicating that node 2 should maintain its original settings. For node 3 under SF8 (with T being t_3^{SF8}), no intersections are found with $t_i^{SF7}(i = 1, 2, 3)$, suggesting the time slot under SF7 is available. Thus, NS sends frame information containing the adjusted SF value and the new communication interval to node 3 during its receive window. Node 3 then resets its parameters accordingly. The updated communication timetable T_i^{SF} is displayed in Table 5.

Table 5. The time communication table T_i^{SF} after the update.

T_i^{SF}	$i = 1$	$i = 2$	$i = 3$	$i = 4$
SF7	t_1^{SF7}	t_2^{SF7}	t_3^{SF7}	t_4^{SF7}
SF8	t_1^{SF8}	t_2^{SF8}		

Next, the parameter calculation and cyclic part of algorithm 2 are discussed. Initially, the algorithm calculates the minimum SNR required ($SNR_{required}$) for the current SF, and computes the average SNR (SNR_{avg}) from the last 20 frames of data. Then, it adjusts the SF based on the SNR_{avg}. Then, the algorithm calculates the SNR margin (SNR_{margin}), which is the difference between the average SNR and the required SNR, minus the device margin (*device_margin*). This SNR margin is then divided by 3 to determine the number of adjustment steps (*nsteps*). Then, the loop is entered; if *nsteps* is greater than 0, indicating that the SNR is higher than required, the algorithm attempts to reduce the TP to save energy. For each reduction in TP (by 3 dB each time), *nsteps* is decreased by 1, until TP reaches the minimum value or *nsteps* becomes 0. If there are remaining *nsteps* after reducing TP, the algorithm tries to decrease the SF. It first checks if lowering the SF would cause a conflict with the communication schedule of other nodes. If there is no conflict, SF is reduced, and *nsteps* is set to 0. If there is a conflict, the algorithm further decreases SF (by 1 each time), and for each decrease in SF, TP is increased by 3 dB, until a conflict-free configuration is found or SF is lowered to its minimum value. If the original nsteps is less than 0, indicating that the SNR is lower than required, the algorithm attempts to increase TP to improve signal quality. For each increase in TP (by 3 dB each time), *nsteps* is incremented by 1, until TP reaches its maximum value or nsteps becomes 0. If nsteps is still less than 0 after increasing TP, the algorithm tries to increase SF. Similarly, it checks for conflicts with the schedule after increasing SF, and if there is a conflict, it continues to increase SF (by 1 each time), and for each increase in SF, TP is decreased by 3 dB, until a conflict-free configuration is found or SF is increased to its maximum value.

It is important to emphasize that our proposed TA-ADR algorithm is implemented on the NS end. All logic and computational operations are performed internally within the NS, sparing the LoRa nodes any additional burden. The NS, after processing through Algorithm 2, sends the optimized SF, TP, transmission time, and the addresses of the specific nodes needing optimization to the gateway. The gateway then conveys this information to the respective nodes, which adjust their parameters and transmission times upon receipt. Additionally, to achieve time synchronization among the nodes, we make the LoRaWAN gateways periodically broadcast time stamp updates to ensure all LoRa nodes are precisely synchronized.

4. Simulation and Results

In this section, we divide our discussion into two parts: simulation parameter settings and result analysis. In the simulation parameter settings section, we present the path loss model used in the LoRaWAN network, as well as the topology, simulation range, and simulation parameters. The result analysis section provides comparative graphs of three algorithms in the LoRaWAN network and elaborates on the advantages of the TA-ADR algorithm.

4.1. Simulation Parameter Settings

In this network simulation, we used the log-normal shadowing path loss model [15] to simulate the path loss caused by attenuation and shadowing when the signal propagates through the air. The mathematical model is defined as follows:

$$L_p(d_i) = L_p(d_0) + 10\gamma lg(d_i/d_0) + X_\sigma \tag{11}$$

where $L_p(d_0)$ represents the average path loss at the reference distance d_0, measured in dB. d_i is the distance from node i to the gateway. γ is the path loss exponent. X_σ (dB) is a zero-mean Gaussian random variable with standard deviation σ.

The received signal power $P_{r,i}(d)$ can be obtained by subtracting the path loss $P_{t,i}$ from the transmit power $L_p(d)$ of node i [21]:

$$P_{r,i}(d) = P_{t,i} - L_p(d_i) \tag{12}$$

All LoRa nodes in the simulation are initialized in Class A transmission mode. The region parameters and path loss parameters are given in Reference [14], where the simulation area size is 480 m × 480 m and the path loss parameters d_0, $L_p(d_0)$, γ, and σ are provided in Table 6.

Table 6. Path loss model parameters in urban scenarios.

Scene	d_0(m)	$L_p(d_0)$(dB)	γ	σ
City	40	127.41	2.08	3.57

According to the predefined network parameters in Table 7, we conducted simulations of LoRa networks using the FLoRa framework in the OMNeT++ platform. We evaluated the performance of three different ADR algorithms in an urban scenario. The LoRa network adopts a star network topology with the gateway placed at the center, as shown in Figure 5. The nodes are uniformly distributed around the gateway.

Table 7. Simulation parameters.

Parameter	Value
Carrier frequency (f)	868 MHz
Bandwidth (BW)	125 KHz
Coding rate (CR)	4/5
Spreading factor (SF)	[7, 12]
Initial SF of nodes	12
Transmission power (TP)	2 − 14 dBm
Initial TP of nodes	14 dBm
Payload (byte)	23 bytes
Simulation time	24 h

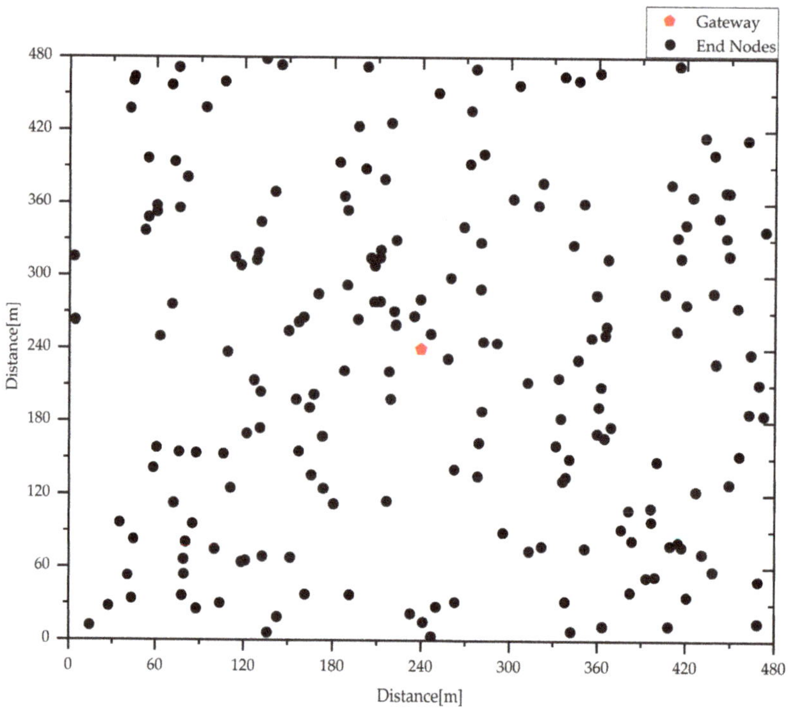

Figure 5. Simulation network with 200 nodes.

Each simulation runs for a duration of 24 h. In the network, each node sends only one data packet at a time (with a payload size of 23 bytes), and nodes with the same SF will wait for a time interval of $2 * T_{SF}$ before sending again. All LoRa nodes periodically send a round of information. The energy consumption of LoRa nodes comes from three states (send, receive, and sleep). Node transmission power consumption depends on node level and instantaneous current value during transmission. The current of the node in receive and sleep mode was obtained from the Semtech SX1272 data manual, and the operating voltage was 3.3 V [23].

Finally, we assessed the performance of the three schemes based on the following three parameters:

Energy consumption (mJ): Defined as the total energy consumed by all nodes in the LoRaWAN network divided by the total number of data packets successfully received by the gateway.

Packet delivery rate (%): Defined as the total number of data packets successfully received by the LoRaWAN network server divided by the total number of data packets sent by all nodes.

Throughput (bps): Defined as the amount of data successfully transmitted per second in the LoRaWAN network.

4.2. Interpretation of Result

From Figure 6a,b, it can be observed that as the number of nodes in the LoRaWAN network increases, the energy efficiency and the packet delivery rate decreases. It is evident that the TA-ADR algorithm performs better compared to the other two algorithms, and its performance advantage becomes even more significant as the number of nodes increases.

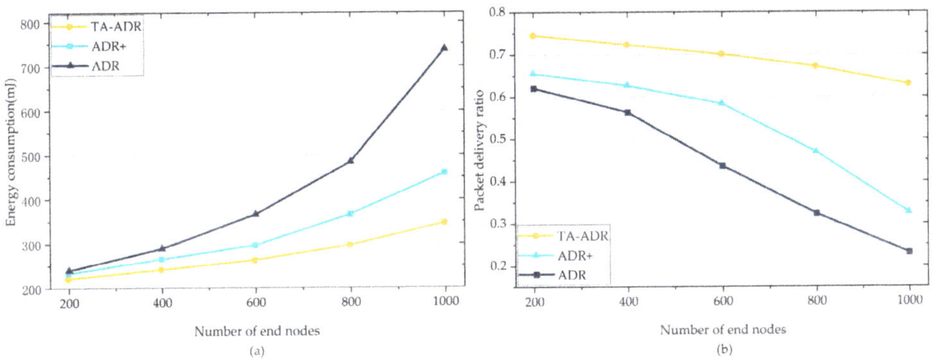

Figure 6. Simulation comparison with different numbers of nodes: (**a**) energy consumption (mJ); (**b**) packet delivery rate.

At a node count of 200, the ADR+ algorithm reduces energy consumption by approximately 2.73% compared to the ADR algorithm, while the TA-ADR algorithm reduces energy consumption by approximately 7.63% compared to the ADR algorithm, and by approximately 5.03% compared to ADR+. At a node count of 1000, the ADR+ algorithm reduces energy consumption by approximately 37.74% compared to the ADR algorithm, while the TA-ADR algorithm reduces energy consumption by approximately 53.04% compared to the ADR algorithm, and by approximately 24.57% compared to ADR+. This is because the ADR algorithm, which selects the maximum SNR value for SF decoding, is overly optimistic. In a noisy channel, the ADR algorithm is prone to selecting a high SNR value for decoding, resulting in a lower SF being decoded. In contrast, the ADR+ algorithm uses the average SNR value as a reference, resulting in more accurate SF decoding. However, in the subsequent adjustment steps, both algorithms prioritize assigning lower spreading factors to nodes. As a result, in the simulation scenario, most nodes under these algorithms transmit data with smaller SF, leading to data collisions and reduced data delivery. Nodes that fail to transmit trigger retransmissions, further increasing energy consumption.

In Figure 7, we present a statistical analysis of the final SF allocation for 1000 LoRa nodes under the three ADR algorithms. The results show that under the ADR algorithm, 643 nodes are assigned SF7 and 269 nodes are assigned SF8. Under the ADR+ algorithm, 503 nodes are assigned SF7 and 406 nodes are assigned SF8. In contrast, the TA-ADR algorithm assigns nodes almost equally in decreasing order of SFs, with an approximately 50% reduction in the number of nodes for each SF. Based on the propagation time of each SF in the 125 kHz bandwidth channel, as calculated in Table 2, when a node uses a higher SF to transmit a data packet, approximately two nodes using lower SFs can transmit data consecutively within that time frame. Therefore, when the spreading channel is fully utilized, the number of nodes using lower SFs should be approximately twice the number of nodes using higher SFs.

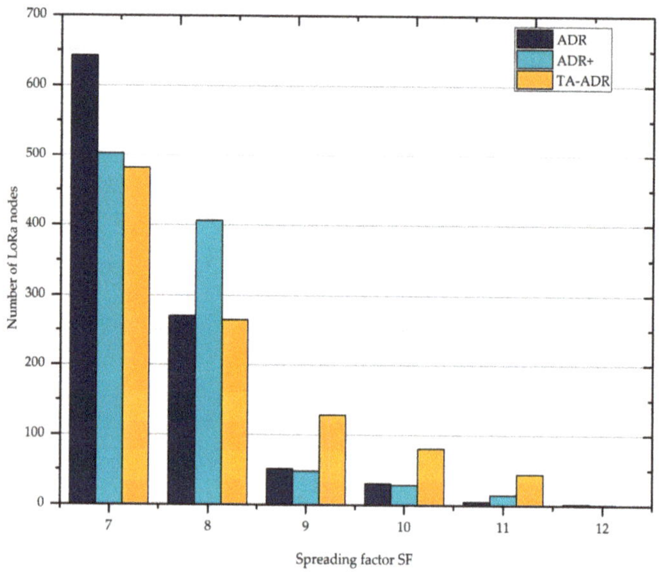

Figure 7. Number of LoRa nodes with different SFs is as follows in urban scenario with 1000 LoRa nodes.

Although the TA-ADR algorithm, like the other two algorithms, prioritizes assigning lower spreading factors to LoRa nodes in subsequent SF adjustments, it allocates nodes with the same SF to different time slots for data transmission. When the low spread spectrum is fully allocated, it continues to allocate LoRa nodes to unused higher spreading channels and correspondingly reduces the TP of that node. This ensures that multiple LoRa nodes with the same SF do not transmit data in the same time slot, reducing the probability of data collisions and improving data delivery rates while reducing the number of node retransmissions. The results in Figure 6b further demonstrate that the LoRaWAN network under the TA-ADR algorithm exhibits superior packet delivery ratio (PDR) performance. On average, the PDR under the TA-ADR algorithm is approximately 30.35% higher than that under the ADR+ algorithm and approximately 59.54% higher than that under the ADR algorithm.

Finally, in order to more intuitively reflect the ability of nodes in the LoRaWAN network to transmit data under the TA-ADR algorithm, we selected the statistical data of the network throughput changes over time during the period from 16 h to the end of the simulation to display in Figure 8, and calculated the average value of throughput, which is given in Table 8. The average network throughput of the TA-ADR algorithm is about 31.25% higher than that of the ADR+ algorithm, and 48.65% higher than that of the ADR algorithm, which further proves the advantages of the TA-ADR algorithm.

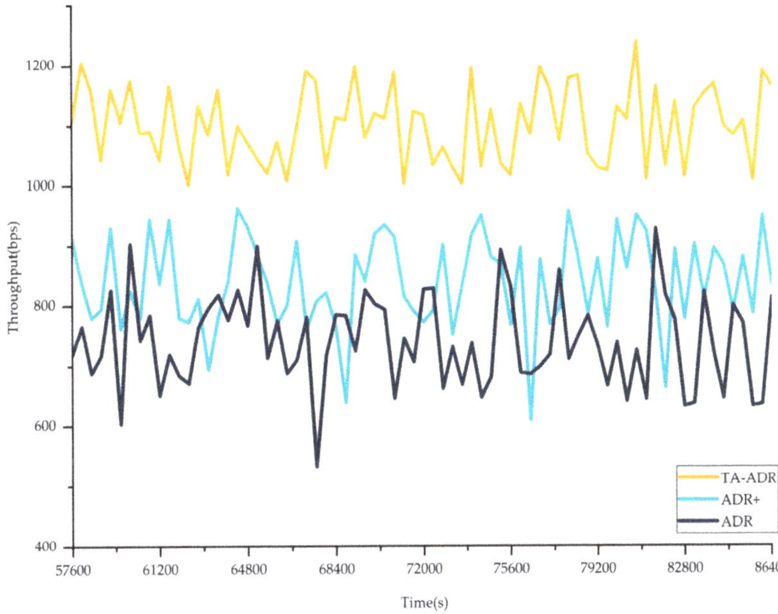

Figure 8. Throughput of 1000 LoRa nodes in 8 h.

Table 8. Average throughput 1000 nodes in 8 h.

Scheme	Average Throughput [bps]
TA-ADR	1115.29
ADR+	849.70
ADR	750.28

5. Conclusions and Prospects

5.1. Conclusions

In this study, we propose an NS ADR algorithm for dynamically adjusting the SF and TP of LoRa nodes in dense LoRaWAN networks. The algorithm introduces the concept of time intervals, denoted as t_i^{SF}, for node transmissions. Its objective is to allocate independent time intervals to each node as much as possible, thereby mitigating data collision issues in densely populated scenarios. Through network simulations of LoRaWAN networks, we evaluated the performance of this algorithm and compared it with other algorithms. The results demonstrate that our proposed TA-ADR algorithm outperforms the comparison algorithms in terms of energy consumption, packet delivery rate, and throughput.

5.2. Deficiencies and Prospects

The optimal application scenario for the TA-ADR algorithm is primarily limited to network environments with deterministic and periodic traffic patterns. This limitation stems from the core principle of the TA-ADR algorithm, which involves pre-planning communication schedules for nodes within the network. In environments characterized by non-deterministic or non-periodic traffic patterns, the communication behavior of nodes may be random or unpredictable. Under such circumstances, the pre-planned communication schedules may not accurately reflect the actual communication needs of the nodes, leading to reduced communication efficiency. With the increase in the noise level in the environment, the effectiveness of the TA-ADR algorithm will become weaker and weaker, but it is still better than the other two algorithms. Therefore, in future work, exploring ways to improve the TA-ADR algorithm to better adapt to diverse traffic patterns will be an important research direction.

Author Contributions: Conceptualization, K.W. (Kunzhu Wang); methodology, K.W. (Kunzhu Wang); software, K.W. (Kunzhu Wang) and K.W. (Kun Wang); data analysis, K.W. (Kunzhu Wang) and K.W. (Kun Wang); investigation, K.W. (Kunzhu Wang) and K.W. (Kun Wang); data curation, K.W. (Kunzhu Wang) and K.W. (Kun Wang); writing and editing manuscript, K.W. (Kunzhu Wang) and Y.R.; writing—review and editing, Y.R.; visualization, K.W. (Kunzhu Wang). All authors have read and agreed to the published version of the manuscript.

Funding: This research received no external funding.

Data Availability Statement: Data are contained within the article.

Conflicts of Interest: The authors declare no conflicts of interest.

Abbreviations

The following abbreviations are used in this manuscript:

LPWAN	Low-Power Wide-Area Networks
IOT	Internet of Things
ADR	Adaptive Data Rate
SF	Spreading Factor
TP	Transmission Power
NS	Network Server
SNR	Signal-to-Noise-Ratio
RSSI	Received Signal Strength Indicator
PDR	Packet Delivery Rate

References

1. Chen, S.; Xu, H.; Liu, D.; Hu, B.; Wang, H. A Vision of IoT: Applications, Challenges, and Opportunities with China Perspective. *IEEE Internet Things J.* **2014**, *1*, 349–359. [CrossRef]
2. URaza, R.; Kulkarni, P.; Sooriyabandara, M. Low Power Wide Area Networks: An Overview. *IEEE Commun. Surv. Tutor.* **2017**, *19*, 855–873.
3. Asad Ullah, M.; Iqbal, J.; Hoeller, A.; Souza, R.D.; Alves, H. K-Means Spreading Factor Allocation for Large-Scale LoRa Networks. *Sensors* **2019**, *19*, 4723. [CrossRef] [PubMed]
4. Kufakunesu, R.; Hancke, G.P.; Abu-Mahfouz, A.M. A Survey on Adaptive Data Rate Optimization in LoRaWAN: Recent Solutions and Major Challenges. *Sensors* **2020**, *20*, 5044. [CrossRef] [PubMed]
5. Lodhi, M.A.; Wang, L.; Farhad, A. ND-ADR: Nondestructive adaptive data rate for LoRaWAN Internet of Things. *Int. J. Commun. Syst.* **2022**, *35*, e5136. [CrossRef]
6. Ragnoli, M.; Esposito, P.; Stornelli, V.; Barile, G.; Santis, E.D.; Sciarra, N. A LoRa-based Wireless Sensor Network monitoring system for urban areas subjected to landslide. In Proceedings of the 2023 8th International Conference on Cloud Computing and Internet of Things (CCIOT 2023), Okinawa, Japan, 22–24 September 2023; ACM: New York, NY, USA, 2023. 10p.
7. Andrić, I.; Vrsalović, A.; Perković, T.; Aglić Čuvić, M.; Šolić, P. IoT approach towards smart water usage. *J. Clean. Prod.* **2022**, *367*, 133065. [CrossRef]
8. Cho, H.; Kim, S.W. Mobile Robot Localization Using Biased Chirp-Spread-Spectrum Ranging. *IEEE Trans. Ind. Electron.* **2010**, *57*, 2826–2835.
9. Kim, J.; Song, J. A Secure Device-to-Device Link Establishment Scheme for LoRaWAN. *IEEE Sens. J.* **2018**, *18*, 2153–2160. [CrossRef]
10. Sornin, N.; Yegin, A. *LoRaWAN 1.1 Specification Version 1.1. LoRa Alliance*; LoRa Alliance Technical Committee: Beaverton, OR, USA, 2017; pp. 10–12.
11. Augustin, A.; Yi, J.; Clausen, T.; Townsley, W.M. A study of LoRa: Long range & low power networks for the internet of things. *Sensors* **2016**, *16*, 1466. [PubMed]
12. Beltramelli, L.; Mahmood, A.; Österberg, P.; Gidlund, M. LoRa beyond ALOHA: An Investigation of Alternative Random Access Protocols. *IEEE Trans. Ind. Inform.* **2021**, *17*, 3544–3554. [CrossRef]
13. Slabicki, M.; Premsankar, G.; Di Francesco, M. Adaptive configuration of lora networks for dense IoT deployments. In Proceedings of the NOMS 2018–2018 IEEE/IFIP Network Operations and Management Symposium, Taipei, Taiwan, 23–27 April 2018; pp. 1–9.
14. Babaki, J.; Rasti, M.; Aslani, R. Dynamic Spreading Factor and Power Allocation of LoRa Networks for Dense IoT Deployments. In Proceedings of the 2020 IEEE 31st Annual International Symposium on Personal, Indoor and Mobile Radio Communications, London, UK, 31 August–3 September 2020; pp. 1–6.

15. Jiang, C.; Yang, Y.; Chen, X.; Liao, J.; Song, W.; Zhang, X. A New-Dynamic Adaptive Data Rate Algorithm of LoRaWAN in Harsh Environment. *IEEE Internet Things J.* **2022**, *9*, 8989–9001. [CrossRef]
16. Al-Gumaei, Y.A.; Aslam, N.; Aljaidi, M.; Al-Saman, A.; Alsarhan, A.; Ashyap, A.Y. A Novel Approach to Improve the Adaptive-Data-Rate Scheme for IoT LoRaWAN. *Electronics* **2022**, *11*, 3521. [CrossRef]
17. Marini, R.; Cerroni, W.; Buratti, C. A Novel Collision-Aware Adaptive Data Rate Algorithm for LoRaWAN Networks. *IEEE Internet Things J.* **2021**, *8*, 2670–2680. [CrossRef]
18. Jeon, W.S.; Jeong, D.G. Adaptive Uplink Rate Control for Confirmed Class A Transmission in LoRa Networks. *IEEE Internet Things J.* **2020**, *7*, 10361–10374. [CrossRef]
19. Anwar, K.; Rahman, T.; Zeb, A.; Khan, I.; Zareei, M.; Vargas-Rosales, C. RM-ADR: Resource Management Adaptive Data Rate for Mobile Application in LoRaWAN. *Sensors* **2021**, *21*, 7980. [CrossRef] [PubMed]
20. Cuomo, F.; Campo, M.; Caponi, A.; Bianchi, G.; Rossini, G.; Pisani, P. EXPLoRa: Extending the performance of LoRa by suitable spreading factor allocations. In Proceedings of the 2017 IEEE 13th International Conference on Wireless and Mobile Computing, Networking and Communications (WiMob), Rome, Italy, 9–11 October 2017; pp. 1–8.
21. Reynders, B.; Meert, W.; Pollin, S. Power and spreading factor control in low power wide area networks. In Proceedings of the 2017 IEEE International Conference on Communications (ICC), Paris, France, 21–25 May 2017; pp. 1–6.
22. Alliance, L. LoRaWAN™ 1.0.3 Regional Parameters. 2018. Available online: https://lora-alliance.org/wp-content/uploads/2020/11/lorawan-regional-parameters-v1.1ra.pdf (accessed on 24 September 2022).
23. Semtech Corporation. *SX1272/73 Datasheet*; Version 4; Semtech Corporation: Camarillo, CA, USA, 2019.

electronics

MDPI

Article

Novel Hybrid SOR- and AOR-Based Multi-User Detection for Uplink M-MIMO B5G Systems

Yung-Ping Tu [1,*], Pei-Shen Jian [1] and Yung-Fa Huang [2,*]

[1] Department of Electronic Engineering, National Formosa University, Yunlin 632301, Taiwan; 11160120@nfu.edu.tw

[2] Department of Information and Communication Engineering, Chaoyang University of Technology, Taichung 413310, Taiwan

* Correspondence: duhyp@gs.nfu.edu.tw (Y.-P.T.); yfahuang@cyut.edu.tw (Y.-F.H.)

Abstract: The Internet of Things (IoT) is one of the most important wireless sensor network (WSN) applications in 5G systems and requires a large amount of wireless data transmission. Therefore, massive multiple-input multiple-output (M-MIMO) has become a crucial type of technology and trend in the future of beyond fifth-generation (B5G) wireless network communication systems. However, as the number of antennas increases, this also causes a significant increase in complexity at the receiving end. This is a challenge that must be overcome. To reduce the BER, confine the computational complexity, and produce a form of detection suitable for 4G and B5G environments simultaneously, we propose a novel multi-user detection (MUD) scheme for the uplink of M-MIMO orthogonal frequency division multiplexing (OFDM) and universal filtered multi-carrier (UFMC) systems that combines the merits of successive over-relaxation (SOR) and accelerated over-relaxation (AOR) named mixed over-relaxation (MOR). Herein, we divide MOR into the initial and collaboration stages. The former will produce the appropriate initial parameters to improve feasibility and divergence risk. Then, the latter achieves rapid convergence and refinement performance through alternating iterations. The conducted simulations show that our proposed form of detection, compared with the BER performance of traditional SOR and AOR, can achieve 99.999% and 99.998% improvement, respectively, and keep the complexity at $\mathcal{O}(N^2)$. It balances BER performance and complexity with fewer iterations.

Keywords: massive multiple-input multiple-output (M-MIMO); beyond fifth-generation (B5G); successive over-relaxation (SOR); accelerated over-relaxation (AOR); mixed over-relaxation (MOR)

Citation: Tu, Y.-P.; Jian, P.-S.; Huang, Y.-F. Novel Hybrid SOR- and AOR-Based Multi-User Detection for Uplink M-MIMO B5G Systems. *Electronics* **2024**, *13*, 187. https://doi.org/10.3390/electronics13010187

Academic Editors: Eleftherios Anastasiadis and Dionisis Kandris

Received: 6 December 2023
Revised: 28 December 2023
Accepted: 28 December 2023
Published: 31 December 2023

1. Introduction

International mobile telecommunications (IMT) [1–3] have formulated the architecture and goals in the future for fifth-generation (5G) systems [4–6], such as enhanced mobile broadband (eMBB), ultra-reliable and low-latency communications (URLLC), and massive machine-type communications (mMTC). They bring users a better experience and inject new vitality into the fields of the Internet of Things (IoT) [7–9], industrial automation, telemedicine, and driverless driving. Among them, the IoT is one of the critical application technologies for 5G wireless communication [10,11], and wireless transmission services have become an influential means of transmitting IoT messages. To meet various applications of the IoT, wireless communication transmission of large amounts of information to data collection centers and extension to big data analysis is an essential requirement for eMBB and URLLC (i.e., beyond 5G (B5G) technology [12–14] will be forced to bear massive amounts of data while being more time-saving than the previous systems).

Apart from the demand increase in data rate and spectral efficiency in the evolution of wireless communication systems, single-carrier modulation technology has disadvantages such as poor resistance to channel delays, a high bit error rate (BER), and significant bandwidth demand. Therefore, it is insufficient for most current applications. Thus, many researchers have developed multi-carrier modulation technology to cut a bandwidth into many subchannels and use multiple subcarriers to transmit signals and combat the above shortcomings. Orthogonal frequency-division multiplexing (OFDM) is one of the most popular technologies among multiple-carrier modulation schemes [15,16]. Although the spectrum overlaps, its subcarriers are orthogonal. Therefore, each subcarrier will not affect one another. Furthermore, its robustness to channel delay and resistance to inter-symbol interference (ISI) is proven [17]. Unfortunately, OFDM still has some disadvantages that are not conducive to B5G [18–20], such as strict synchronization requirements, high sidelobe losses, and inter-carrier interference (ICI), which need to be improved. To approach the needs of B5G systems simultaneously, it is necessary to find new multi-carrier waveforms to combat the shortcomings of OFDM and support higher data rates, low latency, and looser synchronization techniques. Universal filtered multi-carrier (UFMC) [21,22] is a feasible candidate multi-carrier waveform that combines the advantages of OFDM and filter bank multi-carrier (FBMC) [23,24], is resistant to ICI, and has less out-of-band emissions (OOBMs) to reduce sidelobe losses. In addition, UFMC is compatible with the same architecture as OFDM regarding the channel model [25–27].

Regarding the challenges of B5G wireless communication systems, to further stimulate the advantages of UFMC and provide better spectral efficiency, previous researchers have proposed the massive multiple-input multiple-output (M-MIMO) architecture as an essential type of technology for advanced wireless communication [28,29] which provides better link reliability and higher spectral efficiency. Due to coherent combination [30], the transmit power is inversely proportional to the number of transmit antennas. Thus, as the number of transmit antennas increases, the energy efficiency, signal throughput diversity gain, array gain, capacity gain, and beamforming gain will also be improved and can be obtained efficiently [31,32].

As for the current standard optimum detectors, maximum likelihood (ML) [33] has the best BER performance, but many researchers are distressed and discouraged due to its complexity. The main reason for this is that the complexity grows exponentially with the number of antennas, severely impacting hardware costs, while traditional linear detector methods, such as zero forcing (ZF) [34] and the minimum mean square error (MMSE) [35], have BER performance levels that are only inferior to ML. Regrettably, they still involve the calculation of the inverse matrix, which keeps the complexity high. To deal with the hazards of inverse matrices, many researchers have dedicated themselves to proposed methods based on iterative algorithms to avoid the annoying inverse matrices in mathematical operations, such as the Neumann series (NS) method, Gauss–Seidel (GS) method, Jacobi (JA) method, successive over-relaxation (SOR) method, and accelerated over-relaxation (AOR) method proposed by Liu et al. [36], Wu et al. [37], Kong et al. [38], Gao et al. [39], and Hadjidimos et al. [40], respectively. These detectors avoid the inverse matrix operation by linear iteration and then reduce the complexity from $\mathcal{O}(N^3)$ to $\mathcal{O}(N^2)$. However, their performance has yet to reach the required level of the current day and still needs improvement. Given this, Ning et al. [41] and Hu et al. [42] proposed a symmetric successive over-relaxation (SSOR) method and a symmetric accelerated over-relaxation (SAOR) method based on SOR and AOR, respectively, which utilize two similar symmetric matrices for iteration so that the performance can be better than the previous SOR method and AOR method. Even so, their performance results are unsatisfactory. Therefore, Yu et al. [43] and the authors in our previous work [44], through the two-stage structure proposed SOR method and AOR method combined with the Chebyshev algorithm, namely Chebyshev successive over-relaxation (CSOR) and Chebyshev accelerated over-relaxation (CAOR) techniques, respectively, produced efficient performance.

To reduce the high complexity of the linear detector caused by the increased transmitter antenna of M-MIMO and still maintain its BER performance, this study proposes a more advanced novel detection method combining the merits of the SOR and AOR methods to promote a balance between calculated complexity and BER performance. Simultaneously, we provide one option for disparate needs. Herein, we divide the proposed algorithm into two parts to improve the feasibility and reduce the risk of divergence. The first part, named the initial stage, involves preprocessing the parameters required for the proposed detection, such as the iteration matrix, matched filter (MF) compensation vector, and initial estimation signal. In the second part, called the collaboration stage, the parameters processed in the first part are mixed with the respective characteristics of SOR and AOR through the collaborative architecture to speed up convergence and refine performance. It is worth noting that this study cooperates with the SOR and AOR methods to achieve efficient performance through the joint architecture, which offsets their shortcomings and provides the effect of each compensating the other's performance. Therefore, we named it mixed over-relaxation (MOR). The simulation results show that MOR detection only needs moderate computational complexity and good BER performance and is achievable with uplink multi-user M-MIMO OFDM and UFMC systems simultaneously. These are merits and features that other previous works do not present.

The rest of this paper is organized as follows. Section 2 introduces the system model adopted in this paper. Section 3 reviews some traditional iterative methods and the novel MOR detection method proposed in this study. The simulation results, complexity analysis, and verification of the proposed method are given in Section 4. Section 5 provides a concluding remark to summarize the paper.

2. System Model

This section will illustrate the architecture of the OFDM [17,26,45] and UFMC [21,23,26] systems and then briefly describe the M-MIMO channel model [17,28,46], channel estimation method, and standard MMSE detector [47], which acts as a benchmark for comparing BER performance.

2.1. OFDM Systems

As shown in Figure 1, the input data are first modulated by quadrature amplitude modulation (QAM) and inserted into pilot tones to generate a QAM symbol signal. The signal of the QAM symbol is serial-to-parallel (S/P) conversion and performs an N-point inverse fast Fourier transform (IFFT). Finally, parallel-to-serial (P/S) conversion is used to generate OFDM signals. It can be expressed as follows [26]:

$$x_{OFDM}[n] = \sum_{k=0}^{N-1} X[k]e^{\frac{j2\pi kn}{N}}, 0 \le n \le (N-1), \quad (1)$$

where $X[k]$ is the QAM symbol signal and N is the number of subcarriers.

To better combat inter-symbol interference (ISI), adding a cyclic prefix (CP) to the OFDM signal can effectively avoid the occurrence of ISI, and according to [48], the length of the CP has been adopted and proven to be a quarter of the number of subcarriers. Before the signal is transmitted to the channel, the OFDM signal is converted from baseband to a radio frequency (RF) through a process called upconversion and sent to the receiving end through the channel.

At the receiving terminal, the channel's RF signal output must first be converted into a baseband signal by the downconversion function. Then, the serial signal is altered to its parallel form using the S/P converter. Because a CP is added to the OFDM signal at the transmitter, the receiver must remove its CP component first and then perform an N-point fast Fourier transform (FFT). Herein, the signal includes two parts: the pilot and data parts, which must be separated. The pilot part is provided to the channel estimator to estimate the channel matrix, and the channel matrix mixes the data part into the detector to obtain the complex signal transmitted by the transmitter. Finally, the output data are produced through P/S conversion and QAM demodulation.

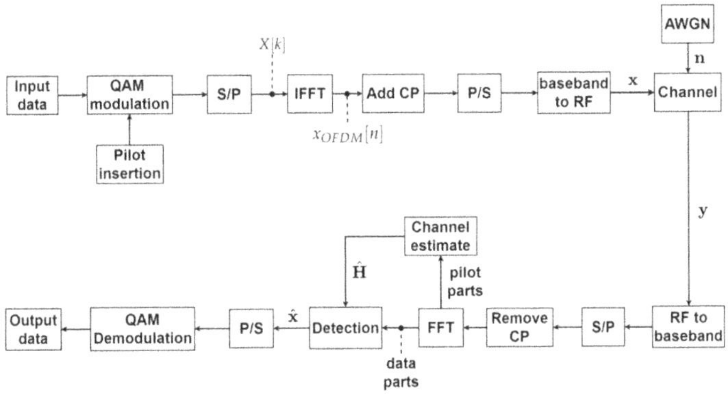

Figure 1. A system block diagram of the OFDM transceiver.

2.2. UFMC Systems

The UFMC system is a B5G candidate waveform extended by the OFDM system architecture to obtain a prompted spectrum efficiency. As shown in Figure 2, the input data are first modulated by QAM and inserted into pilot tones to generate QAM symbol signals. At this time, the QAM symbol signals are converted into S/P form and divided into B sub-bands, each with M QAM sub-symbol signals. To allow vector operations to be performed on the grouped sub-bands, each sub-band needs zero padding to the subcarrier length N. Then, each sub-band is subjected to N-point IFFT and multiplied by a finite impulse response (FIR) filter of a length L individually. Finally, vector addition and P/S conversion of all sub-band signals are performed to obtain the UFMC signal, whose vector length is $(N+L-1)$. The UFMC signal $x_{UFMC}[n]$ can be expressed as follows [26]:

$$x_{UFMC}[n] = \sum_{b=1}^{B}\sum_{l=0}^{L-1}\sum_{m=0}^{N-1} X[b,m]e^{\frac{j2\pi mn}{N}} f_b(l), 0 \le n \le (N+L-1), \tag{2}$$

where $X[b,m]$ is the QAM symbol signal after zero padding, its length is $N \times 1$, and $f_b(l)$ is the FIR filter for each sub-band. In this study, for the UFMC system, $f_b(l)$ is adopted based on the Dolph–Chebyshev filter, which can be written as [49,50]

$$f_b(l) = h_b(l)e^{\frac{j2\pi}{N}\left(\frac{N-N_{ZG}}{2}+\left(b-\frac{1}{2}\right)n+\frac{N}{2}\right)l}, 0 \le l \le (L-1), \tag{3}$$

where N_{ZG} is the zero padding for each sub-band and $h_b(l)$ is the Dolph–Chebyshev prototype FIR filter, whose equation is [49,50]

$$h_b(l) = (-1)^l \frac{\cos\left[N\cos^{-1}\left[\mu\cos\left(\frac{\pi l}{N}\right)\right]\right]}{\cosh\left[N\cosh^{-1}(\mu)\right]}, \tag{4}$$

in which $\mu = \cos\left[\frac{1}{N}\cosh^{-1}(10^{\alpha})\right]$, α is the attenuation parameter, which is a positive real number and determines the relative sidelobe attenuation of the filter. In front, the signal is transmitted to the channel as the OFDM mentioned earlier, and the UFMC signal is upconverted and sent to the receiving end through the channel.

Similar to the aforementioned OFDM receiver, the UFMC channel output signal is downconverted and then S/P converted. At this time, when performing a $2N$-point FFT, it is necessary to first zero fill the received signal to twice the subcarrier length and then downsample it to recover the signal, which is performed to extract the odd-numbered elements after the FFT [25–27]. As in the previous subsection, separating the pilot and data parts is necessary. The pilot part is provided to the signal estimator to estimate the channel matrix, and the channel matrix mixes the data part into the detector to obtain the complex signal transmitted by the transmitter. Finally, the output data are obtained through P/S conversion and QAM demodulation.

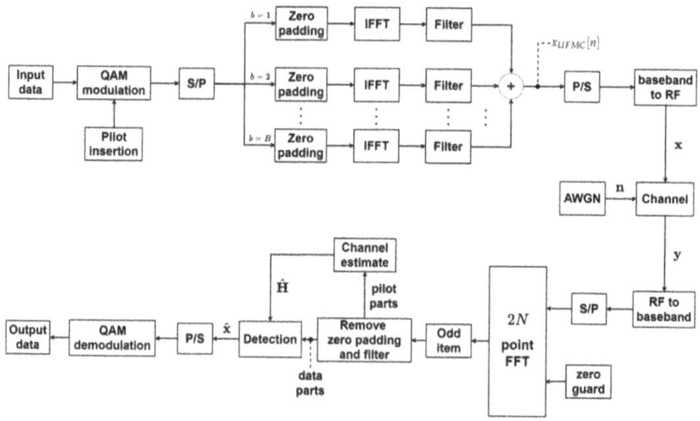

Figure 2. A system block diagram of the UFMC transceiver.

2.3. *Multi-User M-MIMO Channel Model*

For the uplink M-MIMO scenario [28,29] in this article, we assumed that there was a total of KN_t user antennas, denoted as N_T, and mounted N_R antennas at the base station, where K is the number of users and each user has N_t transmission antennas. In terms of setting the number of antennas, to ensure optimal detector performance and minimize thermal noise interference and channel estimation bias, we set the number for N_R to be much larger than the total number of user antennas N_T. Moreover, the signal transmitted into the channel and the received signal can be denoted as $\mathbf{x} = \left[x_1, x_2, \ldots, x_{N_T}\right]^T$ and $\mathbf{y} = \left[y_1, y_2, \ldots, y_{N_R}\right]^T$, respectively. Furthermore, we use bold lowercase letters and bold capital letters to represent vectors and matrices, respectively, to make them easier to read. Therefore, the channel model can be expressed as follows [51,52]:

$$\mathbf{y} = \mathbf{H}\mathbf{x} + \mathbf{n}, \tag{5}$$

where $\mathbf{H} \in \mathbb{C}^{N_R \times N_T}$ and $\mathbf{n}$ is the $N_R \times 1$ noise vector. To be clear, we have rewritten this as

$$\begin{bmatrix} y_1 \\ \vdots \\ y_{N_R} \end{bmatrix} = \begin{bmatrix} h_{11} & \cdots & h_{1N_T} \\ \vdots & \ddots & \vdots \\ h_{N_R1} & \cdots & h_{N_RN_T} \end{bmatrix} \begin{bmatrix} x_1 \\ \vdots \\ x_{N_T} \end{bmatrix} + \begin{bmatrix} n_1 \\ \vdots \\ n_{N_R} \end{bmatrix}. \tag{6}$$

Now, we will use the Rayleigh fading channel matrix [53] to simulate the outdoor environment for the above matrix $\mathbf{H}$. Without loss of generality, considering that the realistic environment has many external factors, we set up a channel with two independent and identically distributed (i.i.d.) paths and a Gaussian distribution that obeyed unit variance and a zero mean to align with the compatible actual environment [44]. The noise vector adopted additive white Gaussian noise (AWGN) that conformed to an i.i.d. and complex Gaussian distribution.

Here, we used a comb-type pilot structure [54] to measure the Rayleigh fading channel at the base station, which periodically inserted pilot tones into the subcarriers. Furthermore, we estimated the channel matrix with the least squares (LS) channel estimation method [55] with the pilot tones, and the result can be expressed as

$$\hat{\mathbf{H}}_{LS} = \left(\mathbf{X}_p^H \mathbf{X}_p\right)^{-1} \mathbf{X}_p^H \mathbf{Y}_p = \mathbf{X}_p^{-1} \mathbf{Y}_p, \tag{7}$$

where $\mathbf{X}_p$ and $\mathbf{Y}_p$ denote the transmitted and received signals' pilot tones, respectively.

As for the detector, it estimates the transmitted signal through the estimated channel matrix and the matched filter (MF). In light of this, according to [35,47,52], the traditional linear MMSE detection algorithm has been proven to be nearly optimal for uplink MIMO systems, and the estimation of its transmitted signal can be expressed as

$$\hat{\mathbf{x}}_{MMSE} = \left(\hat{\mathbf{H}}^H\hat{\mathbf{H}} + \sigma^2\mathbf{I}_{N_T}\right)^{-1}\hat{\mathbf{H}}^H\mathbf{y} = \mathbf{W}^{-1}\mathbf{y}^{MF}, \tag{8}$$

where σ^2 denotes the noise variance, $\hat{\mathbf{H}}$ is the channel estimation matrix, $\mathbf{W}$ is the filter matrix of MMSE, which is equal to $\left(\hat{\mathbf{H}}^H\hat{\mathbf{H}} + \sigma^2\mathbf{I}_{N_T}\right)$, and $\mathbf{y}^{MF}$ is MF's output, which is equivalent to $\hat{\mathbf{H}}^H\mathbf{y}$.

3. Proposed Scheme

To clarify our proposed method, we will briefly describe the conventional SOR [39,56] and AOR [40,44] methods, including their convergence behavior. Immediately afterward, we will introduce our proposed MOR method, which allows improved BER performance and complexity balance with fewer iterations. Moreover, we will briefly discuss and derive convergence in Appendix A.

3.1. Overview of the Conventional SOR Method

According to [56], we consider a linear system whose mathematical equation can be expressed as

$$\mathbf{A}\mathbf{x} = \mathbf{b}, \tag{9}$$

where $\mathbf{A}$ is a symmetric positive definite matrix, $\mathbf{x}$ is an arbitrary complex vector, and $\mathbf{b}$ the MF output of the received signal $\mathbf{y}$ after channel estimation. We can denote $\mathbf{b}$ as $\mathbf{b} = \hat{\mathbf{H}}^H\mathbf{y} = \mathbf{y}^{MF}$, and it is a nonzero complex vector (i.e., $\mathbf{b} \in \mathbb{C}^n \setminus \{0\}$). Moreover, $\mathbf{A}$ can be decomposed into

$$\mathbf{A} = \mathbf{D} - \mathbf{L} - \mathbf{U}, \tag{10}$$

in which $\mathbf{D}$, $-\mathbf{L}$, and $-\mathbf{U}$ are $\mathbf{A}$'s diagonal, lower, and upper triangular matrices, respectively. Then, the SOR iteration equation can be expressed as

$$\mathbf{x}^{(i+1)} = (\mathbf{D} - \omega\mathbf{L})^{-1}\{[(1-\omega)\mathbf{D} + \omega\mathbf{U}]\mathbf{x}^{(i)} + \omega\mathbf{b}\}, \tag{11}$$

where ω is the relaxation parameter. Herein, we replace $(\mathbf{D} - \omega\mathbf{L})^{-1}$ and $[(1-\omega)\mathbf{D} + \omega\mathbf{U}]$ with $\mathbf{M}_{SOR}$ and $\mathbf{N}_{SOR}$, respectively, and further define the matrix $\mathbf{G}_{SOR}$, called the iteration matrix of SOR, which is expressed as

$$\mathbf{G}_{SOR} = \mathbf{M}_{SOR}\mathbf{N}_{SOR}. \tag{12}$$

Then, Equation (11) can be simplified and written as

$$\mathbf{x}^{(i+1)} = \mathbf{G}_{SOR}\mathbf{x}^{(i)} + \mathbf{d}_{SOR}, \tag{13}$$

where $\mathbf{d}_{SOR}$, the MF compensation vector of SOR, comes from $\mathbf{M}_{SOR}$ multiplied by $\mathbf{d}$, where $\mathbf{d}$ is $\omega\mathbf{b}$.

The linear iterative algorithm judges its convergence using the spectral radius $\rho(\mathbf{G})$ of matrix $\mathbf{G}$ as the criterion, which is defined as follows [44,57]:

$$\rho(\mathbf{G}) \triangleq \max_{\lambda \in \rho(\mathbf{G})} |\lambda|, \tag{14}$$

where λ is the eigenvalue of $\mathbf{G}$. Moreover, Equation (11) will converge if $\rho(\mathbf{G}_{SOR})$ satisfies

$$\rho(\mathbf{G}_{SOR}) = \max_{1<c<N_T} |\lambda_c| < 1. \tag{15}$$

According to [56], the SOR iterative algorithm has been proven to converge when it satisfies $0 < \omega < 2$.

3.2. Overview of the Conventional AOR Method

The conventional AOR iterative algorithm [40] is an extended version of the SOR iterative algorithm that, through a combination of relaxation parameters ω and acceleration parameters γ, obtains better performance, and the equation is as follows:

$$\mathbf{x}^{(i+1)} = (\mathbf{D} - \gamma\mathbf{L})^{-1}\{[(1-\omega)\mathbf{D} + (\omega - \gamma)\mathbf{L} + \omega\mathbf{U}]\mathbf{x}^{(i)} + \omega\mathbf{b}\}, \tag{16}$$

Similar to the SOR method, to simplify Equation (16), we define the matrix $\mathbf{G}_{AOR}$, called the iteration matrix of AOR, as

$$\mathbf{G}_{AOR} = \mathbf{M}_{AOR}\mathbf{N}_{AOR}, \tag{17}$$

where $\mathbf{M}_{AOR} = (\mathbf{D} - \gamma\mathbf{L})^{-1}$ and $\mathbf{N}_{AOR} = [(1-\omega)\mathbf{D} + (\omega - \gamma)\mathbf{L} + \omega\mathbf{U}]$. Then, Equation (16) can be written as

$$\mathbf{x}^{(i+1)} = \mathbf{G}_{AOR}\mathbf{x}^{(i)} + \mathbf{d}_{AOR}, \tag{18}$$

where $\mathbf{d}_{AOR}$ is the MF compensation vector of AOR, equal to $\mathbf{M}_{AOR}$ multiplied by $\mathbf{d}$. From Equation (14), we see that Equation (16) will converge if it satisfies

$$\rho(\mathbf{G}_{AOR}) = \max_{1<c<N_T} |\lambda_c| < 1. \tag{19}$$

According to [44], when satisfying $0 < \omega < 2$, $0 < \gamma < 2$, and $\omega = \gamma$, the AOR iterative algorithm has been proven to converge.

3.3. Proposed MOR Method

After reviewing the previous subsections, we developed a novel detection combining the advantages of conventional SOR and AOR to improve BER performance and balance the complexity, which we call mixed over-relaxation (MOR). As shown in Figure 3, the first part is the initial stage, and some parameters required by the iterative algorithm must be processed. In the first step, we define the iteration matrix of MOR according to the iteration matrices $\mathbf{G}_{SOR}$ and $\mathbf{G}_{AOR}$ mentioned earlier in Sections 3.1 and 3.2, respectively, which are written as follows:

$$\mathbf{G}_{MOR} = \mathbf{G}_{AOR}\mathbf{G}_{SOR}. \tag{20}$$

Aside from that, we performed a mixed operation on the MF compensation vectors in the SOR and AOR iterative equations for better BER performance. Here, the MOR MF compensation signal $\mathbf{d}_{MOR}$ can be described as follows:

$$\mathbf{d}_{MOR} = (\mathbf{G}_{AOR}\mathbf{M}_{SOR} + \mathbf{M}_{AOR})\mathbf{d}. \tag{21}$$

Next, to obtain appropriate initial ω and γ values, we must first calculate the spectral radius of $\mathbf{G}_{MOR}$. The definition of the spectral radius $\rho(\mathbf{G}_{MOR})$ is the same as in Equation (14) in the previous subsection, and it can be written as

$$\rho(\mathbf{G}_{MOR}) \triangleq \max_{\lambda \in \rho(\mathbf{G}_{MOR})} |\lambda|, \tag{22}$$

Therefore, we can find the mathematical equations [44] for ω and γ:

$$\omega = \frac{1}{\sqrt{1-\mu^2}}, \tag{23}$$

$$\gamma = \frac{2}{1+\sqrt{1-\mu^2}}, \tag{24}$$

where μ is $\rho(\mathbf{G}_{MOR})|_{\omega=1,\gamma=0}$.

The second part we call the collaboration phase. As shown in the MOR algorithm block (collaboration stage) proposed in Figure 3, we joined the relaxation characteristics of the SOR iteration algorithm and the acceleration ability of the AOR iteration algorithm. Through the collaborative architecture, SOR and AOR assist each other in estimating the better signal and apply the appropriate initialization ω and γ to obtain the MOR iteration matrix $\mathbf{G}_{MOR}$ and MF compensation vector $\mathbf{d}_{MOR}$. The experimental results prove that the proposed MOR method has a faster convergence speed and better BER performance. Moreover, its iteration equation can be simplified as follows:

$$\mathbf{x}^{(i+1)} = \mathbf{G}_{MOR}\mathbf{x}^{(i)} + \mathbf{d}_{MOR}. \tag{25}$$

It is worth noting that the parameters generated in the initial stage in Figure 3 only need to be calculated once, which include the iterative matrix $\mathbf{G}_{MOR}$, MF compensation vector $\mathbf{d}_{MOR}$, and initial estimate signal $\mathbf{x}^{(0)}$, which are provided to the collaboration stage for iterative calculation. The procedure of the proposed MOR detection method is shown in Algorithm 1. As for its convergence of MOR, we will derive this in detail in Appendix A. Therefore, we know that MOR will converge when $0 < \omega < 2$ and $0 < \gamma < 2$. Aside from that, we can compare the convergence conditions of SOR and AOR to MOR and find that MOR does not increase the convergence difficulty and also does not limit $\omega = \gamma$ as AOR does, which means that MOR has higher flexibility in choosing ω and γ.

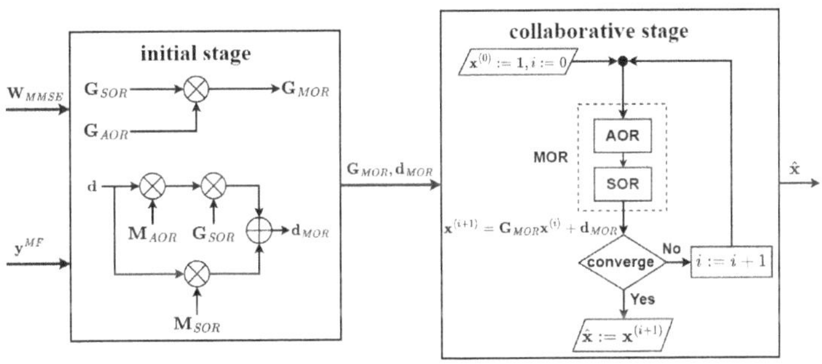

Figure 3. A block diagram of the proposed detection scheme.

Algorithm 1 Proposed MOR detection algorithm.

Receiver signal input:

1. $\mathbf{W}_{MMSE} = \hat{\mathbf{H}}^H\hat{\mathbf{H}} + \sigma^2\mathbf{I}_{N_T} \triangleq \mathbf{A}$, and $\mathbf{A} = \mathbf{D} + \mathbf{L} + \mathbf{U}$
2. $\mathbf{y}^{MF} = \hat{\mathbf{H}}^H\mathbf{y} \triangleq \mathbf{b}$

The first part: (initial stage)

1. $\mathbf{M}_{SOR} = (\mathbf{D} - \omega\mathbf{L})^{-1}, \mathbf{N}_{SOR} = [(1-\omega)\mathbf{D} + \omega\mathbf{U}]$
2. $\mathbf{M}_{AOR} = (\mathbf{D} - \gamma\mathbf{L})^{-1}, \mathbf{N}_{AOR} = [(1-\omega)\mathbf{D} + (\omega-\gamma)\mathbf{L} + \omega\mathbf{U}]$
3. $\mathbf{G}_{SOR} = \mathbf{M}_{SOR}\mathbf{N}_{SOR}, \mathbf{G}_{AOR} = \mathbf{M}_{AOR}\mathbf{N}_{AOR}$
4. $\mathbf{G}_{MOR} = \mathbf{G}_{AOR}\mathbf{G}_{SOR}$
5. $\mathbf{d} = \omega\mathbf{b}, \mathbf{d}_{MOR} = (\mathbf{G}_{AOR}\mathbf{M}_{SOR} + \mathbf{M}_{AOR})\mathbf{d}$
6. $\omega = \dfrac{1}{\sqrt{1-\mu^2}}, \gamma = \dfrac{2}{1+\sqrt{1-\mu^2}}, \mu = \rho(\mathbf{G}_{MOR})|_{\omega=1,\gamma=0}$
7. **Set** $i := 0$ and $\mathbf{x}^{(0)} := \mathbf{1}$

The second part: (collaborative stage)
While not converging, **do**

1. $\mathbf{x}^{(i+1)} = \mathbf{G}_{MOR}\mathbf{x}^{(i)} + \mathbf{d}_{MOR}$
2. $i := i + 1$

End
Set $\hat{\mathbf{x}} := \mathbf{x}^{(i+1)}$
Receiver signal output: The estimate of the transmitted signal vector $\hat{\mathbf{x}}$

---

4. Simulation Results and Complexity Analysis

4.1. Simulation Results and Discussion

In this section, some numerical simulations will be performed to evaluate and verify the performance of our proposed novel receiver. Moreover, we use the famous Matlab(Version R2022a) mathematical software tool to simulate the numerical results and graphics, execute Monte Carlo 500,000 for each graph, and use Microsoft Excel for calculations and statistical tables. Considering the OFDM multi-carrier technology, UFMC multi-carrier technology, and $N_R \times N_T$ uplink multi-user M-MIMO environment described in Section 2, as shown in Table 1, we fed the same total data volume of 512 to experiment with the two systems equitably, and the pilot tone insertion interval was 0.05. Therefore, each symbol inserted 25 pilot tones and 1024 QAM modulation. For the particular parameters of OFDM in the 4G environment and UFMC in the B5G environment, the former needed to set a cyclic prefix (CP) whose length was one quarter of the subcarrier (i.e., 128), and the volume of the data was equal to the number of subcarriers. In the latter, the size of subcarrier N was 1024, the number of sub-bands B was 16, and each sub-band was allocated a data volume of 32 (i.e., M). It is worth noting that the product of the two could not be greater than the number of subcarriers N (i.e., the total data amount needed to be less than or equal to the number of subcarriers), in which the number of zero padding was the subcarrier minus the data amount and would be divided into the starting and tailing of the data vector (i.e., 256). According to [58–60], the UFMC waveform adopted the Chebyshev FIR filter, where the filter length L was 43 and the side attenuation was 40. In addition, we assumed that the channel was a two-multipath flat Rayleigh fading channel with AWGN, and the receiver could obtain channel state information (CSI) through the least squares (LS) estimation scheme. We chose SOR [39], AOR [40], SSOR [41], SAOR [42], CSOR [43], and CAOR [44] as the BER performance and complexity comparison objects of the proposed MOR scheme and used the MMSE as the BER performance benchmark. Moreover, we briefly describe the features of previously published works and our proposed MOR scheme in Table 2.

Table 1. Enumerate parameters used in simulation scenarios.

Parameter	Value
Common parameters	
Modulation scheme	1024 QAM
Volume of data	512
Amount of pilot data in one symbol	25
The maximum SNR (dB)	50
Channel type	Rayleigh fading channel
Number of users K	8
Number of transmission antennas in one user N_t	2
Number of channel taps	2
Noise	AWGN
Channel estimation	LS
The number of experiments for Monte Carlo (times)	500,000
OFDM specific parameters	
CP length	128
UFMC specific parameters	
Number of subcarriers N	1024
Number of sub-bands B	16
Number of subcarriers in each sub-band M	32
Amount of zero padding in each sub-band N_{ZG}	256
Filter type	Chebyshev FIR filter
Filter length L	43
Filter sidelobe attenuation (dB)	40

Table 2. Brief descriptions of our proposed scheme and previously published works.

Scheme	Brief Description
SOR [39]	SOR is a linear iterative method, and it was derived from adding relaxation parameters ω to the Gauss–Seidel iterative algorithm.
AOR [40]	The AOR iterative algorithm is an extension of the SOR iterative algorithm. It is a linear iterative method derived through the relaxation parameter ω and the newly added acceleration parameter γ.
SSOR [41]	SSOR combines two SOR sweeps in a semi-iterative architecture to produce an iterative matrix similar to a symmetric matrix.
SAOR [42]	SAOR combines two AOR sweeps in a semi-iterative architecture to produce an iterative matrix similar to a symmetric matrix.
CSOR [43]	The CSOR method combines the SOR iterative algorithm and the recursive characteristics of the Chebyshev polynomials.
CAOR [44]	The CAOR method combines the AOR iterative algorithm and the recursive characteristics of the Chebyshev polynomials.
MOR	Our proposed MOR method joins the characteristics and abilities of both the SOR and AOR iterative algorithms to accelerate iterative convergence and obtain efficient BER performance through a collaborative architecture.

To verify and roughly observe the characteristics between the OFDM and UFMC waveforms regarding BER performance and the power spectral density (PSD), we conducted some experiments to demonstrate their disparity, as shown in Figures 4 and 5. In Figure 4, we can see that the UFMC effectively improved the BER performance when $N_R \times N_T = 64 \times 16$ and utilized the MMSE detector. Especially when the SNR level was 35 dB, the UFMC and OFDM BER performance values were 2.061×10^{-5} and 9.886×10^{-4}, respectively, which could be improved by approximately 97.916%. As for the PSD, which is an agreeable index of the impact of OOBMs, as shown in Figure 5, we can observe that under the same environment and data volume, the UFMC had better OOBM resistance than OFDM, which means it resisted intercarrier interference (ICI) better [22,61]. By combining the BER performance and PSD simulation results, we know that the UFMC had better BER performance and could achieve better OOBM resistance, which is expected in future B5G wireless communication systems.

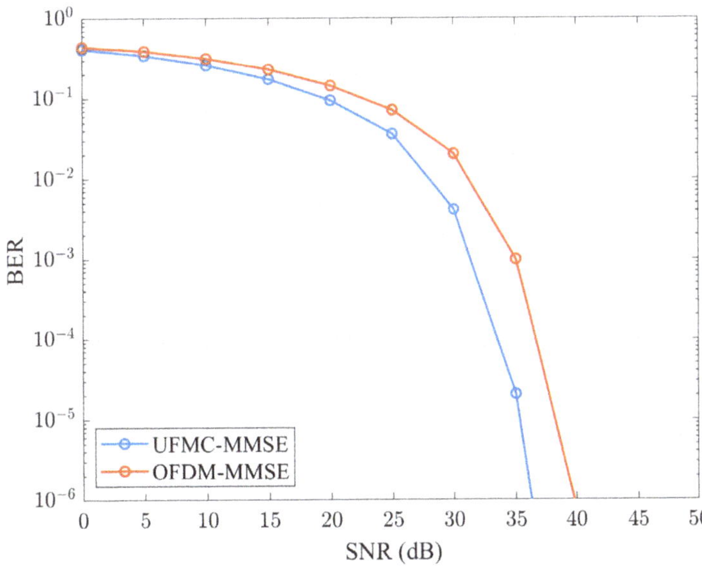

Figure 4. An MMSE BER performance comparison for OFDM vs. UFMC with $N_R \times N_T = 64 \times 16$ and $K = 8$.

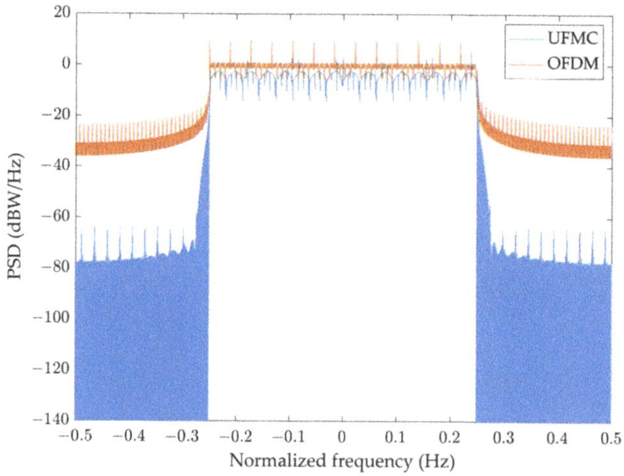

Figure 5. A PSD comparison for OFDM vs. UFMC.

Next, we will explore MOR detectors that can be applied to 4G and B5G multi-carrier technologies. To obtain the appropriate relaxation parameter ω and acceleration parameter γ, we first used Equations (23) and (24) to obtian the preliminary relaxation parameter ω and acceleration parameter γ. Apart from this, as shown in Figures 6 and 7, the acceleration parameter γ and relaxation parameter ω are depicted for different relaxation parameter ω values and acceleration parameter γ values for our proposed method in the OFDM and UFMC, respectively, comparing the BER performance graphs when the iteration number i was 4, $N_R \times N_T$ was 64×16, and the SNR was at 35 dB. We can observe in Figure 6a,b that if γ was the curve of 1.1, the BER performance would improve, whereas the BER performance would decrease, and the best BER performance was when ω was equal to 1.2. Similarly, we also observed the same values of ω and γ in Figures 6 and 7. In light of this, choosing a γ value close to 1.1 and ω value close to 1.2 would have the best estimate. Meanwhile, the experimental simulation data in Figures 6 and 7 were consistent with the theoretical calculation values of Equations (23) and (24). It is worth noting that regardless of whether the MOR operated in the OFDM or UFMC, the optimal values of ω and γ were the same, which is ideal. This means that the MOR-optimized BER performance could still be obtained without recalculation if applied to 4G or B5G systems, substituting ω with 1.1 and γ with 1.2.

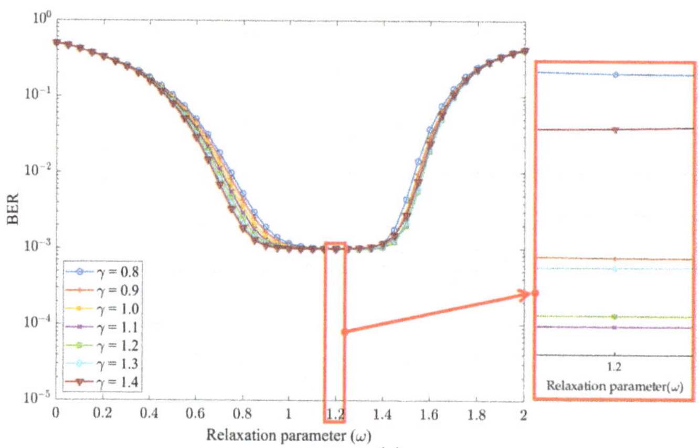

(**a**)

Figure 6. *Cont.*

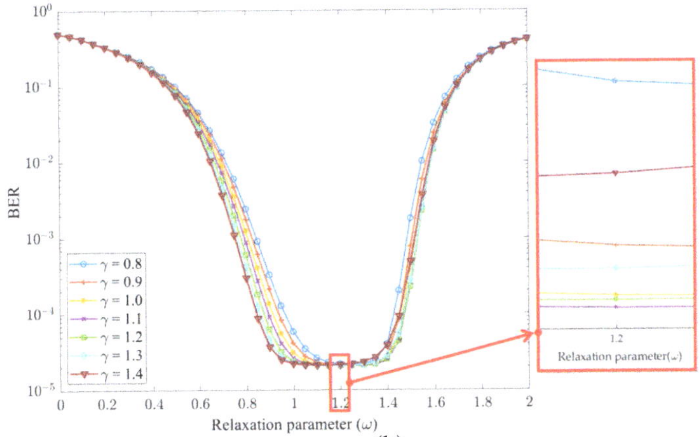

(**b**)

Figure 6. BER performance of MOR method relative to ω with SNR = 35 dB, $N_R \times N_T = 64 \times 16$, $K = 8$, and number of iterations i of 4 for (**a**) OFDM and (**b**) UFMC.

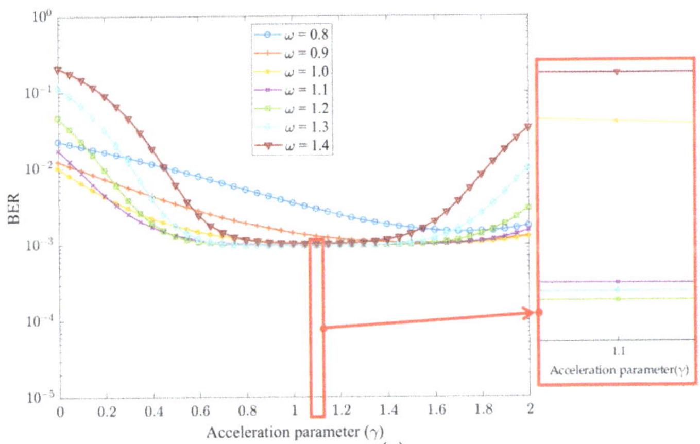

(**a**)

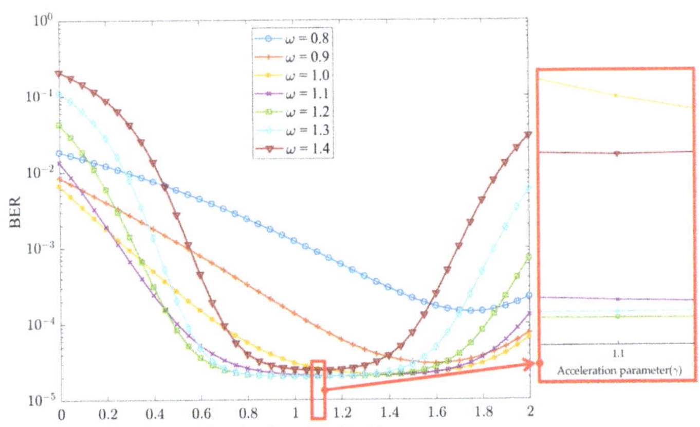

(**b**)

Figure 7. BER performance of MOR method relative to γ with SNR = 35 dB, $N_R \times N_T = 64 \times 16$, $K = 8$, and number of iterations i of 4 for (**a**) OFDM and (**b**) UFMC.

Figures 8 and 9 depict the BER performance comparison of different detection methods when the iteration number i was 3 and 4, respectively. Among them, Figures 8a and 9a apply to the OFDM system, and Figures 8b and 9b apply to the UFMC system. In Figure 8a,b, we can observe that when the iteration number i was equal to 3, our proposed method was already close to the performance of the MMSE. Moreover, when the SNR was at 40 dB, we compared the BER performance of MOR and related it to that of the CAOR, and for the OFDM system, it was approximately 1.395×10^{-6} and 8.563×10^{-3}, respectively, an improvement of 99.984%. The UFMC system's results were roughly 3.878×10^{-8} and 7.773×10^{-3}, respectively, an improvement of 99.999%. As shown in the numerical values, the BER performance of our new method was significantly improved compared with CAOR, let alone other iterative detection methods. In addition, as shown in Figure 9a,b, when the iteration number i was equal to 4, the BER performance of our method overlapped with the MMSE method.

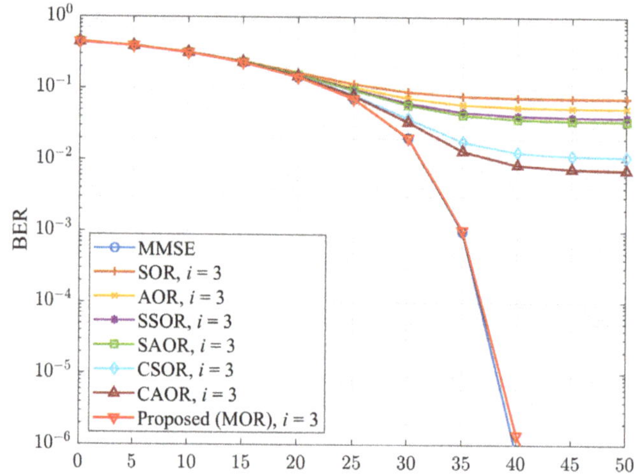

(**a**)

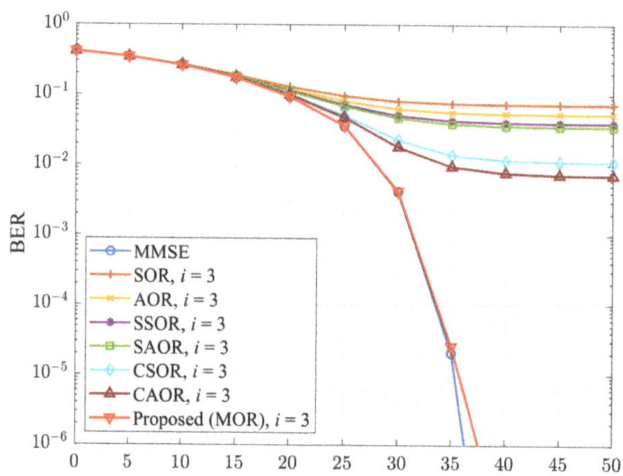

(**b**)

Figure 8. BER performance comparison for different detection methods with $N_R \times N_T = 64 \times 16$, $K = 8$, and number of iterations i of 3, for (**a**) OFDM and (**b**) UFMC.

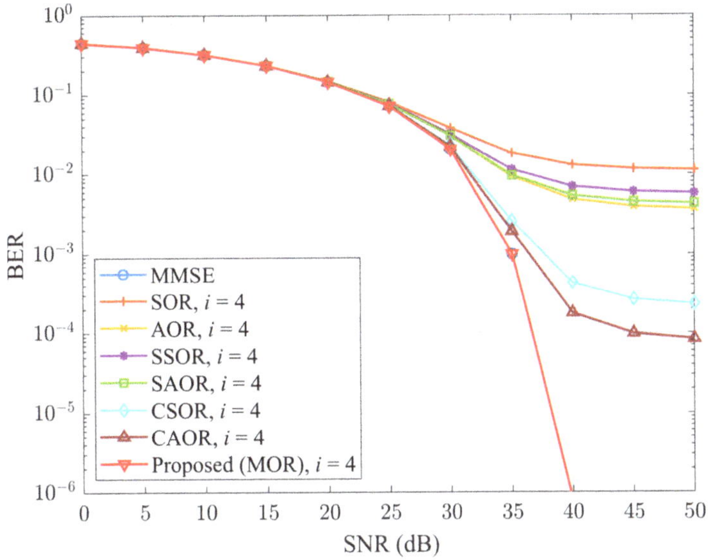

(**a**)

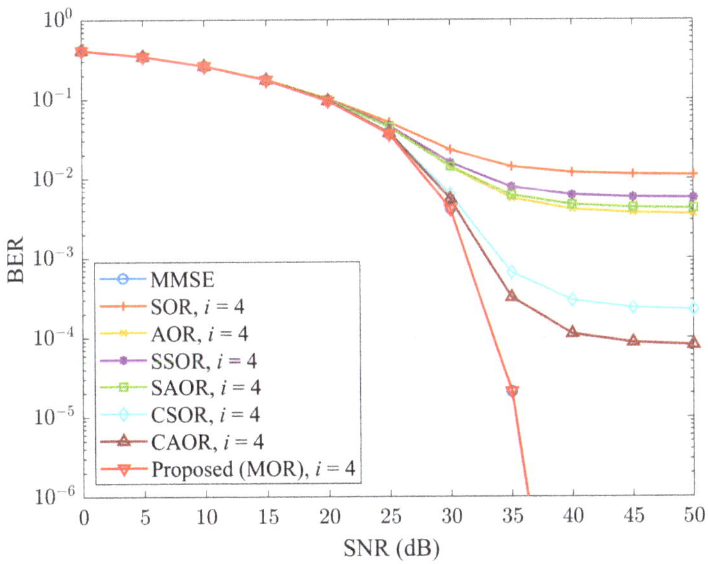

(**b**)

Figure 9. BER performance comparison for different detection methods with $N_R \times N_T = 64 \times 16$, $K = 8$, and number of iterations i of 4 for (**a**) OFDM and (**b**) UFMC.

Following this, we will observe when the number of base station antennas N_R increases, as shown in Figures 10 and 11, which depict the BER performance comparison of different detection methods run in OFDM and UFMC systems when the number of base station antennas N_R was 128 and 256, respectively, while the fixed iteration number i was 2. In Figure 10a, we can observe that when the SNR was at 35 dB and N_R was 128, the MOR and CAOR BER performance of the OFDM system were approximately 1.838×10^{-5} and 6.030×10^{-3}, respectively, an improvement of 99.695%; In Figure 10b, the MOR and CAOR BER performance of the UFMC system were approximately 6.208×10^{-7} and 4.955×10^{-3}, respectively, an improvement of 99.987%. In Figure 11, when N_R kept increasing to 256, whether the OFDM or UFMC system was used, MOR only needed two time iterations, and the BER performance already overlapped with the MMSE.

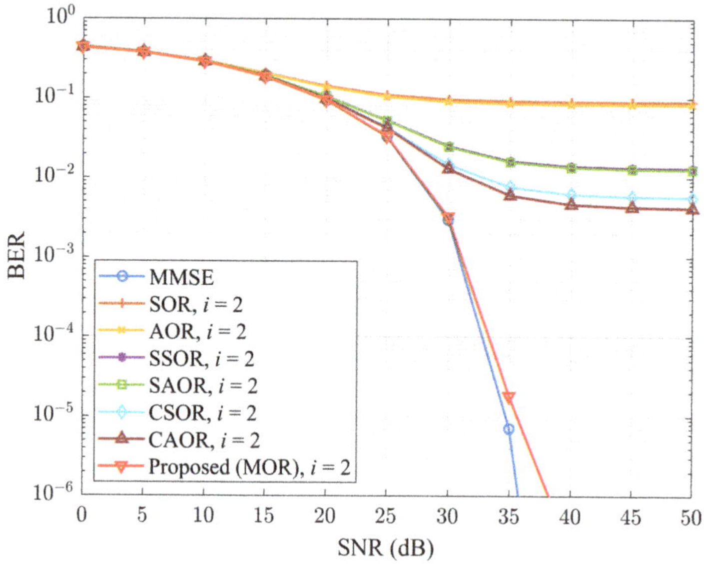

(**a**)

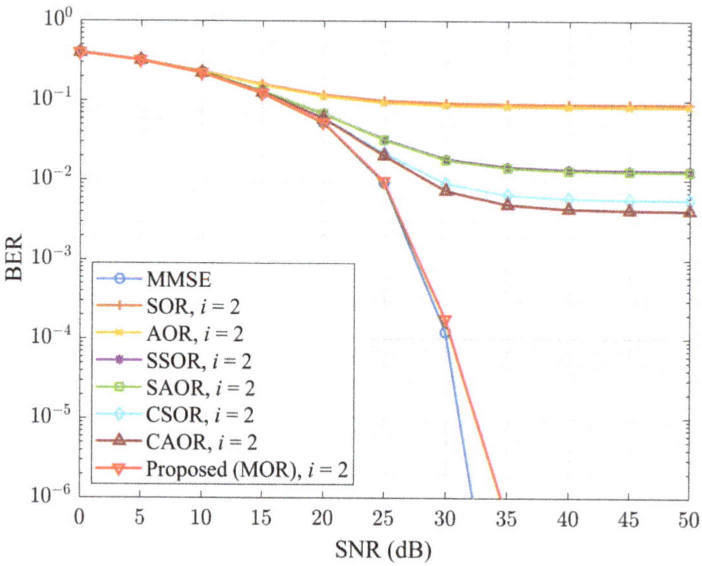

(**b**)

Figure 10. BER performance comparison for different detection methods with $N_R \times N_T = 128 \times 16$, $K = 8$, and number of iterations i of 2 for (**a**) OFDM and (**b**) UFMC.

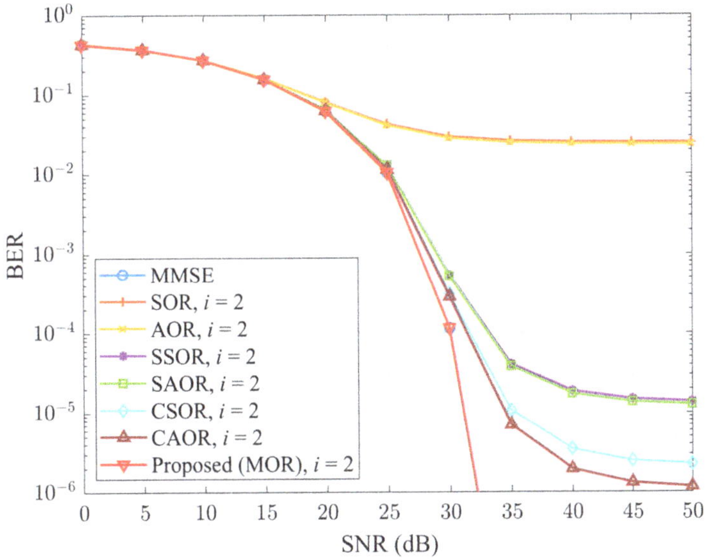

(**a**)

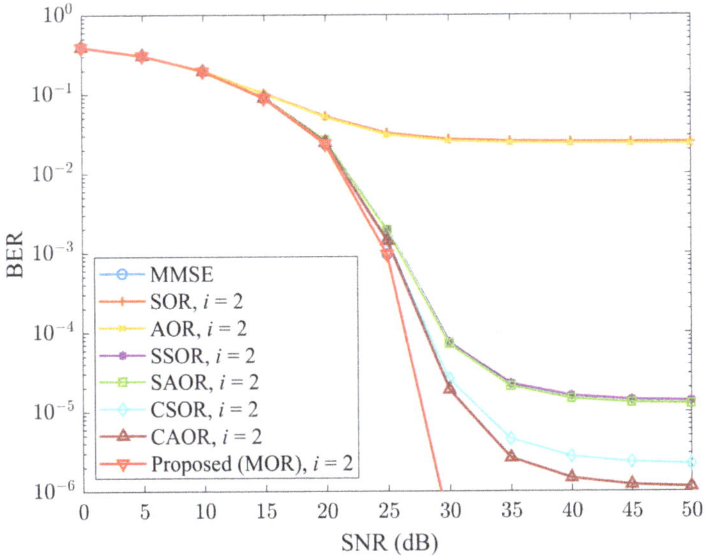

(**b**)

Figure 11. BER performance comparison for different detection methods with $N_R \times N_T = 256 \times 16$, $K = 8$, and number of iterations i of 2 for (**a**) OFDM and (**b**) UFMC.

To analyze the BER performance of different detector methods varies with the ratio of N_R to the total number of user antennas N_T, we denoted this as β, which is called the antenna ratio [52]. Figures 12 and 13 show the variations in BER performance of different detector methods for OFDM and UFMC systems when the iteration number i was 2 and 3, respectively. From Figures 12 and 13, we took some samples to look at the BER improvement, as shown in Tables 3 and 4, respectively. For example, when β was equal to 16, the MOR and CAOR of the OFDM system were approximately 1.174×10^{-4} and 2.862×10^{-4}, respectively, an improvement of 58.979%.

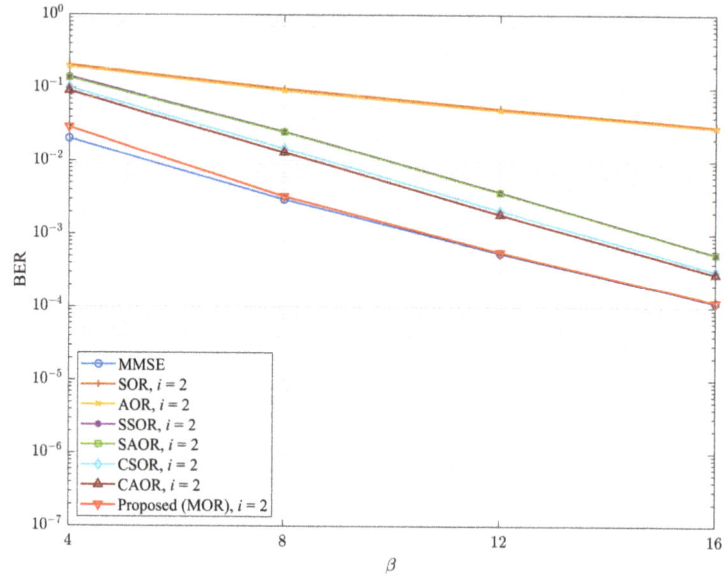

(a)

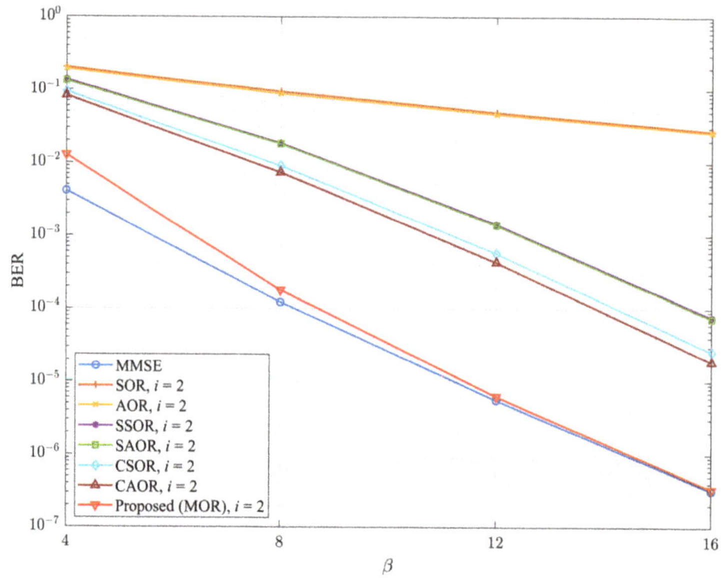

(b)

Figure 12. (a) OFDM and (b) UFMC, BER vs. β for different detection schemes when the number of iterations i was 2 and the SNR was 30 dB.

Table 3. Comparison of BER improvement of MOR vs. previous works at $\beta = 16$ for the OFDM system.

Number of Iterations	CAOR [44]	CSOR [43]	SAOR [42]	SSOR [41]	AOR [40]	SOR [39]
$i = 2$	58.979%	62.589%	77.826%	78.116%	99.589%	99.608%
$i = 3$	2.026%	2.310%	4.290%	4.315%	51.695%	57.018%

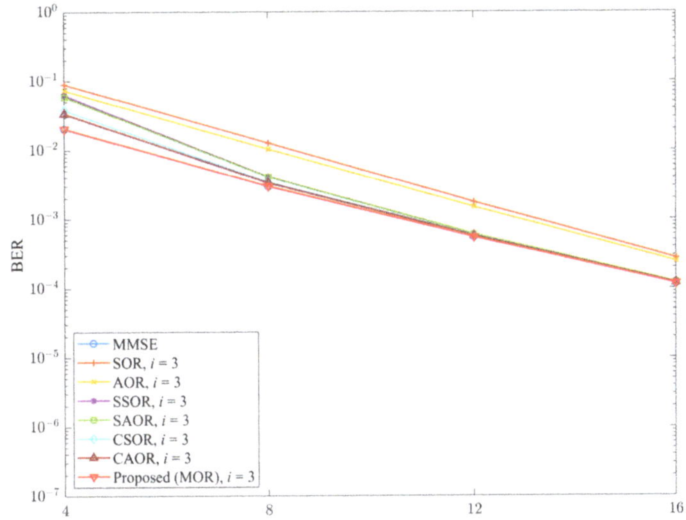

(**a**)

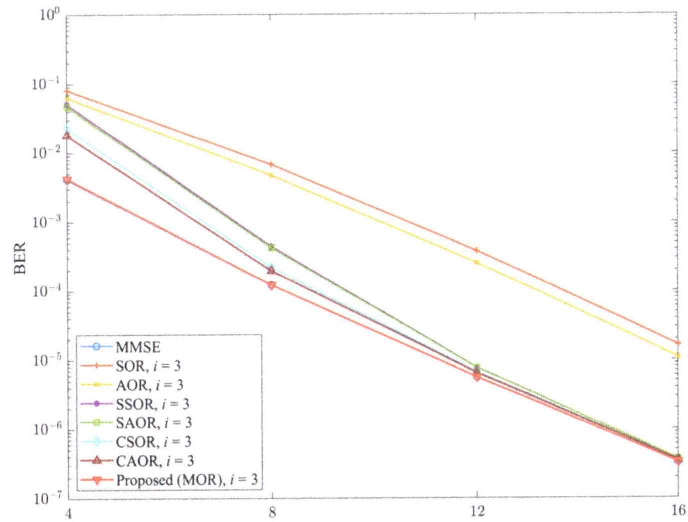

(**b**)

Figure 13. (**a**) OFDM and (**b**) UFMC, BER vs. β for different detection schemes when the number of iterations i was 3 and the SNR was 30 dB.

Table 4. Comparison of BER improvement of MOR vs. previous works at $\beta = 16$ for the UFMC system.

Number of Iterations	CAOR [44]	CSOR [43]	SAOR [42]	SSOR [41]	AOR [40]	SOR [39]
$i = 2$	98.214%	98.690%	99.542%	99.563%	99.998%	99.999%
$i = 3$	5.717%	6.395%	9.784%	9.958%	96.957%	97.998%

As shown by the numerical values, we could find that the gap in BER performance of all iterative methods had a decreasing trend. Undoubtedly, as the β values grew, BER performance was improved with more antennas due to the spatial diversity gain [62,63]. Above all, our proposed method achieved the best BER performance compared with the abovementioned detection methods regardless of the value of β. Simultaneously, as the number of antennas kept increasing, the required iterations were also relatively reduced.

To further illustrate the impact of the numerical antenna ratio β on BER performance, in Tables 5 and 6, we compare the degree to which the BER performance of each iterative method was close to the MMSE when the number of iterations i was 2 and 3 in the form of a logarithm value (i.e., to obtain a more demarcated numerical comparison, we took the logarithm operation $\log(\cdot)$ and denoted it as a separation rate between the MMSE) of the BER distance between each detection method and the traditional MMSE, and the data came from the simulation data in Figures 12 and 13. We can observe that in Tables 5 and 6, under different β conditions, the distance between MOR and the traditional MMSE was the smallest. For instance, in the OFDM and UFMC, when β was 16 and i was 2, the BER separation rates between MOR and the MMSE were 0.0061 and 0.0215, respectively. Aside from that, when β was 8 and i was 3, the BER separation rate of MOR and the MMSE was 0.0001 and 0.0005, respectively, which means the proposed method was already extremely close to the MMSE. Furthermore, when β was 16 and i was 3, the BER separation rate between MOR and the MMSE in the OFDM and UFMC was 0 and -0.0007, respectively. As shown in the numerical value, the distance in the UFMC system was already negative, which means that the BER performance distance ratio between MOR and the MMSE was less than one. Hence, the value became a negative value after the logarithm operation $\log(\cdot)$. In light of this, the separation rate value was smaller, and the BER performance of the detector was closer to the MMSE.

Table 5. Comparison of BER performance separation rates between all detectors and MMSE in different β under an iteration number i of 2 and SNR at 30 dB for (a) OFDM and (b) UFMC.

Scheme	$\beta = 4$	$\beta = 8$	$\beta = 12$	$\beta = 16$
		(a)		
SOR [39]	1.0008	1.5153	1.9799	2.4129
AOR [40]	0.9777	1.4873	1.9552	2.3921
SSOR [41]	0.8439	0.9307	0.8355	0.6660
SAOR [42]	0.8276	0.9239	0.8284	0.6603
CSOR [43]	0.6947	0.7013	0.5836	0.4330
CAOR [44]	0.6482	0.6449	0.5303	0.3930
Proposed MOR	0.1522	0.0399	0.0138	0.0061
		(b)		
SOR [39]	1.6925	2.8858	3.9460	4.9298
AOR [40]	1.6674	2.8555	3.9195	4.9080
SSOR [41]	1.5252	2.1803	2.4161	2.3807
SAOR [42]	1.5076	2.1693	2.3989	2.3610
CSOR [43]	1.3578	1.8727	2.0138	1.9043
CAOR [44]	1.3017	1.7794	1.8911	1.7695
Proposed MOR	0.4893	0.1636	0.0528	0.0215

To understand the convergence of different detectors under different numbers of base station antennas N_R, Figures 14–16 show the relationship between the iteration number i and BER performance in OFDM and UFMC systems when N_R was 64, 128, and 192, respectively. While N_R was 64 and the SNR was 37 dB, as shown in Figure 14, MOR almost converged at four iterations, which can also be verified by Figure 9. Moreover, at the same iteration count of four, when comparing MOR to CAOR, the BER performance in the OFDM and UFMC systems improved by 80.540% and 99.705%, respectively. Similarly, as shown in Figure 15, when N_R increased to 128, the SNR level was 33 dB, and MOR nearly converged while only needing three iterations. In other words, at the same iteration count of three, when comparing MOR to CAOR, the BER performance in the OFDM and UFMC systems improved by 35.220% and 88.542%, respectively. As for N_R increasing to 192 and the SNR being at 31 dB, as shown in Figure 16, we found that the MOR method could approach convergence in only two iterations for either the OFDM or UFMC systems. Compared with CAOR, their BER performance increased by 84.196% and 99.776%, respectively. Echoing the previous antenna ratio β analysis, M-MIMO would enhance the BER performance when increasing the number of base station antennas N_R. Moreover, our proposed MOR algorithm had the best BER performance and the fastest convergence speed among the abovementioned detectors.

Table 6. Comparison of BER performance separation rates between all detectors and MMSE in different β under an iteration number i of 3 and SNR at 30 dB for (a) OFDM and (b) UFMC.

Scheme	$\beta = 4$	$\beta = 8$	$\beta = 12$	$\beta = 16$
		(a)		
SOR [39]	0.6406	0.6283	0.5046	0.3667
AOR [40]	0.5523	0.5364	0.4307	0.3160
SSOR [41]	0.4857	0.1397	0.0415	0.0192
SAOR [42]	0.4594	0.1354	0.0410	0.0191
CSOR [43]	0.2707	0.0643	0.0220	0.0102
CAOR [44]	0.2197	0.0514	0.0186	0.0089
Proposed MOR	0.0026	0.0001	0	0
		(b)		
SOR [39]	1.2935	1.7443	1.8282	1.6979
AOR [40]	1.1828	1.5852	1.6505	1.5160
SSOR [41]	1.0907	0.5603	0.1396	0.0448
SAOR [42]	1.0539	0.5384	0.1365	0.0440
CSOR [43]	0.7417	0.2591	0.0766	0.0280
CAOR [44]	0.6407	0.2048	0.0633	0.0248
Proposed MOR	0.0102	0.0005	0.0001	−0.0007

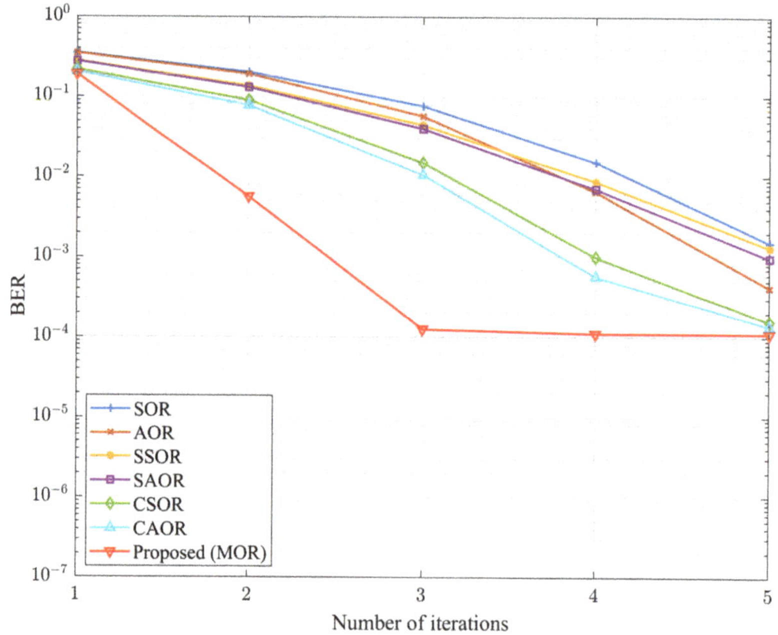

(**a**)

Figure 14. *Cont.*

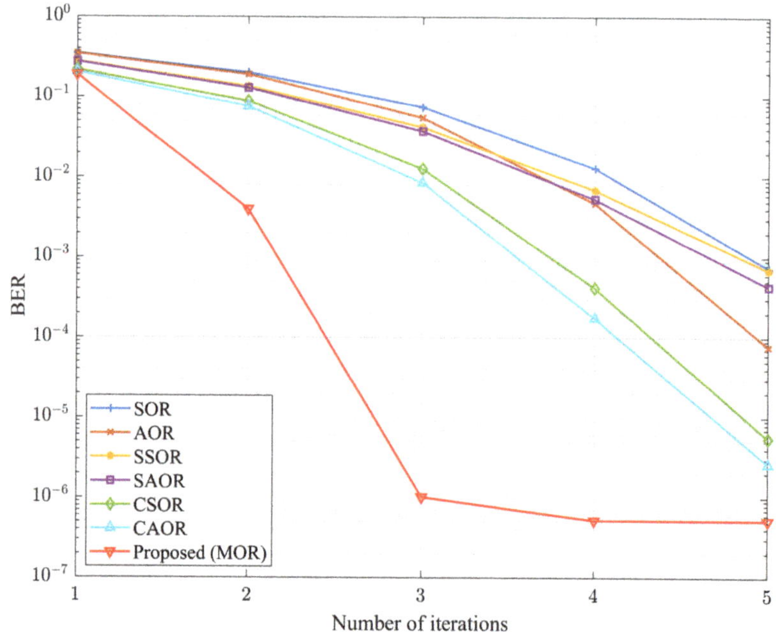

(**b**)

Figure 14. BER performance vs. number of iterations with $N_R \times N_T = 64 \times 16$, $K = 8$, and SNR = 37 dB for (**a**) OFDM and (**b**) UFMC.

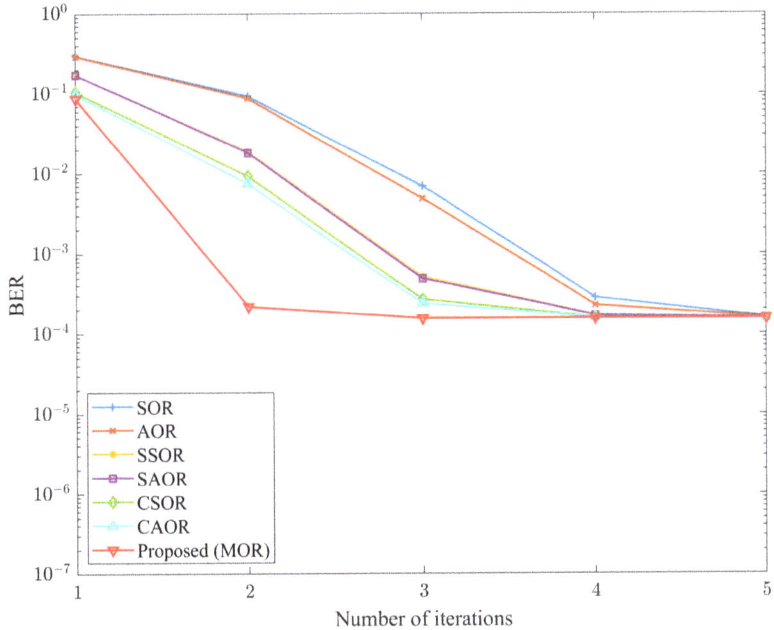

(**a**)

Figure 15. *Cont.*

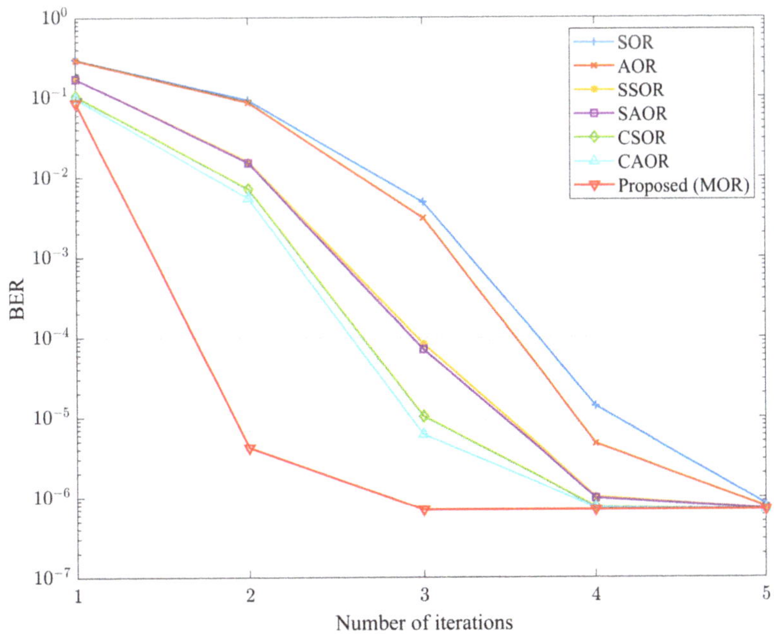

(**b**)

Figure 15. BER performance vs. number of iterations with $N_R \times N_T = 128 \times 16$, $K = 8$, and SNR = 33 dB for (**a**) OFDM and (**b**) UFMC.

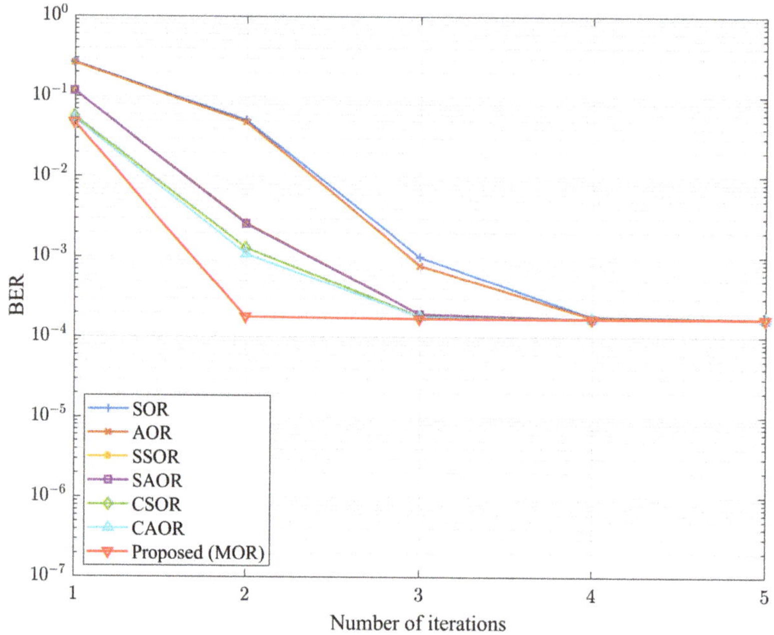

(**a**)

Figure 16. *Cont.*

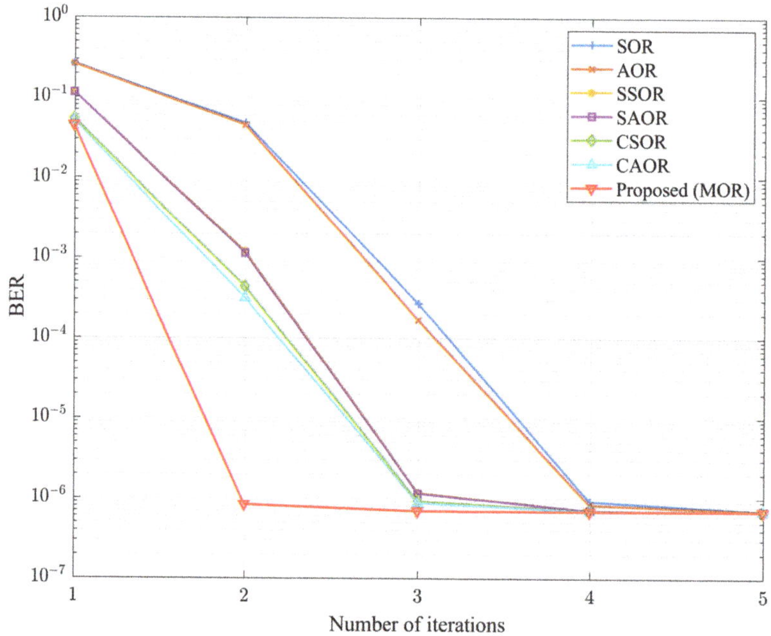

(**b**)

Figure 16. BER performance vs. number of iterations with $N_R \times N_T = 192 \times 16$, $K = 8$, and SNR = 31 dB for (**a**) OFDM and (**b**) UFMC.

To more clearly observe the numerical evolution of the number of iterations and BER performance under different numbers of antennas, where N_R was 128 and 192, as well as the progress of each method approaching the MMSE, we organized the data into Tables 7 and 8 from Figures 15 and 16, respectively. Similar to Tables 5 and 6, to obtain a clear numerical comparison, we took the logarithm operation $\log(\cdot)$ of the BER distance between each detector and the MMSE at iteration numbers i from 2 to 5. In Table 7a,b, we observe that MOR was quite near the MMSE in only three iterations when N_R was 128 and the SNR was at 36 dB. Compared with CAOR, the separation rate with the MMSE in the OFDM system was 0.0006 and 0.1891, respectively, being shortened by 0.1885, while in the UFMC system, the values were 0.0020 and 0.9429, respectively, being compressed by 0.9409. In Table 8, N_R increased to 192. The MOR scheme was extremely close to the MMSE and only needed two iterations. Moreover, there was a significant gap with other iteration methods. It is worth noting that in Tables 7b and 8b, when the iteration number i reached 5 and 4, respectively, the BER separation rate between MOR and the MMSE was already a negative value. To summarize Tables 7 and 8, the MOR detector had the fastest convergence and was closest to the optimal BER performance compared with thte other methods, whether in the OFDM or UFMC systems. Moreover, the experimental results in Figures 14–16 show that in addition to Appendix A theoretically proving the convergence of the MOR scheme, it is also verified convergence from the experimental data.

Finally, to verify M-MIMO affecting the capability of the OFDM and UFMC systems, Table 9 shows the improvement range of various iterative methods in the OFDM and UFMC as N_R increased, which can be referred to in Figure 12. We know that no matter whether the OFDM or UFMC system was used, when the number of antennas increased in the M-MIMO environment, all schemes could obtain significantly improved BER performance, of which the amount of gain in the UFMC was slightly higher than that of the OFDM system, which means that the UFMC system had better adaptability to M-MIMO. Also, the proposed MOR detector is entirely compatible with M-MIMO environments in both OFDM and UFMC systems and has better spatial diversity gain. It is worth noting that MOR can be used seamlessly for the 4G environment of the OFDM system and B5G environment of the UFMC system. Therefore, the proposed MOR detector is highly competitive in the 4G and B5G environments.

Table 7. Comparison of BER performance separation rates between all detectors and MMSE in different iteration numbers when $N_R \times N_T = 128 \times 16$ and SNR level was 33 dB for (**a**) OFDM and (**b**) UFMC.

Scheme	$i = 2$	$i = 3$	$i = 4$	$i = 5$
		(a)		
SOR [39]	2.7875	1.6570	0.2604	0.0162
AOR [40]	2.7573	1.5006	0.1624	0.0089
SSOR [41]	2.0874	0.5203	0.0370	0.0037
SAOR [42]	2.0766	0.5002	0.0355	0.0036
CSOR [43]	1.7833	0.2392	0.0104	0.0003
CAOR [44]	1.6917	0.1891	0.0070	0.0002
Proposed MOR	0.1517	0.0006	0	0
		(b)		
SOR [39]	5.1212	3.8485	1.2989	0.0649
AOR [40]	5.0897	3.6505	0.8284	0.0304
SSOR [41]	4.3528	2.0726	0.1591	0.0101
SAOR [42]	4.3387	2.0076	0.1460	0.0091
CSOR [43]	4.0115	1.1673	0.0344	0.0010
CAOR [44]	3.8973	0.9429	0.0238	0.0010
Proposed MOR	0.7785	0.0020	0.0002	−0.0001

Table 8. Comparison of BER performance separation rates between all detectors and MMSE in different iteration numbers when $N_R \times N_T = 192 \times 16$ and SNR level was 31 dB for (**a**) OFDM and (**b**) UFMC.

Scheme	$i = 2$	$i = 3$	$i = 4$	$i = 5$
		(**a**)		
SOR [39]	2.4801	0.7712	0.0305	0.0011
AOR [40]	2.4547	0.6671	0.0219	0.0006
SSOR [41]	1.1957	0.0596	0.0039	0.0004
SAOR [42]	1.1855	0.0587	0.0039	0.0004
CSOR [43]	0.8873	0.0321	0.0009	0
CAOR [44]	0.8124	0.0269	0.0006	0
Proposed MOR	0.0208	0	0	0
		(**b**)		
SOR [39]	4.8419	2.5898	0.1252	0.0044
AOR [40]	4.8152	2.3851	0.0834	0.0015
SSOR [41]	3.2391	0.2311	0.0077	0.0003
SAOR [42]	3.2191	0.2232	0.0077	0.0003
CSOR [43]	2.8017	0.1264	0.0035	0.0003
CAOR [44]	2.6612	0.1017	0.0022	0.0003
Proposed MOR	0.0286	0.0008	−0.0001	−0.0001

Table 9. BER improvement rates of different detection schemes when iteration number i was 2 and SNR was 33 dB for $N_R = 128$, 192, and 256 for (**a**) OFDM and (**b**) UFMC.

Scheme	$N_R = 128$	$N_R = 192$	$N_R = 256$
	(**a**)		
SOR [39]	52.373%	74.435%	85.341%
AOR [40]	52.902%	74.528%	85.265%
SSOR [41]	82.208%	97.369%	99.613%
SAOR [42]	81.815%	97.369%	99.612%
CSOR [43]	85.210%	97.312%	99.689%
CAOR [44]	85.543%	97.923%	99.685%
Proposed MOR	88.750%	98.050%	99.795%
	(**b**)		
SOR [39]	53.632%	75.869%	96.501%
AOR [40]	54.283%	75.998%	86.430%
SSOR [41]	86.601%	98.955%	99.944%
SAOR [42]	86.397%	98.954%	99.944%
CSOR [43]	90.299%	99.392%	99.973%
CAOR [44]	91.097%	99.478%	99.977%
Proposed MOR	98.600%	99.951%	99.997%

4.2. Computational Complexity Analysis

In this subsection, we evaluate the computational complexity of the proposed detection method in terms of the number of complex multiplications and additions (CMAs) compared with other mentioned detection methods in this article [39–44]. Table 10 shows the algebraic expressions of the computational complexity of different iterative methods, where i and N_T represent the number of iterations and the number of user antennas, respectively. Furthermore, iterative methods for inverse matrices require $(2N_T^2 - N_T)$ CMAs [64]. Relatively, the iterative procedure of our proposed MOR method utilizes Equation (25), where $\mathbf{G}_{MOR} = \mathbf{M}_{SOR}\mathbf{N}_{SOR}\mathbf{M}_{AOR}\mathbf{N}_{AOR}$ and $\mathbf{d}_{MOR} = (\mathbf{M}_{AOR} + \mathbf{G}_{AOR}\mathbf{M}_{SOR})\mathbf{d}$ require $(6N_T^2 - 2N_T)$ and $3N_T^2$ CMAs, respectively. Because there are two stages within our proposed MOR algorithm, with the first being the initial stage that involves spending $(9N_T^2 - 2N_T)$ CMAs for initialization calculation and the second being the collaboration stage to perform the iterative work by Equation (25), requiring $i(2N_T^2)$ CMAs. Therefore, the total complexity of our proposed method is $(9N_T^2 - 2N_T) + i(2N_T^2)$. In addition, Table 11 shows the numerical complexity of each detector when N_T is 16 and the number of iterations is from 2 to 5. Furthermore, to be more straightforward, the numerical complexity is presented using a bar chart in Figure 17.

Here, considering both the BER performance and complexity factors under discussion, we observe from Table 11 and Figure 8 that in the case of three iterations, although the complexity of our proposed method was about 15.546% slightly higher than CAOR, we found that MOR could significantly surpass CAOR in the OFDM and UFMC systems, not to mention other detectors. On the other hand, it can be observed that when the iteration number i was 3, the MOR complexity only required 3808 CMAs, which could outperform the CAOR BER performance with an iteration number i of 5 (needing 4848 CMAs). When further observing the impact of increasing the number of base station antennas N_R on the BER performance and complexity, it can be found from Figures 15 and 16 that as N_R increased, the number of iterations required by the iterative method gradually decreased, and the proposed MOR algorithm especially only required three iterations and two iterations, respectively. In light of the above discussion, we know that as the number of antennas N_R increased, it could arrive at convergence using a small amount of iterations, simultaneously reducing the complexity. Therefore, overall, the computational complexity of MOR was lower than that of other detectors, and it had good BER performance.

Table 10. Algebraic expressions of computational complexity for different detectors.

Iteration Methods	Complex Multiplications and Additions (CMAs)
SOR [39]	$\frac{1}{2}(5N_T^2 + N_T) + i(2N_T^2 + N_T)$
AOR [40]	$3N_T^2 + 3iN_T^2$
SSOR [41]	$(5N_T^2 + N_T) + 2i(2N_T^2 + N_T)$
SAOR [42]	$6N_T^2 + 6iN_T^2$
CSOR [43]	$\frac{1}{2}(5N_T^2 + N_T) + i(2N_T^2 + 3N_T)$
CAOR [44]	$3N_T^2 + 3i(N_T^2 + N_T)$
Proposed MOR	$(9N_T^2 - 2N_T) + i(2N_T^2)$

Table 11. Numerical complexity comparison for different detectors with $N_T = 16$.

Iteration Methods	CMAs $i = 2$	CMAs $i = 3$	CMAs $i = 4$	CMAs $i = 5$
SOR [39]	1704	2232	2760	3288
AOR [40]	2304	3072	3840	4608
SSOR [41]	3408	4464	5520	6576
SAOR [42]	4608	6144	7680	9216
CSOR [43]	2280	3096	3912	4728
CAOR [44]	2400	3216	4032	4848
Proposed MOR	3296	3808	4320	4832

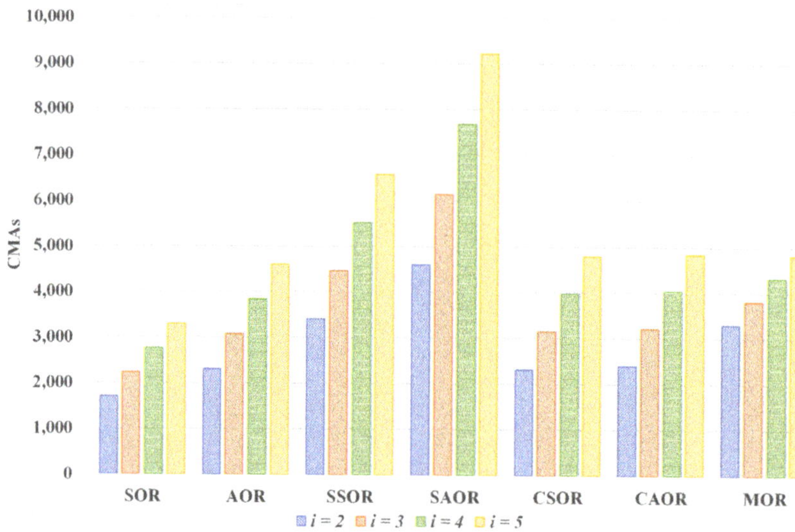

Figure 17. Bar chart of computational complexity for different detectors with $N_T = 16$.

5. Conclusions

This paper proposes a novel collaborative architecture receiver that mixes the relaxation characteristics of the SOR iteration algorithm and the acceleration ability of the AOR iteration algorithm to improve the convergence rate and obtain significant BER performance compared with the other iterative methods. Of course, combining the best convergence merits and complementarity of AOR and SOR is crucial to achieving such excellent BER performance. The numerical results verified that compared with the BER performance of different detection methods under the same environment, it outperformed other detection methods and was simultaneously close to the performance of the MMSE. For the complexity issue, although the proposed method adds a little computational load compared with CSOR and CAOR detectors under consistent iteration numbers, fortunately, due to our proposed MOR detector only needing a small amount of iteration to convergence, simultaneously, the BER performance approached the MMSE the most. In other words, our proposed method can achieve outstanding BER performance and only needs moderate complexity compared with other detectors that require more iterations. In addition, by applying MOR to 4G and B5G environments through experiments, we can verify that it can be ideally used and realize its merit.

Finally, the B5G system is an essential driver of advanced wireless sensor networks. Applications, such as the AIoT face numerous computing and transmission challenges. Therefore, it will be an inevitable trend to develop technologies that meet the requirements of eMMB, URLLC, and mMTC. We propose that the MOR algorithm be applied to M-MIMO systems, which possess lower complexity and BERs, contributing to the demand for large-scale transmission, low latency, and high accuracy in this field. Simultaneously, it is an algorithm worth looking forward to in further development.

Author Contributions: Conceptualization, Y.-P.T., P.-S.J. and Y.-F.H.; methodology, Y.-P.T. and P.-S.J.; software, P.-S.J.; validation, Y.-P.T.; formal analysis, Y.-P.T.; investigation, Y.-P.T. and Y.-F.H.; resources, Y.-P.T. and P.-S.J.; data curation, Y.-P.T. and P.-S.J.; writing—original draft preparation, Y.-P.T. and P.-S.J.; writing—review and editing, Y.-P.T. and Y.-F.H.; visualization, P.-S.J.; supervision, Y.-P.T.; project administration, Y.-P.T.; funding acquisition, Y.-F.H. and Y.-P.T. All authors have read and agreed to the published version of the manuscript.

Funding: This research received no external funding.

Data Availability Statement: All research data are listed in the article, and no additional source data are required.

Conflicts of Interest: The authors declare no conflicts of interest.

Abbreviations

The following abbreviations are used in this manuscript:

4G	fourth-generation
5G	fifth-generation
AOR	accelerated over-relaxation
AWGN	additive white Gaussian noise
B5G	beyond fifth-generation
BER	bit error rate
CAOR	Chebyshev accelerated over-relaxation
CMAs	complex multiplications and additions
CP	cyclic prefix
CSI	channel state information
CSOR	Chebyshev successive over-relaxation
eMBB	enhanced mobile broadband
FBMC	filter bank multi-carrier
FFT	fast Fourier transform
FIR	finite impulse response
GS	Gauss–Seidel
i.i.d.	independent and identically distributed
ICI	inter-carrier interference
IFFT	inverse fast Fourier transform
IMT	international mobile telecommunications
IoT	Internet of Things
ISI	inter-symbol interference
JA	Jacobi
LS	least squares
M-MIMO	massive multiple-input multiple-output
MF	matched filter
ML	maximum likelihood
MMSE	minimum mean square error
mMTC	massive machine-type communications
MOR	mixed over-relaxation
MUD	multi-user detection
NS	Neumann series
OFDM	orthogonal frequency division multiplexing
OOBM	out-of-band emission
QAM	quadrature amplitude modulation
RF	radio frequency
S/P	serial to parallel
SAOR	symmetric accelerated over-relaxation
SE	spectral efficiency
SOR	successive over-relaxation
SSOR	symmetric successive over-relaxation
P/S	parallel to serial
PSD	power spectral density
UFMC	universal filtered multi-carrier
URLLC	ultra-reliable and low-latency communications
WSN	wireless sensor network
ZF	zero forcing

Appendix A

In this subsection, we will deduce whether the MOR iteration equation converges and its convergence conditions.

As in Equation (22), when the spectral radius of the iteration matrix $\rho(\mathbf{G})$ is less than one (i.e., the eigenvalue of the iteration matrix in the iteration equation is less than one), the iteration equation can be proven to converge. On the other hand, the MOR iteration matrix $\mathbf{G}_{MOR}$ is as shown in Equation (20), and its expansion is

$$\mathbf{G}_{MOR} = (\mathbf{D}-\gamma\mathbf{L})^{-1}[(1-\omega)\mathbf{D}+(\omega-\gamma)\mathbf{L}+\omega\mathbf{U}](\mathbf{D}-\omega\mathbf{L})^{-1}[(1-\omega)\mathbf{D}+\omega\mathbf{U}]. \quad \text{(A1)}$$

Assuming that λ is the eigenvalue of $\mathbf{G}_{MOR}$, according to the eigenvalue theorem [65], we can obtain

$$\begin{aligned}\mathbf{G}_{MOR}\mathbf{x} &= (\mathbf{D}-\gamma\mathbf{L})^{-1}[(1-\omega)\mathbf{D}+(\omega-\gamma)\mathbf{L}+\omega\mathbf{U}](\mathbf{D}-\omega\mathbf{L})^{-1}[(1-\omega)\mathbf{D}+\omega\mathbf{U}]\mathbf{x} \\ &= \lambda\mathbf{x},\end{aligned} \quad \text{(A2)}$$

where moving $(\mathbf{D}-\gamma\mathbf{L})^{-1}$ and $(\mathbf{D}-\omega\mathbf{L})^{-1}$ to the right of the equal side yields

$$[(1-\omega)\mathbf{D}+(\omega-\gamma)\mathbf{L}+\omega\mathbf{U}][(1-\omega)\mathbf{D}+\omega\mathbf{U}]\mathbf{x} = (\mathbf{D}-\gamma\mathbf{L})(\mathbf{D}-\omega\mathbf{L})\lambda\mathbf{x}. \quad \text{(A3)}$$

We can simplify Equation (A3) as follows:

$$(\mathbf{D}-\omega\mathbf{D}+\omega\mathbf{L}-\gamma\mathbf{L}+\omega\mathbf{U})(\mathbf{D}-\omega\mathbf{D}+\omega\mathbf{U})\mathbf{x} = \left(\mathbf{D}^2-\omega\mathbf{DL}-\gamma\mathbf{DL}+\omega\gamma\mathbf{L}^2\right)\lambda\mathbf{x}, \quad \text{(A4)}$$

Then, we have

$$\begin{aligned}&\left[\left(\mathbf{D}^2-\omega\mathbf{D}^2+\omega\mathbf{DU}\right)+\left(-\omega\mathbf{D}^2+\omega^2\mathbf{D}^2-\omega\mathbf{U}\right)+\left(\omega\mathbf{LD}-\omega^2\mathbf{LD}+\omega^2\mathbf{LU}\right)\right. \\ &\left.+(-\gamma\mathbf{LD}+\omega\gamma\mathbf{LD}-\omega\gamma\mathbf{LU})+\left(\omega\mathbf{UD}-\omega^2\mathbf{UD}+\omega^2\mathbf{U}^2\right)\right]\mathbf{x} \\ &= \left(\mathbf{D}^2-\omega\mathbf{DL}-\gamma\mathbf{LD}-\omega\gamma\mathbf{L}^2\right)\lambda\mathbf{x},\end{aligned} \quad \text{(A5)}$$

and

$$\begin{aligned}&\left[\mathbf{D}^2\left(1-2\omega+\omega^2-\lambda\right)+\mathbf{DU}\left(\omega-\omega^2\right)+\mathbf{LD}\left(\omega-\omega^2-\gamma+\omega\gamma+\gamma\lambda\right)\right. \\ &\left.+\mathbf{LU}\left(\omega^2-\omega\gamma\right)+\mathbf{UD}\left(\omega-\omega^2\right)+\omega^2\mathbf{U}^2+\omega\mathbf{DL}\lambda-\omega\gamma\mathbf{L}^2\lambda\right]\mathbf{x}=0.\end{aligned} \quad \text{(A6)}$$

Now, we multiply Equation (A6) by $\mathbf{x}^T$ such that

$$\begin{aligned}&\mathbf{x}^T\left[\mathbf{D}^2\left(1-2\omega+\omega^2-\lambda\right)+\mathbf{DU}\left(\omega-\omega^2\right)+\mathbf{LD}\left(\omega-\omega^2-\gamma+\omega\gamma+\gamma\lambda\right)\right. \\ &\left.+\mathbf{LU}\left(\omega^2-\omega\gamma\right)+\mathbf{UD}\left(\omega-\omega^2\right)+\omega^2\mathbf{U}^2+\omega\mathbf{DL}\lambda-\omega\gamma\mathbf{L}^2\lambda\right]\mathbf{x}=0,\end{aligned} \quad \text{(A7)}$$

and transpose Equation (A7):

$$\begin{aligned}&\mathbf{x}\left[\mathbf{D}^2\left(1-2\omega+\omega^2-\lambda\right)+\mathbf{DL}\left(\omega-\omega^2\right)+\mathbf{UD}\left(\omega-\omega^2-\gamma+\omega\gamma+\gamma\lambda\right)\right. \\ &\left.+\mathbf{UD}\left(\omega^2-\omega\gamma\right)+\mathbf{LD}\left(\omega-\omega^2\right)+\omega^2\mathbf{L}^2+\omega\mathbf{DU}\lambda-\omega\gamma\mathbf{U}^2\lambda\right]\mathbf{x}^T=0.\end{aligned} \quad \text{(A8)}$$

Then, we add Equation (A7) to Equation (A8) to obtain Equation (A9) as follows:

$$\begin{aligned}&\mathbf{x}^T\Big[2\mathbf{D}^2\Big(1-2\omega+\omega^2-\lambda\Big)+\mathbf{D}(\mathbf{L}+\mathbf{U})\Big(\omega-\omega^2\Big)+(\mathbf{UL}+\mathbf{LU})\Big(\omega^2-\omega\gamma\Big)\\&+(\mathbf{U}+\mathbf{L})\mathbf{D}\Big(\omega-\omega^2-\gamma+\omega\gamma+\gamma\lambda\Big)+(\mathbf{L}+\mathbf{U})\mathbf{D}\Big(\omega-\omega^2\Big)+\omega^2\Big(\mathbf{L}^2+\mathbf{U}^2\Big)\\&+\omega\mathbf{D}(\mathbf{U}+\mathbf{L})\lambda-\omega\gamma\Big(\mathbf{U}^2+\mathbf{L}^2\Big)\lambda\Big]\mathbf{x}=0.\end{aligned}\tag{A9}$$

We can simplify Equation (A9) to be

$$\begin{aligned}&\mathbf{x}^T\Big[2\mathbf{D}^2\Big(1-2\omega+\omega^2-\lambda\Big)+\mathbf{D}(\mathbf{L}+\mathbf{U})\Big(3\omega-3\omega^2-\gamma+\omega\gamma+\omega\lambda+\gamma\lambda\Big)\\&+\Big(\mathbf{U}^2+\mathbf{L}^2\Big)\Big(\omega^2-\omega\gamma\lambda\Big)+(\mathbf{UL}+\mathbf{LU})\Big(\omega^2-\omega\gamma\Big)\Big]\mathbf{x}=0,\end{aligned}\tag{A10}$$

because $\mathbf{W}=\mathbf{D}+\mathbf{L}+\mathbf{U}$ after transposition becomes $\mathbf{W}-\mathbf{D}=\mathbf{L}+\mathbf{U}$. We substitute this equation into Equation (A10):

$$\begin{aligned}&\mathbf{x}^T\Big[2\mathbf{D}^2\Big(1-2\omega+\omega^2-\lambda\Big)+\mathbf{D}(\mathbf{W}-\mathbf{D})\Big(3\omega-3\omega^2-\gamma+\omega\gamma+\omega\lambda+\gamma\lambda\Big)\\&+\Big(\mathbf{U}^2+\mathbf{L}^2\Big)\Big(\omega^2-\omega\gamma\lambda\Big)+(\mathbf{UL}+\mathbf{LU})\Big(\omega^2-\omega\gamma\Big)\Big]\mathbf{x}=0,\end{aligned}\tag{A11}$$

and simplify Equation (A11) to

$$\begin{aligned}&\mathbf{x}^T\Big[\mathbf{D}^2\Big(2-7\omega+5\omega^2+\gamma-\omega\gamma-\omega\lambda-\gamma\lambda-2\lambda\Big)+\Big(\mathbf{U}^2+\mathbf{L}^2\Big)\Big(\omega^2-\omega\gamma\lambda\Big)\\&+\mathbf{DW}\Big(3\omega-3\omega^2-\gamma+\omega\gamma+\omega\lambda+\gamma\lambda\Big)+(\mathbf{UL}+\mathbf{LU})\Big(\omega^2-\omega\gamma\Big)\Big]\mathbf{x}=0,\end{aligned}\tag{A12}$$

Since $\mathbf{D}$ and $\mathbf{W}$ are symmetric positive definite matrices, both $\mathbf{x}^T\mathbf{D}^2\mathbf{x}$ and $\mathbf{x}^T\mathbf{DWx}$ are more than zero [66], and the following equalities can be written:

$$\begin{aligned}&\Big(2-7\omega+5\omega^2+\gamma-\omega\gamma-\omega\lambda-\gamma\lambda-2\lambda\Big)>0,\\&\lambda<\frac{2-7\omega+5\omega^2+\gamma-\omega\gamma}{\omega+\gamma+2},\end{aligned}\tag{A13}$$

$$\begin{aligned}&\Big(3\omega-3\omega^2-\gamma+\omega\gamma+\omega\lambda+\gamma\lambda\Big)>0,\\&\lambda<\frac{3\omega-3\omega^2-\gamma+\omega\gamma}{-\omega-\gamma},\end{aligned}\tag{A14}$$

Herein, we assume that $0<\omega<2$ and substitute these values into Equations (A13) and (A14).

For Equation (A13), if $\omega=0$, then we can obtain

$$\lambda<\frac{2+\gamma}{\gamma+2}=1,\tag{A15}$$

and when $\omega=2$, we can obtain

$$\lambda<\frac{8-\gamma}{4+\gamma}.\tag{A16}$$

We hope that $\lambda<1$ meets the convergence conditions. Therefore, we have

$$\frac{8-\gamma}{4+\gamma}=1,\tag{A17}$$

and

$$\gamma=2.\tag{A18}$$

From Equations (A15) and (A18), we can infer that Equation (A13) will converge when $0 < \omega < 2$ and $0 < \gamma < 2$.

Similarly, in Equation (A14), when $\omega = 0$, we can obtain

$$\lambda < \frac{-\gamma}{-\gamma} = 1, \tag{A19}$$

and when $\omega = 2$, we can obtain

$$\lambda < \frac{-6 + \gamma}{-2 - \gamma}. \tag{A20}$$

We hope that $\lambda < 1$ meets the convergence conditions. Therefore, we have

$$\frac{-6 + \gamma}{-2 - \gamma} = 1, \tag{A21}$$

and

$$\gamma = 2. \tag{A22}$$

From Equations (A19) and (A22), we can demonstrate that Equation (A14) will converge when $0 < \omega < 2$ and $0 < \gamma < 2$.

In summary, we deduce that when $0 < \omega < 2$ and $0 < \gamma < 2$, it can be proven that the MOR iterative equation converges when ω and γ are not necessarily equal.

References

1. Henry, S.; Alsohaily, A.; Sousa, E.S. 5G is Real: Evaluating the Compliance of the 3GPP 5G New Radio System with the ITU IMT-2020 Requirements. *IEEE Access* **2020**, *8*, 42828–42840. [CrossRef]
2. Fuentes, M.; Carcel, J.L.; Dietrich, C.; Yu, L.; Garro, E.; Pauli, V.; Lazarakis, F.I.; Grøndalen, O.; Bulakci, O.; Yu, J.; et al. 5G New Radio Evaluation Against IMT-2020 Key Performance Indicators. *IEEE Access* **2020**, *8*, 110880–110896. [CrossRef]
3. Kim, Y.; Park, S. Calculation Method of Spectrum Requirement for IMT-2020 eMBB and URLLC With Puncturing Based on M/G/1 Priority Queuing Model. *IEEE Access* **2020**, *8*, 25027–25040. [CrossRef]
4. Agiwal, M.; Roy, A.; Saxena, N. Next Generation 5G Wireless Networks: A Comprehensive Survey. *IEEE Commun. Surv. Tutorials* **2016**, *18*, 1617–1655. [CrossRef]
5. Andrews, J.G.; Buzzi, S.; Choi, W.; Hanly, S.V.; Lozano, A.; Soong, A.C.K.; Zhang, J.C. What Will 5G Be? *IEEE J. Sel. Areas Commun.* **2014**, *32*, 1065–1082. [CrossRef]
6. Pirinen, P. A brief overview of 5G research activities. In Proceedings of the 1st International Conference on 5G for Ubiquitous Connectivity, Akaslompolo, Finland, 26–28 November 2014; pp. 17–22. [CrossRef]
7. Shafique, K.; Khawaja, B.A.; Sabir, F.; Qazi, S.; Mustaqim, M. Internet of Things (IoT) for Next-Generation Smart Systems: A Review of Current Challenges, Future Trends and Prospects for Emerging 5G-IoT Scenarios. *IEEE Access* **2020**, *8*, 23022–23040. [CrossRef]
8. Dudhe, P.; Kadam, N.; Hushangabade, R.M.; Deshmukh, M.S. Internet of Things (IOT): An overview and its applications. In Proceedings of the 2017 International Conference on Energy, Communication, Data Analytics and Soft Computing (ICECDS), Chennai, India, 1–2 August 2017; pp. 2650–2653. [CrossRef]
9. Pan, J.; McElhannon, J. Future Edge Cloud and Edge Computing for Internet of Things Applications. *IEEE Internet Things J.* **2018**, *5*, 439–449. [CrossRef]
10. Chettri, L.; Bera, R. A Comprehensive Survey on Internet of Things (IoT) Toward 5G Wireless Systems. *IEEE Internet Things J.* **2020**, *7*, 16–32. [CrossRef]
11. Kar, S.; Mishra, P.; Wang, K.C. 5G-IoT Architecture for Next Generation Smart Systems. In Proceedings of the 2021 IEEE 4th 5G World Forum (5GWF), Montreal, QC, Canada, 13–15 October 2021; pp. 241–246. [CrossRef]
12. Mishra, D.; Natalizio, E. A survey on cellular-connected UAVs: Design challenges, enabling 5G/B5G innovations, and experimental advancements. *Comput. Netw.* **2020**, *182*, 107451. [CrossRef]
13. Morocho-Cayamcela, M.E.; Lee, H.; Lim, W. Machine Learning for 5G/B5G Mobile and Wireless Communications: Potential, Limitations, and Future Directions. *IEEE Access* **2019**, *7*, 137184–137206. [CrossRef]
14. Agiwal, M.; Kwon, H.; Park, S.; Jin, H. A Survey on 4G-5G Dual Connectivity: Road to 5G Implementation. *IEEE Access* **2021**, *9*, 16193–16210. [CrossRef]
15. Khurshid, K.; Khokhar, I.A. Comparison survey of 4G competitors (OFDMA, MC CDMA, UWB, IDMA). In Proceedings of the 2013 International Conference on Aerospace Science & Engineering (ICASE), Islamabad, Pakistan, 21–23 August 2013; pp. 1–7. [CrossRef]
16. Sit, Y.L.; Reichardt, L.; Sturm, C.; Zwick, T. Extension of the OFDM joint radar-communication system for a multipath, multiuser scenario. In Proceedings of the 2011 IEEE RadarCon (RADAR), Kansas City, MI, USA, 23–27 May 2011; pp. 718–723. [CrossRef]

17. Cho, Y.S.; Kim, J.; Yang, W.Y.; Kang, C.G. *MIMO-OFDM Wireless Communications with MATLAB*; John Wiley & Sons: Hoboken, NJ, USA, 2010. [CrossRef]
18. Hammoodi, A.; Audah, L.; Aljumaily, M.S.; Taher, M.A.; Shawqi, F.S. Green Coexistence of CP-OFDM and UFMC Waveforms for 5G and Beyond Systems. In Proceedings of the 2020 4th International Symposium on Multidisciplinary Studies and Innovative Technologies (ISMSIT), Istanbul, Turkey, 22–24 October 2020; pp. 1–6. [CrossRef]
19. Huang, M.; Chen, J.; Feng, S. Synchronization for OFDM-Based Satellite Communication System. *IEEE Trans. Veh. Technol.* **2021**, *70*, 5693–5702. [CrossRef]
20. An, C.; Ryu, H.G. CPW-OFDM(Cyclic Postfix Windowing OFDM) for the B5G (Beyond 5th Generation) Waveform. In Proceedings of the 2018 IEEE 10th Latin-American Conference on Communications (LATINCOM), Guadalajara, Mexico, 14–16 November 2018; pp. 1–4. [CrossRef]
21. Rani, P.N.; Rani, C.S. UFMC: The 5G modulation technique. In Proceedings of the 2016 IEEE International Conference on Computational Intelligence and Computing Research (ICCIC), Chennai, India, 15–17 December 2016; pp. 1–3. [CrossRef]
22. Mukherjee, M.; Shu, L.; Kumar, V.; Kumar, P.; Matam, R. Reduced out-of-band radiation-based filter optimization for UFMC systems in 5G. In Proceedings of the 2015 International Wireless Communications and Mobile Computing Conference (IWCMC), Dubrovnik, Croatia, 24–28 August 2015; pp. 1150–1155. [CrossRef]
23. Durga, V.; Anuradha, S. On Channel Estimation in Universal Filtered Multi-Carrier (UFMC) System. In Proceedings of the 2019 PhotonIcs & Electromagnetics Research Symposium - Spring (PIERS-Spring), Rome, Italy, 17–20 June 2019; pp. 3708–3713. [CrossRef]
24. Schaich, F.; Wild, T. Waveform contenders for 5G—OFDM vs. FBMC vs. UFMC. In Proceedings of the 2014 6th International Symposium on Communications, Control and Signal Processing (ISCCSP), Athens, Greece, 21–23 May 2014; pp. 457–460. [CrossRef]
25. Doré, J.B.; Gerzaguet, R.; Cassiau, N.; Ktenas, D. Waveform contenders for 5G: Description, analysis and comparison. *Phys. Commun.* **2017**, *24*, 46–61. [CrossRef]
26. Ramadhan, A.J. Overview and Comparison of Candidate 5G Waveforms: FBMC, UFMC and F-OFDM. *Int. J. Comput. Netw. Inf. Secur.* **2022**, *14*, 27–38. [CrossRef]
27. Gerzaguet, R.; Bartzoudis, N.; Baltar, L.G.; Berg, V.; Doré, J.B.; Kténas, D.; Font-Bach, O.; Mestre, X.; Payaró, M.; Färber, M. The 5G candidate waveform race: A comparison of complexity and performance. *EURASIP J. Wirel. Commun. Netw.* **2017**, *2017*, 13. [CrossRef]
28. Albreem, M.A.; Juntti, M.; Shahabuddin, S. Massive MIMO Detection Techniques: A Survey. *IEEE Commun. Surv. Tutorials* **2019**, *21*, 3109–3132. [CrossRef]
29. Ali, M.Y.; Hossain, T.; Mowla, M.M. A Trade-off between Energy and Spectral Efficiency in Massive MIMO 5G System. In Proceedings of the 2019 3rd International Conference on Electrical, Computer & Telecommunication Engineering (ICECTE), Rajshahi, Bangladesh, 26–28 December 2019; pp. 209–212. [CrossRef]
30. Pan, X.; Zheng, Z. Cooperative Distributed Antenna Systems Based Secure Communications for Industrial Internet of Things. In Proceedings of the 2020 IEEE 20th International Conference on Communication Technology (ICCT), Nanning, China, 28–31 October 2020; pp. 790–794. [CrossRef]
31. Larsson, E.G.; Edfors, O.; Tufvesson, F.; Marzetta, T.L. Massive MIMO for next generation wireless systems. *IEEE Commun. Mag.* **2014**, *52*, 186–195. [CrossRef]
32. Paulraj, A.; Gore, D.; Nabar, R.; Bolcskei, H. An overview of MIMO communications—A key to gigabit wireless. *Proc. IEEE* **2004**, *92*, 198–218. [CrossRef]
33. Peng, W.; Ma, S.; Ng, T.S.; Wang, J. A novel analytical method for maximum likelihood detection in MIMO multiplexing systems. *IEEE Trans. Commun.* **2009**, *57*, 2264–2268. [CrossRef]
34. Hu, S.; Rusek, F. Modulus Zero-Forcing Detection for MIMO Channels. In Proceedings of the 2018 IEEE Global Communications Conference (GLOBECOM), Abu Dhabi, United Arab Emirates, 9–13 December 2018; pp. 1–7. [CrossRef]
35. Yan, L. Linear Mmse Interference Cancellation Detection for MIMO-OFDM System. In Proceedings of the 2017 9th International Conference on Measuring Technology and Mechatronics Automation (ICMTMA), Changsha, China, 14–15 January 2017; pp. 106–108. [CrossRef]
36. Liu, X.; Zhang, Z.; Wang, X.; Lian, J.; Dai, X. A Low Complexity High Performance Weighted Neumann Series-based Massive MIMO Detection. In Proceedings of the 2019 28th Wireless and Optical Communications Conference (WOCC), Beijing, China, 9–10 May 2019; pp. 1–5. [CrossRef]
37. Wu, Z.; Zhang, C.; Xue, Y.; Xu, S.; You, X. Efficient architecture for soft-output massive MIMO detection with Gauss-Seidel method. In Proceedings of the 2016 IEEE International Symposium on Circuits and Systems (ISCAS), Montreal, QC, Canada, 22–25 May 2016; pp. 1886–1889. [CrossRef]
38. Lee, Y. Decision-aided Jacobi iteration for signal detection in massive MIMO systems. *Electron. Lett.* **2017**, *53*, 1552–1554. [CrossRef]
39. Young, D.M. Convergence Properties of the Symmetric and Unsymmetric Successive Overrelaxation Methods and Related Methods. *Math. Comput.* **1970**, *24*, 793–807. [CrossRef]
40. Hadjidimos, A. Accelerated Overrelaxation Method. *Math. Comput.* **1978**, *32*, 149–157. [CrossRef]

41. Ning, J.; Lu, Z.; Xie, T.; Quan, J. Low complexity signal detector based on SSOR method for massive MIMO systems. In Proceedings of the 2015 IEEE International Symposium on Broadband Multimedia Systems and Broadcasting, Ghent, Belgium, 17–19 June 2015; pp. 1–4. [CrossRef]
42. Hu, Y.; Wu, J.; Wang, Y. SAOR-Based Precoding with Enhanced BER Performance for Massive MIMO Systems. In Proceedings of the 2019 International Conference on Artificial Intelligence in Information and Communication (ICAIIC), Okinawa, Japan, 11–13 February 2019; pp. 512–516. [CrossRef]
43. Berra, S.; Dinis, R.; Shahabuddin, S. Fast matrix inversion based on Chebyshev acceleration for linear detection in massive MIMO systems. *Electron. Lett.* **2022**, *58*, 451–453. [CrossRef]
44. Tu, Y.P.; Chen, C.Y.; Lin, K.H. An Efficient Two-Stage Receiver Base on AOR Iterative Algorithm and Chebyshev Acceleration for Uplink Multiuser Massive-MIMO OFDM Systems. *Electronics* **2022**, *11*, 92. [CrossRef]
45. Armstrong, J. OFDM for Optical Communications. *J. Light. Technol.* **2009**, *27*, 189–204. [CrossRef]
46. Gesbert, D.; Shafi, M.; shan Shiu, D.; Smith, P.; Naguib, A. From theory to practice: an overview of MIMO space-time coded wireless systems. *IEEE J. Sel. Areas Commun.* **2003**, *21*, 281–302. [CrossRef]
47. Eilert, J.; Wu, D.; Liu, D. Implementation of a programmable linear MMSE detector for MIMO-OFDM. In Proceedings of the 2008 IEEE International Conference on Acoustics, Speech and Signal Processing, Las Vegas, NV, USA, 31 March–4 April 2008; pp. 5396–5399. [CrossRef]
48. Firdaus, M.; Moegiharto, Y. Performance of OFDM System against Different Cyclic Prefix Lengths on Multipath Fading Channels. *arXiv* **2022**, arXiv:2207.13045.
49. Manda, R.; Gowri, R. Filter Design for Universal Filtered Multicarrier (UFMC) based Systems. In Proceedings of the 2019 4th International Conference on Information Systems and Computer Networks (ISCON), Mathura, India, 21–2 Novomber 2019; pp. 520–523. [CrossRef]
50. Sidiq, S.; Mustafa, F.; Sheikh, J.A.; Malik, B.A. FBMC and UFMC: The Modulation Techniques for 5G. In Proceedings of the 2019 International Conference on Power Electronics, Control and Automation (ICPECA), New Delhi, India, 16–17 November 2019; pp. 1–5. [CrossRef]
51. Raj, T.; Mishra, R.; Kumar, P.; Kapoor, A. Advances in MIMO Antenna Design for 5G: A Comprehensive Review. *Sensors* **2023**, *23*, 6329. [CrossRef]
52. Albreem, M.A.; Salah, W.; Kumar, A.; Alsharif, M.H.; Rambe, A.H.; Jusoh, M.; Uwaechia, A.N. Low Complexity Linear Detectors for Massive MIMO: A Comparative Study. *IEEE Access* **2021**, *9*, 45740–45753. [CrossRef]
53. Kang, M.; Alouini, M. Capacity of correlated MIMO Rayleigh channels. *IEEE Trans. Wirel. Commun.* **2006**, *5*, 143–155. [CrossRef]
54. Coleri, S.; Ergen, M.; Puri, A.; Bahai, A. Channel estimation techniques based on pilot arrangement in OFDM systems. *IEEE Trans. Broadcast.* **2002**, *48*, 223–229. [CrossRef]
55. Fang, Z.; Shi, J. Least Square Channel Estimation for Two-Way Relay MIMO OFDM Systems. *ETRI J.* **2011**, *33*, 806–809. [CrossRef]
56. Hadjidimos, A. Successive overrelaxation (SOR) and related methods. *J. Comput. Appl. Math.* **2000**, *123*, 177–199. [CrossRef]
57. Liu, D.; Zhou, W. A Low-Complexity Precoding Algorithm Based on Improved SOR Method for Massive MIMO Systems. In Proceedings of the 2019 11th International Conference on Wireless Communications and Signal Processing (WCSP), Xi'an, China, 23–25 October 2019; pp. 1–6. [CrossRef]
58. Tuli, E.A.; Kim, D.S.; Lee, J.M. Performance Enhancement of UFMC Systems using Kaiser Window Filter. In Proceedings of the 2021 International Conference on Information and Communication Technology Convergence (ICTC), Jeju Island, Republic of Korea, 19–21 October 2021; pp. 386–388. [CrossRef]
59. Ravindran, R.; Viswakumar, A. Performance evaluation of 5G waveforms: UFMC and FBMC-OQAM with Cyclic Prefix-OFDM System. In Proceedings of the 2019 9th International Conference on Advances in Computing and Communication (ICACC), Kochi, India, 6–8 November 2019; pp. 6–10. [CrossRef]
60. Vaigandla, K.K.; Siluveru, M.; Karne, R. Study and Comparative Analysis of OFDM and UFMC Modulation Schemes. *J. Electron. Comput. Netw. Appl. Math. (JECNAM)* **2023**, *3*, 41–50. [CrossRef]
61. Vakilian, V.; Wild, T.; Schaich, F.; ten Brink, S.; Frigon, J.F. Universal-filtered multi-carrier technique for wireless systems beyond LTE. In Proceedings of the 2013 IEEE Globecom Workshops (GC Wkshps), Atlanta, GA, USA, 9–13 December 2013; pp. 223–228. [CrossRef]
62. Agarwal, A.; Mehta, S.N. Design and performance analysis of MIMO-OFDM system using different antenna configurations. In Proceedings of the 2016 International Conference on Electrical, Electronics, and Optimization Techniques (ICEEOT), Chennai, India, 3–5 March 2016; pp. 1373–1377. [CrossRef]
63. Malik, W.Q.; Edwards, D.J. Measured MIMO Capacity and Diversity Gain With Spatial and Polar Arrays in Ultra Wideband Channels. *IEEE Trans. Commun.* **2007**, *55*, 2033–2033. [CrossRef]
64. Yu, A.; Zhang, C.; Zhang, S.; You, X. Efficient SOR-based detection and architecture for large-scale MIMO uplink. In Proceedings of the 2016 IEEE Asia Pacific Conference on Circuits and Systems (APCCAS), Jeju Island, Republic of Korea, 25–28 October 2016; pp. 402–405. [CrossRef]
65. Cox, D.A. Stickelberger and the Eigenvalue Theorem. *arXiv* **2020**, arXiv:2007.12573.
66. Nhat Cuong, C.; Thi Hong, T.; Duc Khai, L. Hardware Implementation of the Efficient SOR-Based Massive MIMO Detection for Uplink. In Proceedings of the 2019 IEEE-RIVF International Conference on Computing and Communication Technologies (RIVF), Danang, Vietnam, 20–22 March 2019; pp. 1–6. [CrossRef]

electronics

MDPI

Article

Q-Learning and Efficient Low-Quantity Charge Method for Nodes to Extend the Lifetime of Wireless Sensor Networks

Kunpeng Xu [1], Zheng Li [1], Ao Cui [2], Shuqin Geng [2,*], Deyong Xiao [1], Xianhui Wang [1] and Peiyuan Wan [2]

[1] Beijing Zhixin Microelectronics Technology Co., Ltd., Beijing 100096, China; xukunpeng@sgchip.sgcc.com.cn (K.X.); lizheng1@sgchip.sgcc.com.cn (Z.L.); xiaodeyong@sgchip.sgcc.com.cn (D.X.); wangxianhui1@sgchip.sgcc.com.cn (X.W.)

[2] The Faculty of Information Technology, College of Electronic Science and Technology, Beijing University of Technology, Beijing 100124, China; booker@emails.bjut.edu.cn (A.C.); wanpy@bjut.edu.cn (P.W.)

* Correspondence: gengshuqin@bjut.edu.cn

Abstract: With the rapid development of the Internet of Things (IoT), improving the lifetime of nodes and networks has become increasingly important. Most existing medium access control protocols are based on scheduling the standby and active periods of nodes and do not consider the alarm state. This paper proposes a Q-learning and efficient low-quantity charge (QL-ELQC) method for the smoke alarm unit of a power system to reduce the average current and to improve the lifetime of the wireless sensor network (WSN) nodes. Quantity charge models were set up, and the QL-ELQC method is based on the duty cycle of the standby and active times for the nodes and considers the relationship between the sensor data condition and the RF module that can be activated and deactivated only at a certain time. The QL-ELQC method effectively overcomes the continuous state–action space limitation of Q-learning using the state classification method. The simulation results reveal that the proposed scheme significantly improves the latency and energy efficiency compared with the existing QL-Load scheme. Moreover, the experimental results are consistent with the theoretical results. The proposed QL-ELQC approach can be applied in various scenarios where batteries cannot be replaced or recharged under harsh environmental conditions.

Keywords: wireless sensor networks; node lifetime; charge consumption; Q-learning

Citation: Xu, K.; Li, Z.; Cui, A.; Geng, S.; Xiao, D.; Wang, X.; Wan, P. Q-Learning and Efficient Low-Quantity Charge Method for Nodes to Extend the Lifetime of Wireless Sensor Networks. *Electronics* **2023**, *12*, 4676. https://doi.org/10.3390/electronics12224676

Academic Editors: Dionisis Kandris and Eleftherios Anastasiadis

Received: 3 October 2023
Revised: 27 October 2023
Accepted: 6 November 2023
Published: 17 November 2023

1. Introduction

Internet of Things (IoT) technology enables machines, such as home appliances, medical equipment, and industrial instruments, to interact with users and other machines via the Internet [1]. Wireless sensor networks (WSNs) are a broad category of IoT applications. WSNs can send and receive data via the Internet using a sink node [2,3]. The successful operation of a power system requires the support of communication networks with massive node access and latency-critical two-way reliable transmission [4]. However, power management in WSNs poses a significant challenge when the WSN must operate continuously for sustained periods without a consistent power source. In such contexts, the nodes have specific limitations regarding their memory, processing capacity, radio communication range, and energy supply [5]. One type of node uses batteries that cannot be replaced or recharged under harsh environmental conditions [6].

Although many complex communication protocols and routing algorithms have been proposed for WSNs, disadvantages, such as power dissipation, network complexity, and high costs, must be overcome for hardware and software implementation [7]. For long-term operation, the power-constrained condition is strict and limited, and a back-end circuit system is required to obtain the sensor information and to transmit the acquired data [8]. As the network scales up and the number of nodes increases, certain fundamental problems, such as energy-efficient data transmission, scalability, data gathering, and aggregation, become concerns [9]. Thus, an effective low-power circuit system is indispensable to ensure the long-term operation of WSNs [10–12]. To improve the WSN's lifetime, the high-coverage communication of targets must be ensured before performing a sensor-node duty cycle [13,14]. The Cooperative Medium Access Control (C-MAC) [15] method for improving the duty cycle-based MAC with idle listening has been proposed. However, an additional channel is required to synchronize the nodes which consume additional energy.

Recently, reinforcement learning (RL) has been widely employed to address resource management problems in next-generation wireless networks [16]. The Q-learning technique is an RL approach in which the algorithm continuously learns by interacting with the environment, gathering information to take certain actions and to improve a specific policy [17]. It is based on iterative offline operations that predict the next optimal step based on obtained experience. Hence, the lifetimes of nodes and WSNs have been extended using Q-learning [18,19], and low power consumption has been achieved via energy management [20,21]. A novel Q-learning-based data-aggregation-aware energy-efficient routing algorithm was proposed in [22]. A runtime-decentralized self-optimization framework based on deep RL for configuring the parameters of a multi-hop network was presented in [23]. This maximizes the performance by determining the optimal result from the environment [24,25]. However, in using a Q-learning algorithm that has too many actions or states to control throughout the duty cycle of a WSN, both the storage requirement and dimensions of the problem become intractable for the end node [26]. Furthermore, a systematic literature review revealed that energy consumption is the most fundamental problem in WSNs [27]. However, this has not been sufficiently considered by scholars and practitioners [28]. Therefore, a low-power-consumption method must be designed to improve the long-term operation of nodes in WSNs by considering various performance metrics with relatively few states and actions.

This study proposed a Q-learning, efficient low-quantity charge (QL-ELQC) method with a small number of states and actions to extend the lifetime of a photoelectric smoke end node (PSEN) in the WSN of a power system. Mathematical models were established to describe the relationships between the main parameters and the principal charge consumption. The outcome of the mathematical analysis formed the basis for the measures taken to optimize the PSEN system and to improve its lifetime. Furthermore, Q-learning-based ELQC was applied to self-adjust the standby time of the modules to optimize the duty cycle of the sensor and RF module's standby time to reduce the average current of the node system. The proposed method effectively overcomes the limitations of Q-learning by solving the problem of a continuous state–action space using the state classification method based on the relationship between the sensor data and the threshold. A lifetime testing system for a wireless photoelectric smoke sensor end node is introduced.

The remainder of this paper is organized as follows. In Section 2, we describe the proposed system architecture. In Section 3, we propose an ELQC model. Section 4 presents the proposed QL-ELQC method. The testing of the modules is provided in Section 5, and the experiment on the node system is discussed in Section 6. Finally, the conclusions are presented in Section 7.

2. Architecture

2.1. WSN PSEN-SM System Architecture

As depicted in Figure 1, the WSN smoke and smart meter system has three hierarchy levels and relationships. The first level comprises the PSENs and SMNs, which monitor the smoke, humidity, ambient temperature, and electricity consumption and send the related compressed data to the sink nodes. The PSENs and SMNs receive commands or acknowledgments from the sink nodes. The second level represents the sink nodes (always in an active state), which receive the PSEN and SMN data and send acknowledgments or commands back to them via the radio frequency (RF) module. The sink nodes receive layer commands from the PC via the Internet and simultaneously send related data to the PC via the Internet and alarm signals to the mobile device of an operator. The third level comprises a PC with Internet access and a data server, which receives data from the sink nodes and sends commands back via the Internet. The following section introduces how the PSEN is used. The SMN method is not involved here.

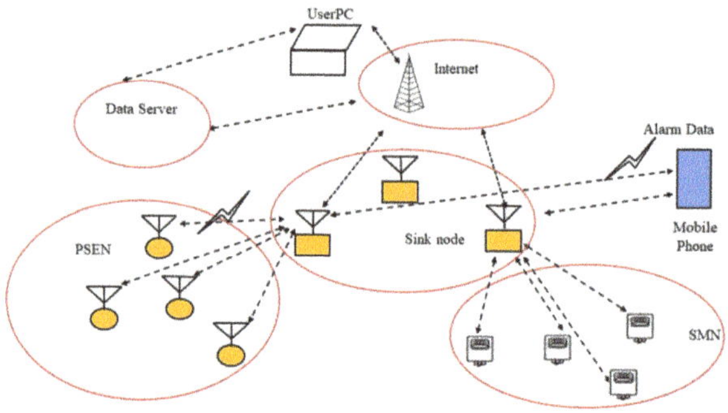

Figure 1. Proposed WSN architecture.

A time-sharing communication protocol is used between the PSENs and the sink node. Each node and module applies a duty-cycling method to reduce charge consumption. Moreover, the sink node of the WSN has high performance, which can reduce the communication time with the PSEN. When all PSENs have a long lifetime, the total lifetime of the WSN can be extended.

2.2. PSEN System Architecture

The PSEN system architecture is illustrated in Figure 2. The system comprises a microcontroller (MCU), an RF module, a power module, and a sensor module. To reduce the charge consumption, each module has a quantity charge model associated with the dominant charge consumer. A component can be regarded as a functional block, and the operational state of various modules is dynamically adapted to the required performance level, which can minimize the power wasted by idle or underutilized components [29]. The PSEN integrates temperature and humidity sensors to detect environmental changes rapidly. For the smoke sensor, we used an ultralow-power photoelectric amplifier with a low supply voltage.

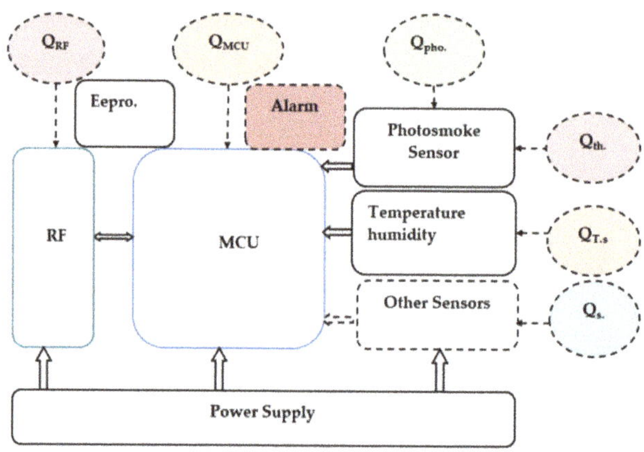

Figure 2. PSEN system architecture.

The PSEN is set to a low power state after the interrupt is initialized and opened. When there is an interrupt signal, the MCU wakes up to execute the interrupt events. Since a node battery's charge is limited, we define three states for the PSEN, namely, the ordinary, warning, and alarm states. The proposed node system can optimize the hardware and software systems, simplify the protocol, and compress signal data.

3. Proposed ELQC Model

Each node in a WSN consists of multiple modules, which can be abstracted as a series, such as 1, 2, ..., m, and each module has multiple states, which can also be seen as a sequence 1, 2, ..., n. We can encode them as $m \times n$ matrices, as expressed in Equation (1). Hence, we can determine the charge consumption of each module in each state.

$$Q_{total} = \begin{bmatrix} Q_{11} & Q_{12} & \ldots Q_{1j} & \ldots Q_{1n} \\ Q_{21} & Q_{22} & \ldots Q_{2j} & \ldots Q_{2n} \\ \ldots & & & \ldots \\ Q_{i1} & Q_{i2} & \ldots Q_{ij} & \ldots Q_{in} \\ \ldots & & & \ldots \\ Q_{m1} & Q_{m2} & \ldots Q_{mj} & \ldots Q_{mn} \end{bmatrix}. \tag{1}$$

The charge consumption Q_{ij} of the i-th module in the j-th state is the time integral of the current, and its average current I_{ij} at quantum time T_{ij} can be represented by the following:

$$Q_{ij} = \int i_{ij} dt = I_{ij} T_{ij}, i = 1, 2 \ldots m; j = 1, 2 \ldots n \tag{2}$$

The node total charge consumption is the time integral of the current (total sum method), which is the sum of the time integrals of the current for each component at different states and can be represented as follows:

$$Q_{total} = \sum_{i=1}^{m} \sum_{j=1}^{n} Q_{ij} = \sum_{i=1}^{m} \sum_{j=1}^{n} \int i_{ij} dt = \sum_{i=1}^{m} \sum_{j=1}^{n} I_{ij} T_{ij}, i = 1, 2 \ldots m, j = 1, 2 \ldots n \tag{3}$$

The average current $I_{total-aver.}$ and time T_{total} matrices for the nodes in various states are represented by the following:

$$I_{total-aver.} = \begin{bmatrix} I_{11} & I_{12} & \ldots I_{1j} & \ldots I_{1n} \\ I_{21} & I_{22} & \ldots I_{2j} & \ldots I_{2n} \\ \ldots & & & \ldots \\ I_{i1} & I_{i2} & \ldots I_{ij} & \ldots I_{in} \\ \ldots & & & \ldots \\ I_{m1} & I_{m2} & \ldots I_{mj} & \ldots I_{mn} \end{bmatrix}, \quad T_{total} = \begin{bmatrix} T_{11} & T_{12} & \ldots T_{1j} & \ldots T_{1n} \\ T_{21} & T_{22} & \ldots T_{2j} & \ldots T_{2n} \\ \ldots & & & \ldots \\ T_{i1} & T_{i2} & \ldots T_{ij} & \ldots T_{in} \\ \ldots & & & \ldots \\ T_{m1} & T_{m2} & \ldots T_{mj} & \ldots T_{mn} \end{bmatrix}. \tag{4}$$

The total charge consumption of the node is the sum of the consumption of each module, and $Q_{Mi} = [Q_{i1} Q_{i2} \cdots Q_{ij} \cdots Q_{in}]$ denotes the charge consumption of the i-th module including n states. These are abbreviated as follows:

$$Q_{total} = \sum_{i=1}^{m} Q_{Mi}, i = 1, 2 \ldots m \tag{5}$$

First, we study the calculations of the i-th module. The charge consumption Q_{Mi} is the sum of the i-th module in the different n states, which is the sum of the corresponding item scores of the two matrices in the i-th row in Equation (4), represented as follows:

$$Q_{Mi} = \sum_{j=1}^{n} Q_{ij} = \sum_{j=1}^{n} \int i_{ij} dt = \sum_{j=1}^{n} I_{ij} T_{ij}, i = 1, 2 \ldots m, j = 1, 2 \ldots n \tag{6}$$

During the period of the i-th module in all states, the average current and period I_{Mi} and T_{Mi} for the i-th module of the node is given by the following:

$$I_{Mi} = \frac{Q_{Mi}}{T_{Mi}} = \frac{\sum_{j=1}^{n} I_{ij} T_{ij}}{\sum_{j=1}^{n} T_{ij}}, T_{Mi} = \sum_{j=1}^{n} T_{ij}, j = 1, 2 \ldots m, j = 1, 2 \ldots n, \tag{7}$$

Similar to real-world node implementations, we divided the states of the *i*-th module into working, idle listening, and standby states. (I_{wi}, T_{wi}), (I_{sti}, T_{sti}), and (I_{li}, T_{li}) are the currents and times corresponding to the working, standby, and idle listening states, respectively. In general, $I_{li} > I_{sti}$. In the sleep state, the current is almost zero and consumes almost no charge; therefore, it is ignored. These can then be represented as follows:

$$I_{Mi} = \frac{I_{wi}T_{wi} + I_{sti}\,T_{sti} + I_{li}T_{li}}{T_{wi} + T_{sti} + T_{li}} = I_{wi} - (I_{wi} - I_{li})R_{li} - (I_{wi} - I_{sti})R_{sti}, i = 1,\ 2\ldots m,$$
$$R_{sti} = T_{sti}/T_{Mi},\ R_{li} = T_{li}/T_{Mi} = 1 - (T_{sti} + T_{wi})/T_{Mi}, T_{Mi} = T_{wi} + T_{sti} + T_{li},\ \ i = 1,2\ldots m, \quad (8)$$

where R_{sti} *and* R_{li} denote the standby and idle listening time duty cycle of the *i*-th module.

If T_{wi} *and* I_{wi} are fixed, R_{sti} $(1 \geq R_{sti} \geq 0)$ and R_{li} $(1 \geq R_{li} \geq 0)$ increase as the standby time T_{sti} and idle listening T_{li} increase. When the other parameters remain unchanged and the standby time is known, then the idle listening duration can be obtained, and vice versa. When R_{sti} and R_{sli} increase, the average current and the charge consumption of the *i*-th module decrease. The average current and period of the module are represented by the following:

$$I_M = \begin{bmatrix} I_{M1} \\ I_{M2} \\ \cdots \\ I_{Mi} \\ \cdots \\ I_{Mm} \end{bmatrix},\ T_M = \begin{bmatrix} T_{M1} \\ T_{M2} \\ \cdots \\ T_{Mi} \\ \cdots \\ T_{Mm} \end{bmatrix}. \quad (9)$$

The node's total average current can be obtained as follows:

$$I_{node-aver} = \sum_{i=1}^{m} I_{Mi}, i = 1,2\ldots m. \quad (10)$$

The total node charge consumption during the battery's lifetime is equal to the available battery charge. The quantity of charge $Q_{battery}$, availability rate η of the battery, and self-discharge rate $R_{self\text{-}dischage}$ can be obtained from the datasheets of the battery. We can then obtain the battery life $T_{batt.life}$ of the WSN node as follows:

$$I_{node-aver.}T_{batt.life} = Q_{battery}\eta\left(1 - R_{self-discharge}\right)^{T_{batt.life}}. \quad (11)$$

For η of 0.72 and $R_{self\text{-}discharge}$ of 3%, the lifetime graph from 0 to 20 years and charge consumption from 950 to 2800 mAh are illustrated in Figure 3. It can be seen that as the current *I* decreases, the lifetime *t* of the node increases, as the yellow color in the figure deepens. As the battery capacity *Q* increases, the allowable current for node with the same lifespan increases, and the light yellow parts in the figure become more numerous.

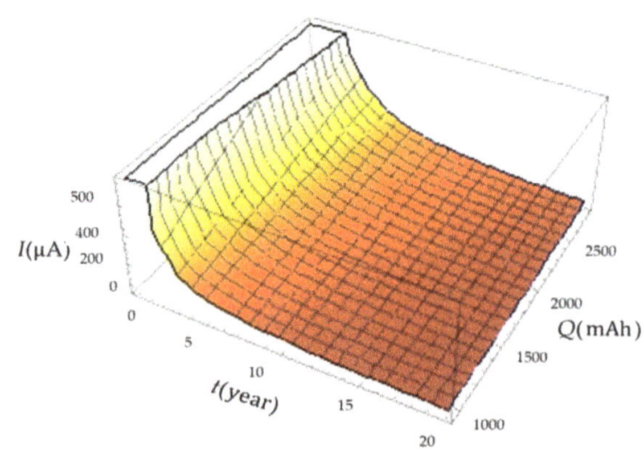

Figure 3. Current of the PSEN for battery charge and lifetime.

4. Proposed QL-ELQC Method

To minimize the average current and to extend the node's life, the designed communication distance is greater than the actual distance, so all nodes can communicate directly with the sink nodes. If other parameters are not changed, when implementing multi-hops between adjacent nodes to the sink node, one data transmission exchanges twice the receiving and transmitting data with the upper and next nodes, but when the node communicates directly with the sink node, it need only exchange once, which can eliminate the charge consumption according to the ELQC model (8).

In special circumstances, some nodes require multi-hops to communicate with the sink node. Since a routing table is used for data transfer, Q-table is used for the next idle listening duration and standby time of a node in WSN. Therefore, this can minimize the time for changing the radio state to RX. The QL-ELQC scheduling method adaptively adjusts the idle listening duration and standby times of the nodes according to the alarm level, which reduces the delay and energy consumption required for data transmission. Here, the QL-ELQC will mainly focus on standby time.

4.1. Proposed QL-ELQC Block Diagram

QL is based on iterative offline operations that predict the next optimal step based on obtained experience. To alert the node in time and to extend its lifetime, we used a QL-ELQC method for duty cycle optimization to determine its operating and propagation strategy in a dynamic environment.

For the proposed QL-ELQC method, the atmospheric sensor data are defined as "state", while the standby time in the entire period is regarded as an "action". The level of alarm and the reduction in the average current are the "reward". In this paper, each node is regarded as an agent that interacts with the environment, calculates the reward, updates the Q-value, self-learns, and selects the optimal state and action, as depicted in Figure 4. Then, the optimal transition between states can send alarm data in time and reduce the quantity of charge consumed to extend the lifetime of the node.

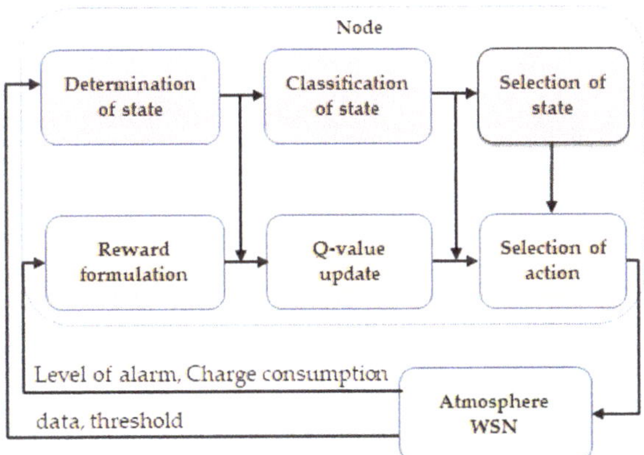

Figure 4. Proposed QL-ELQC block diagram.

Because the state of the environment is significantly large, the state space is also large. Concurrently, the different duty cycles of standby time in the entire period are considered environmental actions. This renders the typical implementation of QL infeasible. To address this problem, the state classification method adopted in this paper aims to limit the acceptable computational overhead and to reduce the energy and time consumption caused by excessive computational complexity. One of its distinctive features compared with other MAC protocols is that the standby time is modified based on the relationship between the atmospheric sensor data and the threshold. The data compression method can be used when the data are in the same state. Simultaneously, carrier detections exist for "listening before transmitting" protocols and transmissions are repeated if the data are not received. This ensures fast point-to-point communication during alarm states.

4.2. Proposed QL-ELQC Model

4.2.1. QL-ELQC of Standby Time Optimization

Owing to the complexity of atmospheric data in WSNs, the duty cycle must be dynamically altered based on the variable sensor data. The node determines the transmission frequency based on state vector $S = (s_1, s_2, \ldots, s_N)$ and sends the results to the RF module. In this paper, a model with three states and one optimal action was created using a self-learning process and interaction with the atmosphere to satisfy the rapid alarm and early warning requirements of the system. This overcomes the problem caused by major atmospheric conditions and action spaces. Based on the relationship between the monitored data values V and thresholds V_{th}, which are set in many experiments, the environmental states are divided into three categories, alarm, warning, and normal states, which can be expressed as follows:

$$S = \begin{cases} s_1, V \geq V_{th}, \ continuous\ 3\ times, \ in\ alarm\ state \\ s_2, V \geq V_{th}, \ 1-2\ times, \qquad in\ warning\ state \\ s_3, V \leq V_{th}, \qquad in\ normal\ state \end{cases}. \tag{12}$$

If the measurement data are larger than the threshold by one time, the node enters the warning state and the sensors immediately increase the monitoring frequency to continuously determine whether it has exceeded the threshold to lessen the error alarm. Subsequently, if one of the data points is still larger than the related threshold, the node system is in an alarm state. The node system then sends the data continuously until the alarm state is cleared, and the sink node system (always active) sends the alarm data to the user PC and the mobile phone of the worker on duty. It continuously reduces the latency through Q-learning training in the alarm state. If the measured data do not exceed the threshold, the node system is in the normal state and the data are processed using the QL-ELQC method to optimize the duty cycle of the node.

In general, the data monitored by sensors do not change significantly in a short period or fluctuate within an allowed range within a certain period. As opposed to continuous monitoring, this can considerably reduce charge consumption. Meanwhile, data aggregation substantially reduces energy consumption compared with transmitting all raw data to the sink node and can reduce traffic and improve the sensing quality for this type of smoke alarm system. The sensors and RF module duty cycles were then optimized using the QL-ELQC method to reduce the charge consumption, considering parameters such as communication distance, operating frequency band, voltage, and current. Therefore, the PSEN with the QL-ELQC quantity charge function to predict the next duration can trigger the alarm in time and can minimize charge consumption.

This policy is crucial for handling the priority relationship between alarms in time and reducing charge consumption. The shorter the standby time, the faster the node system reacts to an alarm state. However, the greater the standby time, the smaller the charge consumption for the node system. The maximum standby time does not exceed the sensitivity requirements of the system. Concurrently, the standby times of the sensor and RF module are not necessarily zero because each module has a minimum time interval. Moreover, in the alarm state, real-time monitoring and communication are superior to the quantity of charge consumed by the smoke alarm system. In an ordinary state, data compression and the duty-cycling algorithm should be prioritized to reduce charge consumption. Based on the sensor data state and policy, QL-ELQC selects an optimal action from the action set $A = [T_{st1}, T_{st2}, T_{st3}, T_{st4}]$. The duty cycle of standby time R_{st} can then be calculated using $R_{sti} = T_{sti}/T_{Mi}$, $i = 1, 2, \ldots m$:

$$R_{st} = [R_{st.al.}, R_{st.war.1}, R_{st.war.2}, R_{st.nor.}], \tag{13}$$

where $R_{st.al.}$, $R_{st.war.1}$, $R_{st.war.2}$, and $R_{st.nor.}$ are the duty cycles for the standby time of the RF module and sensor module during the alarm, warning 1–2 times, and normal action states, respectively.

In this model, reductions in the node's average current and the times that the sensor data continuously exceed the threshold are used as reward values to guide the next steps. The more the average current is reduced, the greater the reward in the normal state. The greater the number of times that the data exceed the threshold, the greater the reward value for the alarm level and the smaller the standby time. Using linear regression and function approximation [26], the reward at time t, R_t, can be determined as follows:

$$R_t = \delta I_t + (1-\delta) l_t + \varnothing, \quad (14)$$

where I_t denotes the average current and l_t indicates the level of alarm of the node at time t and the initialization of t = 0. Furthermore, δ symbolizes the weight of I_t. The reward computed by both the average current and alarm levels ensures an alarm in time and prolongs the lifetime of the node.

The Q function for a node with standby time T_{st} is represented as $Q_t(s_t, R_{st})$, which represents the real value at time t. It is updated based on a dynamic programming concept. If the objective value function Q_{target} at time t is $Q_{target} = R_t + \beta \max_{a\in A} Q_t(s_{t+1}, R_{st+1})$, then β indicates the discount factor of the node. If A represents a set of actions, $\mathrm{Max}_{a\in A} Q_t(s_{t+1}, R_{st+1})$ indicates the largest Q function in the corresponding state s_{t+1} at standby time T_{st+1}. The learning rate α is set as the step size for each update to reduce the difference between the two values; the specific update formula is as follows:

$$Q_{t+1}(s_t, T_{st}) = Q_t(s_t, T_{st}) + \alpha[R_t + \beta \max_{a\in A} Q_t(s_{t+1}, T_{st+1}) - Q_t(s_t, T_{st})]. \quad (15)$$

The node adopts the ε—greedy strategy to optimize its standby time, rather than directly selecting the maximum Q value as the setting. When Q-table converges, selecting action a in any state s to maximize $Q(s,a)$ can yield the optimal control strategy $a^* = \mathrm{argmax}_{a\in A} Q(s,a)$. An optimization control scheme based on Q-learning is presented in the algorithm.

The values of parameters α, β, and ε are crucial for the algorithm to work properly. If α is too small, the convergence speed of the algorithm will slow down; if α is too high, it may prevent the algorithm from converging or it may experience oscillations. These parameter values were selected by initialization, dynamic adjustment, and experimental verification, based on the Algorithm 1's performance and convergence.

Algorithm 1: Standby time optimization control scheme based on QL-ELQC

1: Initialization $\varepsilon = 0.1$, $\alpha = 0.1$, $\beta = 0.9$, $t = 0$, $Q(s, R_{st}) = 0$;
2: Observation sensor data and status s_t Equation (12);
3: Select standby time optimization action value control scheme based on the ε—greedy strategy T_{st};
4: Set the standby time according to policy and calculate R_{st} Equation (13);
5: Obtain the instant reward value Equation (14);
6: Update $Q_t(s_t, T_{st})$, $Q_t(s_{t+1}, T_{st+1})$. according to Equation (15);
7: Determine whether the learning process has ended. If not, set $t = t + 1$ and return to step 2, else end the learning procedure.

4.2.2. Simulation Results

To verify the algorithm, ten nodes were deployed at distances of 30 m using a tree topology. The transmission distance for each node was set to 55 m. One node is a sink node (always active). The other node sensor modules detect the environment and generate data at intervals of 10 s in the normal state and 1 s in the alarm state. As analyzed above, the three different atmospheric states were classified based on the relationship between the data, threshold, and state set $S = [s_1, s_2, s_3]$. The different standby time choices of the RF module and sensor module were considered environmental actions. Action set $A = [a_1, a_2, a_3, a_4]$, initialization at $t = 0$, the node learning rate $\alpha = 0.1$, discount factor $\beta = 0.9$, $\varnothing = 4$, $\delta = 2$, and $\varepsilon = 0.1$ were set. The transmitting, receiving, and standby currents of the RF module were $I_{w.tr} = 16$ mA, $I_{w.rx} = 12.5$ mA, and $I_{st} = 0.68$ μA. Carrier detection exists for "listen before transmit" protocols, and each node sends data based on the allocated time slot to reduce collisions.

As shown in Figure 5, using the two methods, the PSEN's lifetime was compared in the alarm and normal states. In the alarm state, the PSEN's lifetime using the two methods is identical. In the normal state, the PSEN's lifetime under QL-ELQC is longer than that for the QL-Load [26]. This indicates that the QL-ELQC scheme is suitable for the duty cycle of alarm nodes in response to dynamic environmental changes in the WSN. QL-ELQC makes self-adaptive decisions based on the classification of states, actions, and function approximations in a dynamic environment and prolongs the lifetime of the node and the WSN.

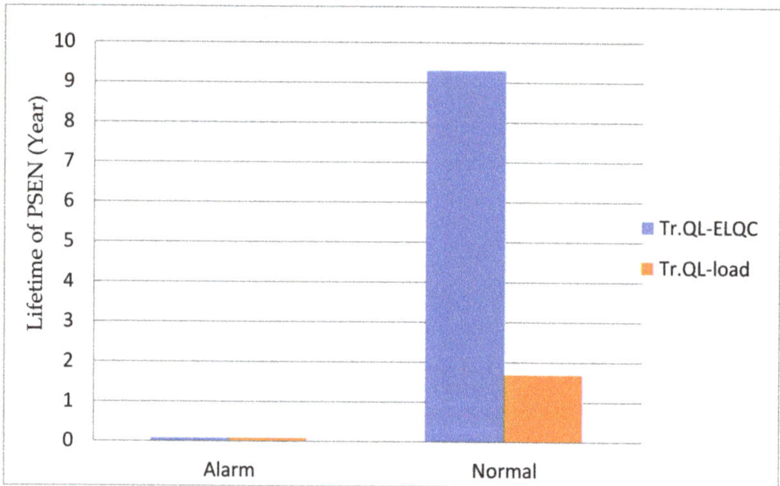

Figure 5. The lifetimes of PSENs.

This study used the data compression method for the case when the sensor data were in the same state. End-to-end latency in packet transmission is occasionally caused by re-transmissions. Due to the carrier detection measures for "listening before transmitting" protocols and the alarm channel, the delay is generally less than 1 s, which is much smaller than that of other QL schemes.

5. Experimentation

According to the ELQC model, when the PSEN is in a different state, the charge consumption is different. For experimental convenience and to verify that the QL-ELQC prolongs the lifetimes of PSENs, we divided the mode of the *i*-th module into two categories, namely, working mode I_{wi}, T_{wi} and standby mode I_{sti}, T_{sti}. Since the current in the sleep state is almost zero, it was ignored. The operating voltage (VCC) was fixed, and the power consumption was calculated from the current in the module connection path. Thus, the low dropout regulator (LDO) fixed the VCC to measure the current of each module circuit using an oscilloscope (RIGOL DS1074). In the figures, the relation between the I_w values in the tables and the voltage values registered by the oscilloscope is 10 mV/mA and 1 mV/μA.

5.1. RF Module

The RF module is integrated via an nRF905 Nordic chip, as depicted in Figure 6; the specifications and measured currents of the RF module are listed in Table 1. $T_{\text{Nor.lt.}}$ indicates the maximum time of the RF module in standby mode under normal environmental conditions, and $T_{al.lt}$ indicates the minimum value of the RF module in the alarm state. This means that the range of standby time $T_{RF.st}$ for the RF module is $0.22 \leq T_{RF.st} \leq 86,400$ s. In this experiment, the TX current was 16 mA, and the transmission time was T_{tx} = 7 ms, while the RX current was 12.5 mA, and the receiving time was T_{rx} = 10 ms. Thus, the working time and average current of the RF module were approximately $T_{tx\text{-}rx}$ = 17 ms and $I_{tx\text{-}rx}$ = 13,941 μA, respectively.

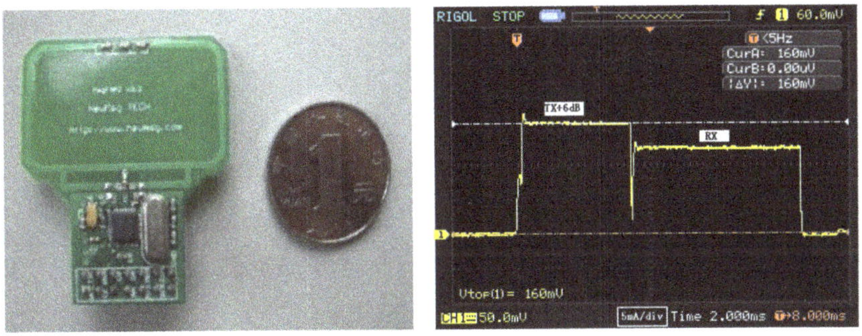

Figure 6. RF module current of the PSEN at TX+6 dBm (5 mA/div).

Table 1. RF module parameters of the PSEN in different states.

State		I_w	Time			Dist.(m)
			T_W (ms)	$T_{Nor\text{-}st}$. (s)	$T_{al.st}$ (ms)	
RX		12.5 mA	10	86,400	220	
TX	+6 dBm	16.0 mA	7	86,400	220	40–55
Standby		0.68 μA	All time		0	0

5.2. Sensor Module

Here, we only list the experimental results for the smoke sensors. The varying current and operating times of the A5303 smoke sensor at different stages were measured in several experiments, as shown in Figure 7. Table 2 lists the varying currents to the smoke sensors. With a large value at the starting point, the signals promptly increased to the maximum value and then gradually slowed down. The average current was 33 μA, which can be calculated using Equation (7), and the operating time was approximately 410 ms when it detected the environment once. With values lower than the threshold, the operational interval is 10 s, and the sensor is in standby mode. When the value exceeds the threshold, the sensor measures the environment three times repeatedly at 1 s intervals. This means that the standby time $T_{sen.st}$ range for the sensor is $1 \leq T_{sen.st} \leq 10$ s.

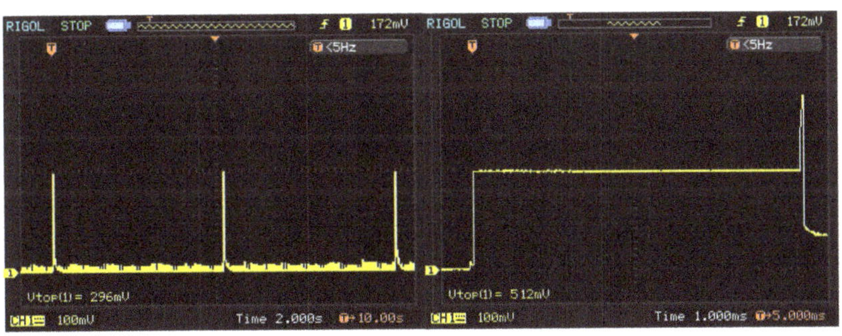

Figure 7. Smoke sensor experimental results (100 μA/div).

Table 2. Smoke sensor experimental data for different states (power: 3.3 V).

State		I_W (μA)	Time T_W (ms)	$I_{w.aver.}$ (μA)
Smoke sensor during one period	Start	300	10	33
	signal MAX.	500	0.2	
	Attenua. meas.	50	100	
		25	100	
		20	100	
		10	100	
Standby		2.6	All time	2.6

5.3. MCU

Microcontrollers are widely used in terminal devices. Therefore, they are listed separately and discussed herein. The PSEN system used a low-power-consumption MCUMSP430 from Texas Instruments. The software uses the interrupts of the MCU to awaken the standby state to execute the QL-ELQC period monitoring, to compress data, to set the alarm, to transmit data, and to receive commands or acknowledgments from the sink node. The clock system is specifically designed for battery-powered applications. Table 3 presents the experimental results for the MSP430 when the PSEN was in different states. Environmental monitoring included monitoring the temperature, humidity, and smoke.

Table 3. MCU experimental data of the PSEN in different states (power: 3.3 V).

State	$I_{w.aver}$	T_w
Low battery detect	420 μA	120 ms
Environment detect	420 μA	120 ms
Environ. detect & RF	500 μA	250 ms
Standby	1.96 μA	10 s

5.4. Power Management

In this study, we used the analog-to-digital converter (ADC) feature of an MCU MSP430F149 to detect the battery voltage periodically (10 s in the normal state and 1 s in the alarm state). The reference voltage of the ADC was 2.5 V, and resistors R1 and R2 were used to distribute the battery voltage. The circuit of the low-battery detector is shown on the left-hand side of Figure 8.

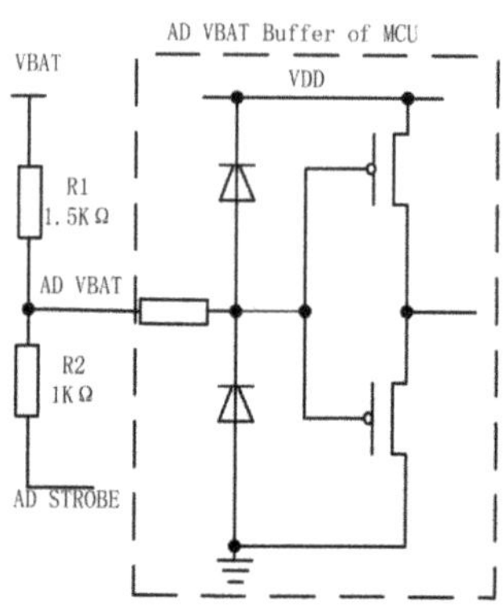

Figure 8. Low-battery circuit and AD VBAT input buffer.

The new battery's voltage is slightly higher than the nominal voltage, and the AD VBAT voltage is greater than the break-over voltage for the clamp diode of the AD VBAT input buffer, which is the circuit in the MCU. In the experiment, when the AD STROBE was set with a high resistance input, the current in decades of μA could be detected through R1. To solve this problem, a new low-battery circuit was designed, as shown in Figure 9.

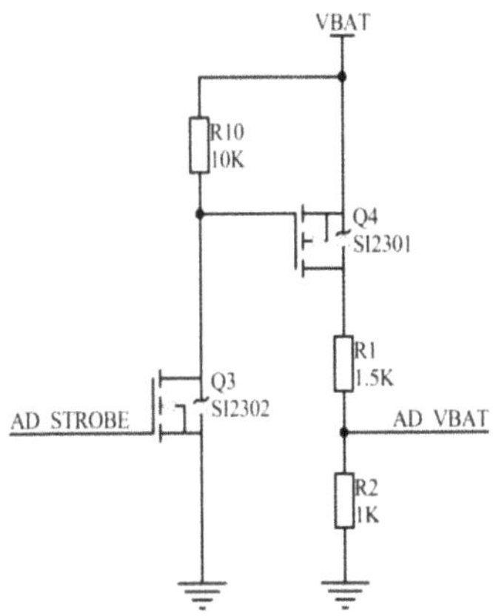

Figure 9. Newly designed low-battery circuit.

When the MCU does not detect the battery voltage, AD_STROBE is low and Q3 is off. Simultaneously, the grid voltage of Q4 becomes high and Q4 is off. When Q3 and Q4 are in the off state, AD_VBAT is reduced by R2. Thus, the detection circuit does not consume charge. When the MCU detects the battery voltage, AD_STROBE is high. In this case, Q3 is turned on, which pulls down the grid voltage of Q4, turning Q4 on. In this instance, R1 and R2 distribute the battery voltage, and the MCU detects the voltage of AD_VBAT to obtain the battery voltage. Table 4 lists the low-voltage detector experimental current and operating time of the PSEN. The average current is 0.0083 μA, which can be calculated using Equation (7).

Table 4. Low-voltage detector experimental data of the PSEN (power: 3.3 V).

Compo.	Stage	I_W (μA)	T_W (ms)	$I_{w.aver.}$ (μA)
Low voltage	I_R	1980	12	0.0083
	AD	900	2	
	MCU	420	10	

6. PSEN System Measurements and Discussion

Table 5 lists the experimental average current for each module and the total average current of the PSEN system. The actual communication time $T_{tx\text{-}rx}$ of the RF module in Table 1 was 17 ms, and the average current of $I_{tx\text{-}rx}$ is 13,941 μA. Note that the redundant RF module's operational time (200 ms) and the current (16,000 μA) were calculated for the average current $I_{total\text{-}ave.}$ and $I_{al.ave.}$, considering the collision and retransmitting. The LDO current has three components, among which 1.54 μA was the standby current, 2.5 μA was the PSEN's current for monitoring the environment, and 11 μA was the PSEN's current for communicating with the sink node. Based on these currents, as well as T_W and T_{st}, the average LDO current under normal and alarm states was determined as 3.14 μA and 5.06 μA using Equation (7), respectively.

From the module measurements, we obtained the average current for different components using Equation (7), and then, the total average standby current of all components was calculated to be 6.92 μA, which is close to the PSEN's total system standby current of 6.8 μA obtained from the experiment. As shown in Table 6, the error between the measurements and calculation with ELQC was 1.73%, which verifies the accuracy of the ELQC model.

Table 5. Experimental data for each module in the normal or alarm state (VCC = 3.3 V).

Module	I_w (μA)	I_{st} (μA)	T_w (s)	$T_{norm\text{-}st/lt}$ (s)	$T_{al.\text{-}st}$ (s)	$I_{total\text{-}ave.}$ (μA)	$I_{al.ave.}$ (μA)
LDO	11 2.5	1.54	0.200 0.721	86,400 10	1 1	3.14	5.06
Low-vol.	2488	0	0.012	3600	1	0.0083	29.50
MCU	420	2	0.120	10	1	6.47	46.79
Smoke	33	2.6	0.410	10	1	3.79	11.44
SHT10	386	0.1	0.103	10	1	4.08	36.14
RF-module	16,000	0.68	0.200	86,400	1	0.717	2667.23
Total			6.8			18.65	2796.16

Table 6. Theoretical and experimental data of the PSEN standby current.

Standby Parameters of the Node System	I (μA)
Experimental	6.8
Theoretical calculation	6.92
Error (%)	1.73

Meanwhile, the PSEN's total normal average current was 18.65 μA, and the total average current was 2.79 mA in the alarm state. As shown in Figure 10, the standby time (86,400 s) set by the QL-ELQC in the normal state was much longer than that (1 s) in the alarm state, and the current in the normal state was approximately 1/150 times lower than that in the alarm state. The advantage is not reflected enough within 10 s, and the longer the standby time, the more obvious the advantage. When power equipment operates normally, the probability of smoke occurrence remains extremely low; therefore, the QL-ELQC method used in the normal state significantly extends the total lifespan of the PSEN.

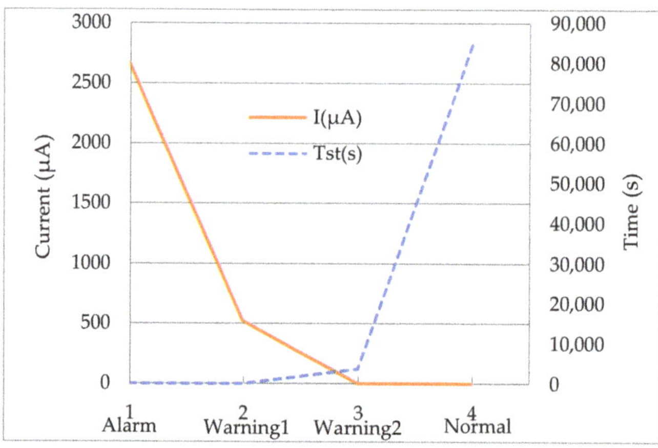

Figure 10. Experimental current of the RF module under different states (each experiment was repeated three times).

The simulation lifetime of the PSEN is 9.2 years for E91 in Figure 5, which is similar to a theoretical lifetime of 9.29 years but so long that we cannot test it for approximately 10 years, based on a practical system current of 18.65 μA. Using Equation (7), we can vary the current and change the lifetime of the PSEN to test the low-quantity charge design method; namely, we can select a small $Q_{battery}$ and shorten the lifetime for the test. Here, we used three E92 (not ordinary E91) small batteries, which have an approximately 950-mAh charge ranging from 1.6 to 1.2 V. The relevant data for the tested system are presented in Table 7. By increasing the communication times of the PSEN to every half second with sensors continuously monitoring the environment, the current of the PSEN can be increased. Thus, the current of the tested system is 5.48 mA (not 18.65 µA) to shorten the lifetime of the test, and the calculation lifetime is 173.35 h. When the voltage decreased to 1.2 V, the practical lifetime of our tested system was 181 h, and the error was 7.64 h, which is approximately 4%. As our proposed method considers redundancy, the tested system ran slightly longer than the calculated lifetime. Through practical experiments and algorithms, we tested the lifetime of a photoelectric smoke node and verified that our method, which is based on the charge quantity, is reasonable. Our proposed approach is general and can be applied to alarm scenarios where the node requires long-term operation.

Table 7. Measurement and theoretical lifetime of the PSEN with increasing transmission times.

Battery Type	E92
Quantity charge from 1.6 to 1.2 V	950 mAh
Tested practical lifetime (h)	181
Calculation lifetime (h)	173.35
Error (%)	4

7. Conclusions

In this paper, a Q-learning and efficient low-quantity charge (QL-ELQC) method is presented for the smoke alarm unit of a power system to reduce the average current and to improve the lifetime of the nodes of wireless sensor networks (WSNs). Analytical functions were derived to describe the behavior of the parameters versus those with which they were compared. The Q-learning-based ELQC method was applied to self-adjust the standby time of the modules to optimize the duty cycle of the sensor and RF modules to prolong the lifetime of the node system. This could effectively overcome the continuous state–action space limitations of Q-learning using the state classification method. Methods were used to extend the lifetime of PSENs in WSNs by reducing the average current in each module and every state, respectively. The simulation results reveal that the proposed scheme significantly improves the lifetime compared with the existing QL-Load scheme. Furthermore, the experimental results are consistent with the theoretical results. The model appears to be accurate for nodes in WSNs. The experimental results show that the proposed QL-ELQC method extends the lifetime of the PSEN, which is capable of long-term operation. We concluded that the QL-ELQC method proposed in this paper can be used for reference to prolong the lifetime of the node in alarm scenarios where batteries cannot be replaced or recharged under harsh environmental conditions.

Author Contributions: Conceptualization, K.X. and S.G.; methodology, Z.L. and D.X.; software, A.C., P.W. and X.W.; validation, S.G.; writing—S.G. and K.X.; supervision, S.G. and K.X.; project administration, K.X. and P.W.; funding acquisition, K.X. and P.W. All authors have read and agreed to the published version of the manuscript.

Funding: This work was supported in part by the National Key Research and Development Program of China (Project No. 2020YFC 1807-903).

Data Availability Statement: No new data were created or analyzed; thus, data sharing does not apply to this paper.

Acknowledgments: This work was derived from Research Project No. 40043001202310 of the Topology Identification Channel Model Simulation and Analog Signal Processing Research Technical Services. In this section, we acknowledge the support provided, which was not covered by the author contributions or funding sections.

Conflicts of Interest: K.X., Z.L., D.X. and X.W. were employed by the Beijing Zhixin Microelectronics Technology Co., Ltd. The remaining authors declare that the research was conducted in the absence of any commercial or financial relationships that could be construed as a potential conflict of interest.

References

1. Bhuiyan, M.N.; Rahman, M.M.; Billah, M.M.; Saha, D. Internet of Things (IoT): A review of its enabling technologies in healthcare applications, standards protocols, security, and market opportunities. *IEEE Internet Things J.* **2021**, *8*, 10474–10498. [CrossRef]
2. Ghayvat, H.; Mukhopadhyay, S.; Gui, X.; Suryadevara, N. WSN- and IOT-based smart homes and their extension to smart buildings. *Sensors* **2015**, *15*, 10350–10379. [CrossRef] [PubMed]
3. Lazarescu, M.T. Design of a WSN platform for long-term environmental monitoring for IoT applications. *IEEE J. Emerg. Sel. Topics Circuits Syst.* **2013**, *3*, 45–54. [CrossRef]
4. Zhang, Y.; Wang, W.; Xie, H.; Du, S.; Ma, M.; Zeng, Q. Wireless multi-node uRLLc B5G/6G networks for critical services in electrical power systems. *Energies* **2022**, *15*, 9437. [CrossRef]
5. Tarighi, R.; Farajzadeh, K.; Hematkhah, H. Prolong network lifetime and improve efficiency in WSNUAV systems using new clustering parameters and CSMA modification. *Int. J. Commun. Syst.* **2020**, *33*, e4324. [CrossRef]
6. Hatime, H.; Namuduri, K.; Watkins, J.M. OCTOPUS: An on-demand communication topology updating strategy for mobile sensor networks. *IEEE Sens. J.* **2011**, *11*, 1004–1012. [CrossRef]
7. Yu, C.M.; Ku, M.L.; Wang, L.C. Balanced Routing Algorithm with Transmission Range Adjustment for Network Lifetime Improvement in WSNs. In Proceedings of the IEEE 13th Annual Ubiquitous Computing, Electronics and Mobile Communication Conference (UEMCON), New York, NY, USA, 26–29 October 2022; pp. 308–312.
8. Wang, W.S.; Huang, H.Y.; Chen, S.C.; Ho, K.C.; Lin, C.Y.; Chou, T.C.; Hu, C.H.; Wang, W.F.; Wu, C.F.; Luo, C.H. Real-time telemetry system for amperometric and potentiometric electrochemical sensors. *Sensors* **2011**, *11*, 8593–8610. [CrossRef]
9. Koo, B.; Shon, T. Implementation of a WSN-based structural health monitoring architecture using 3D and AR mode. *IEICE Trans. Commun.* **2010**, *E93.B*, 2963–2966. [CrossRef]
10. Mahdi Elsiddig Haroun, F.; Mohamad Deros, S.N.; Ahmed Alkahtani, A.; Md Din, N. Towards self-powered WSN: The design of ultra-low-power wireless sensor transmission unit based on indoor solar energy harvester. *Electronics* **2022**, *11*, 2077. [CrossRef]
11. Hassan, A.A.; Shah, W.M.; Habeb, A.-H.H.; Othman, M.F.I.; Al-Mhiqani, M.N. An improved energy-efficient clustering protocol to prolong the lifetime of the WSN-based IoT. *IEEE Access* **2020**, *8*, 200500–200517. [CrossRef]
12. Li, N.; Xiao, M.; Rasmussen, L.K.; Hu, X.; Leung, V.C.M. On resource allocation of cooperative multiple access strategy in energy-efficient industrial Internet of things. *IEEE Trans. Ind. Inf.* **2021**, *17*, 1069–1078. [CrossRef]
13. Panhwar, M.A.; Liang, D.Z.; Memon, K.A.; Khuhro, S.A.; Abbasi, M.A.K.; Noor-ul-Ain, Z.A.; Ali, Z. Energy-efficient routing optimization algorithm in WBANs for patient monitoring. *J. Ambient Intell. Hum. Comput.* **2021**, *12*, 8069–8081. [CrossRef]
14. Xie, J.Z.; Zhang, B.J.; Zhang, C.P. A novel relay node placement and energy efficient routing method for heterogeneous wireless sensor networks. *IEEE Access* **2020**, *8*, 202439–202444. [CrossRef]
15. Liu, S.; Fan, K.W.; Sinha, P. CMAC: An energy-efficient MAC layer protocol using convergent packet forwarding for wireless sensor networks. *ACM Trans. Sens. Netw. (TOSN)* **2009**, *5*, 29. [CrossRef]
16. Khoramnejad, F.; Joda, R.; Sediq, A.B.; Abou-Zeid, H.; Atawia, R.; Boudreau, G.; Erol-Kantarci, M. Delay-aware and energy-efficient carrier aggregation in 5G using double Deep Q-networks. *IEEE Trans. Commun.* **2022**, *70*, 6615–6629. [CrossRef]
17. Wu, Z.; Pan, P.; Liu, J.; Shi, B.; Yan, M.; Zhang, H. Environmental perception Q-learning to prolong the lifetime of poultry farm monitoring networks. *Electronics* **2021**, *10*, 3024. [CrossRef]
18. Tarasia, N.; Swain, A.R.; Roy, S.; Kar, U.N. Improved localized sleep scheduling techniques to prolong WSN lifetime. *Scalable Comput. Pract. Exp.* **2021**, *22*, 81–92. [CrossRef]
19. Yao, Y.-D.; Wang, C.; Li, X.; Zeng, Z.; Zhao, B.; Su, Z.; Li, H. Multihop clustering routing protocol based on improved coronavirus herd immunity optimizer and Q-learning in WSNs. *IEEE Sens. J.* **2023**, *23*, 1645–1659. [CrossRef]
20. Tao, J.; Zhang, R.; Qiao, Z.; Ma, L. Q-Learning-based fuzzy energy management for fuel cell/supercapacitor HEV. *Trans. Inst. Meas. Control* **2022**, *44*, 1939–1949. [CrossRef]
21. Hsu, R.C.; Lin, T.-H.; Su, P.-C. Dynamic energy management for perpetual operation of energy harvesting wireless sensor node using fuzzy Q-learning. *Energies* **2022**, *15*, 3117. [CrossRef]
22. Karunanayake, P.N.; Könsgen, A.; Weerawardane, T.; Förster, A. Q learning based adaptive protocol parameters for WSNs. *J. Commun. Netw.* **2023**, *25*, 76–87. [CrossRef]
23. Hajizadeh, H.; Nabi, M.; Goossens, K. Decentralized configuration of TSCH-based IoT networks for distinctive QoS: A deep reinforcement learning approach. *IEEE Internet Things J.* **2023**, *10*, 16869–16880. [CrossRef]
24. Al-Jerew, O.; Bassam, N.A.; Alsadoon, A. Reinforcement learning for delay tolerance and energy saving in mobile wireless sensor networks. *IEEE Access* **2023**, *11*, 19819–19835. [CrossRef]
25. Redhu, S.; Hegde, R.M. Cooperative network model for joint mobile sink scheduling and dynamic buffer management using Q-learning. *IEEE Trans. Netw. Serv. Manage.* **2020**, *17*, 1853–1864. [CrossRef]
26. Huang, H.Y.; Kim, K.T.; Youn, H.Y. Determining node duty cycle using Q-learning and linear regression for WSN. *Front. Comput. Sci.* **2021**, *15*, 151101. [CrossRef]
27. Shafiq, M.; Ashraf, H.; Ullah, A.; Tahira, S. Systematic literature review on energy efficient routing schemes in WSN—A survey. *Mobile Netw. Appl.* **2020**, *25*, 882–895. [CrossRef]

28. Kamble, A.A.; Patil, B.M. Systematic analysis and review of path optimization techniques in WSN with mobile sink. *Comput. Sci. Rev.* **2021**, *41*, 100412. [CrossRef]
29. Chen, H.; Qin, Y.; Lin, K.; Luan, Y.; Wang, Z.; Yu, J.; Li, Y. PWEND: Proactive wakeup based energy-efficient neighbor discovery for mobile sensor networks. *Ad. Hoc. Netw.* **2020**, *107*, 102247. [CrossRef]

electronics

MDPI

Article

Investigation of the Information Interaction of the Sensor Network End IoT Device and the Hub at the Transport Protocol Level

Viacheslav Kovtun *, Krzysztof Grochla and Konrad Połys

Internet of Things Group, Institute of Theoretical and Applied Informatics Polish Academy of Sciences, Bałtycka 5, 44-100 Gliwice, Poland; kgrochla@iitis.pl (K.G.); kpolys@iitis.pl (K.P.)
* Correspondence: kovtun_v_v@vntu.edu.ua

Abstract: The study examines the process of information transfer between the sensor network end IoT device and the hub at the transport protocol level focused on using the 5G platform. The authors interpreted the researched process as a semi-Markov (focused on the dynamics of the size of the protocol sliding window) process with two nested Markov chains (the first characterizes the current size of the sliding window, and the second, the number of data blocks sent at the current value of this characteristic). As a result, a stationary distribution of the size of the sliding window was obtained both for the resulting semi-Markov process and for nested Markov chains, etc. A recursive approach to the calculation of the mentioned stationary distribution is formalized. This approach is characterized by linear computational complexity. Based on the obtained stationary distribution of the size of the sliding window, a distribution function is formulated that characterizes the bandwidth of the communication channel between the entities specified in the research object. Using the resulting mathematical apparatus, the Window Scale parameter of the TCP Westwood+ protocol was tuned. Testing has shown the superiority of the modified protocol over the basic versions of the BIC TCP, TCP Vegas, TCP NewReno, and TCP Veno protocols in conditions of data transfer between two points in the wireless sensor network environment.

Keywords: information and communication technologies; data transfer; transport protocol; end IoT device; hub; sliding window size; bandwidth

Citation: Kovtun, V.; Grochla, K.; Połys, K. Investigation of the Information Interaction of the Sensor Network End IoT Device and the Hub at the Transport Protocol Level. *Electronics* **2023**, *12*, 4662. https://doi.org/10.3390/electronics12224662

Academic Editors: Dionisis Kandris and Eleftherios Anastasiadis

Received: 31 October 2023
Revised: 9 November 2023
Accepted: 13 November 2023
Published: 15 November 2023

1. Introduction

A key role in managing modern network traffic belongs to transport layer protocols (in particular the TCP (Transmission Control Protocol) [1–4]). By controlling network connections at a point-to-point level, the main algorithms of these protocols form both quantitative and qualitative characteristics of bidirectional packet flow following the physical characteristics and level of congestion of a network route used. Therefore, it is a property of transport protocols that make a decisive contribution to ensuring the reliability, stability and performance of data networks. The latter makes the task of modeling the behavior and analyzing the performance of transport protocols especially the TCP protocol very relevant. We separately, note that the study of the properties of the TCP protocol in various application scenarios is relevant, including because more than 95% of all data flows in the world are controlled by this protocol [5–7].

The Internet of Things (IoT) is built on existing network infrastructure, technologies and protocols currently used in homes, offices and the Internet. This means that most IoT runs on existing TCP/IP networks. TCP/IP uses a four-layer model with specific protocols at each layer. The diagram below (Figure 1) shows a comparison of the protocols currently in use and those most likely to be used for IoT [8–10] in the future.

Modern Internet protocols	Current and expected IoT protocols
Application layer	
HTTP (FTP, SMTP, IMAP)	MQTT (COAP, AMQP)
Transport layer	
TCP, UDP	TCP, UDP
Network layer	
IPv4, IPv6	IPv6, IPv4
Communication layer	
Ethernet, Wi-Fi, GSM	Ethernet, Wi-Fi, GSM, LTE-M, Lora, SigFox
Protocol	
TCP/IP model Internet and IoT protocols	

Figure 1. Analysis of the most current protocols used in the TCP/IP model to support Internet and IoT data flows.

Figure 1 shows that most changes will occur at the communication and application layers, while the network and transport layers are likely to remain unchanged. Thus, by solving the problems of optimizing the TCP protocol for the specific needs of IoT, we are engaged in promising research, the results of which will be in demand not only in the present but also in the future.

Note that TCP creates end-to-end connections on top of the inherently unreliable and best-effort IP packet service through a unique procedure known as a "three-way handshake". During this process, the client sends three TCP segments to the server, and the server responds, establishing a connection. However, due to the nature of IP routing networks, packets containing the TCP segment requesting a new connection and the server's response can sometimes get lost, leading to uncertainty for the communicating hosts. The third message in the sequence enhances the overall reliability of the connection. TCP employs distinct terminology for its connection establishment process. It utilizes a solitary bit known as the SYN (SYNchronization) bit to signify a connection request. This single bit is encapsulated within a comprehensive 20-byte (typically) TCP header, and additional data, including the Initial Sequence Number (ISN) for segment tracking, is transmitted to the receiving host. ACKnowledgments for connections and data segments are confirmed using the ACK bit, while a request to conclude a connection is conveyed through the FIN (FINal) bit. It's important to highlight that when transmitting a single request and response pair within segments, TCP necessitates the creation of an additional seven packets. This results in a substantial packet overhead, and the entire procedure tends to be sluggish when operating over high-latency (delayed) connections. This is a contributing factor to the growing popularity of UDP, especially as networks continue to improve their reliability.

However, let us focus on the network and transport layers in more detail. At the network layer, IPv6 will dominate in the long term. It is unlikely that IPv4 will be used, but it may play a role in the initial stages. Most IoT devices for the home, such as smart light bulbs, currently use IPv4. TCP dominates the transport layer on the Internet. It is used in both HTTP and many other popular Internet protocols (SMTP, POP3, IMAP4, etc.). MQTT (Message Queuing Telemetry Transport) protocol is expected to become one of the main application layer protocols for messaging. Currently, MQTT uses TCP. However, there is an opinion [8] that in the future, due to lower overhead costs, the share of using UDP to serve IoT needs will grow (MQTT-SN operating on top of UDP will probably become more widespread). At the same time, the share of IoT devices that simultaneously operate both on the Internet and on IoT is growing. This statement is confirmed by data on protocol support for IoT platforms: Microsoft Azure (MQTT, AMQP, HTTP and HTTPS), AWS (MQTT, HTTPS, MQTT over WebSockets), IBM Bluemix (MQTT, HTTPS, MQTT), Thingworx (MQTT, HTTPS, MQTT, AMQP). Thus, the trinity of the Internet, IoT, and

TCP will be stable and inseparable in the future. At the same time, we will mention the main disadvantages of TCP, namely, the difficulty in setting up and managing specific use cases; the protocol does not guarantee the delivery of data packets. It is the optimization of TCP protocol parameters to eliminate these shortcomings in the context of the interaction between the "sensor network end IoT device" and the "hub" that is the motivation for the research presented below.

2. State-of-the-Art

The chosen subject area is characterized by a high intensity of research. As confirmation of this fact, we mention both highly cited [1–9] and new research works [10–15].

In [10], T. Toprasert and W. Lilakiataskun introduce a Markov Decision Process (MDP) aimed at improving congestion avoidance. The researchers argue that the MIMD mechanism surpasses both TCP-Illinois and TCP-Scalable in managing window size congestion. Drawing from their findings, they presented a new iteration of the TCP protocol named TCP-Siam. This protocol incorporates a coefficient designed to optimize the congestion window (cWnd), enhancing performance when packets are lost over loss-prone links in a WMN.

In [11], Hurni and colleagues investigated methods to enhance TCP performance. They delved into the impacts of distributed caching coupled with local retransmission techniques. In this approach, every intermediary node stores TCP segments and retransmits the segment if its ACK (acknowledgement) is notably delayed, determined by RTT (round-trip time) estimates. They incorporated their solution into the Contiki OS's uIP stack with a module they named "caching and congestion control" (cctrl). This was then tested across several radio-duty cycling MAC (medium access control) protocols using a real-world testbed of seven TelosB motes. While experiments revealed the cctrl module boosted TCP throughput in numerous settings, its efficacy largely hinged on the specific RDC MAC protocols employed.

Kim et al., as mentioned in [12], undertook an experimental analysis of TCP performance over RPL within an IPv6-driven testbed. This testbed, a low-power and lossy network (LLN), incorporated 30 TelosB devices operating on the TinyOS BLIP stack, complemented by one LBR (LoWPAN border router) and a Linux server. Their observations revealed that TCP displayed notable throughput disparities across nodes in multi-hop LLNs. Moreover, RPL's lack of consideration for traffic load balancing could potentially deteriorate TCP performance.

In an effort to rectify the TCP fairness issues among LLN endpoints, Park and Paek introduced TAiM (TCP assistant in the middle) in [13]. This system intervenes mid-way in TCP communication, specifically at the LBR, and adjusts the RTT of ongoing flows. TAiM holds onto packets and deliberately introduces a delay before forwarding them. This means a flow with reduced throughput experiences a briefer delay, while a flow with increased throughput faces a more extended delay. Experiments utilizing the BLIP stack indicated that TAiM enhanced TCP fairness without compromising the overall throughput.

Gomez et al., in [14], offered guidelines for streamlined TCP implementation, optimized for IoT contexts. Specifically, when considering the RTO algorithm, they advocated for the adoption of CoCoA within TCP.

With the advent of advanced low-power embedded devices boasting greater processing capabilities and increased memory, Kumar et al. demonstrated that a comprehensive TCP can comfortably operate within the CPU and memory limits of contemporary wireless sensor network platforms. They achieved this by implementing a full-scale TCP named TCPlp [15], which draws from the complete capabilities of the TCP in the FreeBSD OS.

As the main results in most of the mentioned works, there are various estimates of the stationary mathematical expectation of the bandwidth of the communication channel, which is controlled by the transport protocol (directly or consolidated). There is also a study of the second moment of bandwidth [12–15] for a point stochastic process, specially selected within the original process of information transfer. Such a trend is quite understandable because, among the typical tasks that are solved in the field of information and communication technologies, there is not only the determination of the state of the communication process at an arbitrary moment in time but also the forecasting of its development and the planning of resources "for growth". It is in such problems that the highest points of the distributions of controlled parameters, as well as their quantiles, are of greatest interest.

Most of the mentioned estimates of the characteristic parameters of the information interaction process are formalized in the form of simple algebraic constructs, in which the efficiency and simplicity of calculations are balanced by the generality of the approximation of the obtained values. In particular, general assessments completely ignore several significant features that appear in situations where the process of information exchange in sensor networks is observed. Let us formulate the most obvious of these features:

- service signals regarding the result of the transfer of a packet of data blocks arrive at random moments, which leads to the interpretation of RTT as a stochastic parameter with a known distribution, depending on the size of the sliding window of the transport protocol. This feature represents the basic specificity of the Ultra-Reliable Low Latency Communication (URLLC) technology as a component of the 5G platform;
- when forming the parametric space of the model of the information exchange process, one should take into account the fact that in real information and communication systems, both the size of the sliding window and the bandwidth of the communication channel are large but always finite. This feature represents the basic specificity of the Massive Machine-Type Communications (mMTC) technology as a component of the 5G platform;
- both the distribution of the size of the sliding window and the distribution of bandwidth should be defined in terms of the ratio of RTT and bandwidth of the communication channel. This approach will make it possible to eliminate the potential influence of the speed characteristics of the data transfer channel on the adequacy of the description of the studied process by the created mathematical apparatus. This feature represents the basic specificity of the enhanced Mobile BroadBand (eMBB) technology as a component of the 5G platform.

Taking into account the strengths and weaknesses of the mentioned methods, we will formulate the necessary attributes of scientific research.

Studied object: The object of our research is the transport layer for managing the process of data transfer between the sensor network end IoT device and the hub using the communication capabilities of the 5G platform.

Research subject: the probability theory and mathematical statistics, the stochastic processes theory, the queuing theory, and the experiment planning theory.

The aim of the research: is to formalize the process of information transfer between the sensor network end IoT device and the hub at the transport level with the determination of the essential characteristic parameters of the protocol.

Research objectives:

- to formalize the process of information transfer between the sensor network end IoT device and the hub based on the stochastic processes theory and the queuing theory;
- to formalize in the analytical basis of the researched process the stationary distribution of the size of the sliding window as a characteristic parameter that determines the intensity of the information flow from the addressee;
- to formalize in the analytical basis of the researched process the distribution function of the bandwidth of the communication channel between the entities specified in the research object;
- justify the adequacy of the proposed mathematical apparatus and demonstrate its functionality with an example.

Main contribution. The study examines the process of information transfer between the sensor network end IoT device and the hub at the transport protocol level. In this context, a queuing system with controlled input flow, deterministic service, feedback and an unlimited queue is synthesized. The authors interpreted the research object as a semi-Markov (focused on the dynamics of the size of the protocol sliding window) process with two nested Markov chains (the first characterizes the current size of the sliding window, and the second—the number of data blocks sent at the current value of this characteristic). As a result, a stationary distribution of the size of the sliding window (a parameter that determines the intensity of the information flow from the addressee) was obtained both for the resulting semi-Markov process and for nested Markov chains, etc. A recursive approach to the calculation of the mentioned stationary distribution is formalized. This approach is characterized by linear computational complexity. Based on the obtained stationary distribution of the size of the sliding window, a distribution function is formulated that characterizes the bandwidth of the communication channel between the entities specified in the research object.

One of the key algorithms in the family of TCP-like transport protocols is an algorithm conventionally called Additive-Increase/Multiplicative-Decrease (AIMD) [16–18]. The AIMD algorithm is focused on increasing the intensity of the flow of packets generated by the sender if the recipient confirms their successful delivery. The mathematical model presented in Section 3 reflects the impact of the AIMD algorithm on massive traffic between the sensor network end IoT device and the hub. Also in Section 3, an analytical form of bandwidth distribution for such a connection is obtained. Section 4 presents the results of the experiments, describing the equipment used and the technologies used to register the empirical data. This section presents the results of comparing the TCP Westwood+ protocol, the Window Scale parameter of which was determined based on the author's mathematical apparatus, with BIC TCP, TCP Vegas, TCP NewReno, TCP Veno without tuning. In Section 5, conclusions are drawn taking into account the results obtained, and directions for further research are formulated.

3. Materials and Methods

Regardless of the type of operating system, the modern transport protocol is designed to ensure the reliable reception of data blocks sent by the communication channel by the addressee. The TCP protocol uses the sliding window to regulate how many packets are in transit to maximize the transmission throughput assuring the reliability of the communication. A few different algorithms have been proposed to regulate the window size, starting from TCP Reno and NewReno to the TCP BIC and CUBIC used in modern operating systems. Suppose that the size of the sliding window is equal to $l > 0$. The basic mechanism of the regulation of the window size can be modeled as follows: if the sender received confirmation from the addressee about the successful receipt of the data block packet at the current sliding window size, then the sliding window size for the next in line to send the data block packet will be increased to the value $l + \lfloor l/n \rfloor$, where $n \geq 2$, $n \in \mathbb{N}$. Otherwise, the size of the sliding window will be reduced to the value $\lfloor l/n \rfloor$. While this is a simplified model mimicking the behavior of the traditional TCP Reno algorithm, the more advanced window size control algorithms, such as BIC and CUBIC can still be approximated with it.

Let us generalize the probabilities f_i that $i = 1, 2, \ldots$ consecutively sent data blocks will be received by the addressee in the form of a distribution of $\{f_i\}$. We consider the stochastic elements of this set to be independent. Compliance with this condition allows us to classify the entity $\{f_i\}$ as a geometric distribution, the parameter $p \in (0, 1)$ of which characterizes the probability of losing a data block during the transfer process.

We will assume that the circular delay D is constant and equal to one. Under the condition of guaranteed successful transfer of all data blocks sent by n RTT, the protocol will send

$$N(n) = 1 + 2 + \ldots + n = \frac{1}{2}\left(n^2 + n\right) \tag{1}$$

data blocks starting from the size of the sliding window $l = 1$. Therefore, the amount of data equal to $N(n)$ will be sent in $N(t) = O(t^2)$ units of time. The non-linear growing character of the function $N(t)$ prompts the introduction of a parameter whose value will limit this growth from above. As such a parameter, we will use the estimate of the bandwidth limit of the communication channel L.

The estimate L is determined empirically for the communication technology used in the investigated information and communication system and the configuration of the hardware component (which in itself is a non-trivial task). In turn, the circular delay is characterized by a stochastic value γ_l, the value of which depends on the size of the sliding window l, which was relevant at the time of transfer of the corresponding data blocks packet. We denote the distribution function of the stochastic value γ_l as $R_l(t)$ and $D_l = \exists\gamma_l$ (if the latter exists).

We focus our research on the description of the information transfer process between the sensor network end IoT device and the hub. The organization of such a process in modern conditions assumes that the monitoring sender (most often – the end IoT device) always has information for transfer. As already mentioned, in real information and communication systems, the bandwidth of the communication channel is limited (estimate L). This circumstance prompts us to introduce a parameter $l_{\max}$ related to the estimate L, the value of which is the upper limit of the growth of the sliding window size.

Based on the above-formulated features of the information transfer process between the sensor network end IoT device and the hub, the analytical description of this process will be carried out based on the stochastic processes theory and the queuing theory. In this context, our goal is to synthesize a queuing system Σ with controlled input flow, deterministic service, feedback, and unlimited queuing. In the terminology of queuing theory:

- a set of data blocks for transfer is a set of requests;
- a communication channel is a service device;
- a service duration distribution is deterministic with a parameter $t_0 = 1/L$.

The functioning of feedback, which regulates the dynamics of the size of the sliding window, is taken into account by entering the set $W = \{W^+, W^-\}$, where W^+ is a service signal about successful data transfer (positive service signal) and W^- characterizes the reversed situation (negative service signal).

The moment of arrival of a signal of type W is a stochastic value: the probability of the appearance of the signal W^+ is characterized by the parameter $1-p$ and the probability of the appearance of the signal W^- is characterized by the parameter p.

The receipt of requests is regulated by the transport protocol according to the type of service signal W. When a feedback service signal W is received, $k \geq 0$ requests (data blocks) are received from the set of requests to the system Σ. The value k depends on the current size of the sliding window l and the number of facts of receiving negative service signals W^-. Requests available in the system Σ are served sequentially (without regard to priority, in order of arrival). To describe the process of information transfer between the sensor network end IoT device and the hub, it is necessary to analytically characterize the output flow of the system Σ.

We formalize analytically the distribution of the size of the sliding window. Suppose that at the time $t > 0$ the size of the sliding window is determined by the function $l(t)$. Also, let us generalize by the set $\mathrm{T} = \{\tau_i\}$, $i = 1,2,\ldots$, the sequence of moments when the value of the parameter l changed in response to the service signals W. The development of this concept will be the Markov chain $l_i = l(\tau_i)$, $l_i \in X\left\{\overline{2, l_{\max}}\right\}$, $i = 1,2,\ldots$; moreover, the minimum size of the sliding window is $l_{\min} = 2$, which corresponds to the specifics of the investigated process. We define the step process $\{l(t)\}_{t>0}$ as semi-Markov. We characterize the event of a transition of chain $\{l_i\}$ from state u to state v in k steps with probability p^k_{uv}, $u, v \in X$. Based on the above, we write: $p^k_{uv} \underset{k\to\infty}{\to} \pi_v$, $\sum\limits_{l=2}^{l_{\max}} \pi_l = 1$.

Let us generalize the set of probabilities $P\{l(t) = l\}$ $P_l(t) = P\{l(t) = l\}$ and conditional mathematical expectations $\alpha_l = A(\tau_{i+1} - \tau_i | l(\tau_i) = l)$.

By definition: $\varepsilon = \tau_{i+1} - \tau_i > 0$ in which case, either $\tau_{i+1} - \tau_i \geq l/L$ or $\tau_{i+1} - \tau_i \geq \gamma_l$. Accordingly, $\exists\varepsilon : F_l(\varepsilon) \leq 1 - \varepsilon$, where $F_l(\varepsilon)$ are the distribution function of the difference $\tau_{i+1} - \tau_i$ and the conditional mathematical expectation α_l exists if D_l exists. Therefore, if a finite mathematical expectation can be determined for a stochastic quantity γ_l then for a stochastic dependence $P_l(t)$ it is possible to write

$$P_l(t) \underset{t\to\infty}{\to} \alpha_l \pi_l / \sum_{l=2}^{l_{\max}} \alpha_l \pi_l. \tag{2}$$

The logic of reasoning embodied in expression (2) echoes that which is the basis of the ergodic theorem for semi-Markov processes [11].

We define the stationary distribution π_l for the Markov chain $\{l_i\}$ in the form of Chapman's equations:

$$\begin{gathered}\pi_i = f_{i-1}\pi_{i-1} + (1 - f_{2i})\pi_{2i} + \\ +(1 - f_{2i+1})\pi_{2i+1} \forall 2i \leq l_{\max},\end{gathered} \tag{3}$$

$$\pi_i = f_{i-1}\pi_{i-1} \forall l_{\max} < 2i < 2l_{\max}, \tag{4}$$

where $f_i = (1-p)^i$ and

$$\pi_{l_{\max}} = f_{l_{\max}-1}\pi_{l_{\max}-1} + f_{l_{\max}}\pi_{l_{\max}}. \tag{5}$$

In the states π_i defined by Equation (3) the system Σ can enter both under the condition of linear growth (under the condition of receiving positive service signals) and under the condition of gradual decline (under the condition of receiving negative service signals).

In the states π_i defined by Equation (4) the system Σ can enter only under the condition of linear growth (provided positive service signals are received).

Equation (5) describes the situation when the system Σ has reached the upper limit of the size of the sliding window $l_{\max}$ and is in this state before the arrival of a negative service signal.

Obtaining an explicit analytical solution to the system of Equations (3) and (4) taking into account Equation (5) is difficult, and its practical implementation will be accompanied by significant computational costs even with the empirical selection of normalizing constants. So, let us resort to the recurrent representation of π_i taking

$$F_i = \prod_{k=1}^{i} f_k, \; j = \lfloor l_{\max}/2 \rfloor. \tag{6}$$

We will obtain:

$$\pi_i = \pi_j C_i, \tag{7}$$

where the values π_j are determined by the normalization condition and

$$C_i = F_{i-1} \forall j < i < l_{\max}, \tag{8}$$

$$C_{i-1} = \frac{1}{f_{i-1}}(C_i - (C_{2i}(1 - f_{2i}) + C_{2i+1}(1 - f_{2i+1}))), \tag{9}$$

$$C_{l_{\max}} = F_{l_{\max}-1}/(1 - f_{l_{\max}}). \tag{10}$$

If we substitute Expressions (8) and (9) into Equation (7) and take into account notation (6) and expression (10), we will obtain the original system of Equations (3), (4) and Expression (5). Therefore, Expression (7) completely determines the distribution of π_l and the recurrent procedure characterized by Expressions (8)–(10) is characterized by linear complexity $O(l_{\max})$.

Let us combine the entities $l(t)$ (the current size of the sliding window) and $n(t)$ (the number of data blocks sent at the size of the sliding window $l(t)$) into a dual function $\eta(t) = \{l(t), n(t)\}$. We accept $n(t) = 1$ each time when, as a result of receiving service signals W, the size of the sliding window $l(t)$ changes. We denote the moment of sending the i-th data block (the moment of completion of service of the i-th request by the system Σ) as τ_i'. Accordingly, if $\tau_{i_1}' > \tau_{i_2}'$ then $i_1 > i_2$. The sequence $\eta_i = \eta(\tau_i')$ is by definition a Markov chain.

Accepting the notation $\pi_\eta = \pi(l, n)$, we formulate the stationary distribution π_η of the Markov chain $\{\eta_i\}$ based on Chapman's equations:

$$\pi(l,n) = (1-p)\pi(l,n-1)\forall l, 1 < n < l, \tag{11}$$

$$\begin{gathered}\pi(l,1) = (1-p)\pi(l-1,l-1) + p\sum_{j=1}^{2l}\pi(2l,j) + \\ +p\sum_{j=1}^{2l+1}\pi(2l+1,j) = 0\forall 2l \le l_{\max},\end{gathered} \tag{12}$$

$$\pi(l,1) = (1-p)\pi(l-1,l-1)\forall l_{\max} < 2l < 2l_{\max}, \tag{13}$$

$$\begin{gathered}\pi(l_{\max},1) = (1-p)\pi(l_{\max}-1,l_{\max}-1) + \\ +(1-p)\pi(l_{\max},l_{\max}).\end{gathered} \tag{14}$$

Based on Equation (11), we write

$$\pi(l,n) = (1-p)^{n-1}\pi(l,1) = (1-p)^{n-1}\pi_l. \tag{15}$$

The distribution π_η is determined by Expressions (7) and (15).

Let us focus on defining conditional mathematical expectations α_l. If the system Σ has time to process the requests in the queue before the service signal W arrives, then $\tau_{i+1} - \tau_i = \gamma_l$. Otherwise, $\tau_{i+1} - \tau_i = l/L$. Accordingly:

$$\alpha_l = \int_{lt_0}^{\infty} t dR_l(t) + lt_0 R_l(lt_0),\ t_0 = 1/L. \tag{16}$$

The change in the size of the sliding window carried out by the mechanisms of the transport protocol is one of the main sources of the stochastic nature of the output stream of the system Σ. A significant characteristic parameter of this flow is bandwidth B. The stochastic characteristic B is defined as the ratio of the number of outgoing requests of the system Σ for the time interval $[\tau_i, \tau_{i+1})$ to the duration of this interval.

For the effective application of the transport protocol in wireless communication networks of 5G technology, the analytical formalization of the distribution of the stochastic characteristic B is relevant. Using 5G technology, the authors focus on such an area of its application as eMBB. The Industry 5.0 paradigm declares the active use of virtualization with the effect of presence, which is impossible without the stable transfer of high-resolution video streams. Moreover, it is wireless communication channels that are optimal in terms of expectations of industrialists, officials, environmentalists and consumers.

If the continuous right function $N(t)$ characterizes the number of outgoing requests of the system Σ at the time t then the character B can be described by the expression

$$B = \frac{N(\tau_{i+1}) - N(\tau_i)}{\tau_{i+1} - \tau_i}. \tag{17}$$

Note, that the logic of determining the size of the sliding window implemented in the transport protocol assumes: if $l < \gamma_l$ then $\tau_{i+1} - \tau_i = \gamma$ and $B = l/\gamma_l$. If the condition $l < \gamma_l$ is not fulfilled, then, for an a priori positive queue length of the system Σ, equality $B = L$ holds. We have already obtained the analytical characterization of the independent stochastic value $l(t)$ (the size of the sliding window, see above). Using the full probability formula [19] $\forall 0 \leq x < L$, we define the distribution function $F_B(x)$ as

$$F_B(x) = P(B < x) = \sum_{i=2}^{l_{\max}} P(l = i)P(D > i/x), \tag{18}$$

We introduce the limit P_l defined by expression (2) into expression (18):

$$F_B(x) = \sum_{l=2}^{l_{\max}} P_l(1 - R_l(l/x)) = 1 - \sum_{l=2}^{l_{\max}} P_l R_l(l/x). \tag{19}$$

Expression (19) is valid for $\forall (x < L) \cup (B(L) = 1)$.

Therefore, we have analytically formalized both the distribution of the size of the sliding window and the distribution of the bandwidth of the communication channel for the process of information transfer between the sensor network end IoT device and the hub based on the stochastic processes theory and the queuing theory. The study takes into account that:

- service signals regarding the result of the transfer of a packet of data blocks arrive at random moments, which led to the interpretation of RTT as a stochastic parameter with a known distribution, depending on the size of the sliding window;
- when forming the parametric space of the model of the researched process, it is taken into account that in real information and communication systems both the size of the sliding window and the bandwidth of the communication channel is finite;
- both the sliding window size distribution and the bandwidth distribution are defined in terms of the ratio of RTT and bandwidth of the communication channel. This eliminates the potential influence of the speed characteristics of the data transfer channel on the adequacy of the description of the researched process by the created mathematical apparatus.

4. Results and Discussion

Let us start setting up the experiment by specifying the entities mentioned in the research object, namely, "sensor network ends IoT device" and "hub". By the hub, we understand the data processing center [20] in the classical sense of this term. Now let us define the end IoT device as an integral element of the sensor network. A modern trend in the organization of sensor networks is the use of edge computing technology [21–23] for processing the data collected by the sensors to relieve the burden of the connection channel with the hub. The spread of this technology is explained by an objective fact: the scale of sensor networks is constantly growing. The introduction of edge computing makes it possible to smooth out the problem of controllability of the sensor network in the conditions of its expansion. The implementation of edge computing technology on a cloud computing platform has both advantages (scalability, survivability) and disadvantages (complications of ensuring confidentiality and regular subscription costs).

The authors propose to carry out preprocessing of sensor data on a developed set of low-level data processing centers. Technologically, each such low-level data preprocessing center or end IoT device is a Raspberry Pi computer (the authors focus on the 4B and Compute Module 4 models, which combine a powerful quad-core ARM v8 Cortex-A72 processor with a full range of current communication interfaces, including Gigabit Ethernet, Wi-Fi 802.11ac, Bluetooth 5.0). Another important fact in favor of the Raspberry Pi is that this computer can function under the control of both an open license Linux operating system (individual features of the installation process are solved with the help of the NOOBS tool presented by Raspberry Pi and a licensed Windows 10 IOT operating system. With this interpretation of the sensor network end IoT device, it is possible to implement the process of information transfer both under the control of the TCP protocol (Next Generation TCP/IP Stack) and under the control of the TCP BIC, TCP CUBIC, Highspeed TCP, H-TCP,

TCP Hybla, TCP Illinois, TCP Low Priority, TCP Vegas, TCP NewReno, TCP Veno, TCP Westwood+, YeAH-TCP.

The application of the mathematical apparatus presented in the previous section for estimating the size of the sliding window and the bandwidth of the communication channel is implemented through the possibilities defined by the standards RFC1323, RFC2018, and RFC3168, namely:

- TCP Window Scale Option: the ability to vary the size of the sliding window up to the limit value of $2^{30} = 1$ GB,
- TCP selective acknowledgment (SACK) options: the possibility of feedback (receiving positive and negative sensory signals from the addressee),
- Explicit Congestion Notification (ECN): the ability to detect congestion of the communication channel without losing data packets.
- TCP timestamps: the possibility of more accurate measurement of RTT due to Prevention Against Wrapped Sequence ACK numbers (PAWS).

The external resource https://www.speedguide.net/analyzer.php (accessed on 8 November 2023) was used to check the current settings of the transport protocol.

Other settings of the transport protocol regarding the used operating system were carried out according to the advice of colleagues from the Pittsburgh Supercomputing Center: https://www.psc.edu/research/networking/tcp-tune (accessed on 8 November 2023).

The New TTCP utility (nuttcp: https://www.nuttcp.net/Welcome%20Page.html (accessed on 8 November 2023)) was used to directly measure the performance of the TCP/IP stack. The advantages of this utility are:

- a simple and effective method of measuring bandwidth via TCP or UDP,
- cross-platform,
- the ability to check the effectiveness of the local TCP/IP stack (loopback),
- the correct termination of TCP connections, and the ability to work with clients under NAT.

Let us apply the mathematical apparatus obtained in the previous section, generalized by Expression (19), to the task of analyzing real network information transactions. By default, we assume that the function $R_l(t) = R(t)$ does not depend on the size of the sliding window and is a normal distribution with parameters $\mu = 60$ ms and $\sigma = 30$ ms.

Figure 2 visualizes the stationary distribution of the size of the sliding window π_l for different probabilities p of the appearance of the service signal W^- and at $l_{\max} = 120$ ms (typical value for the Internet): $\pi_l = f(l)$, $p_{W^-}, l_{\max} = \text{const}$.

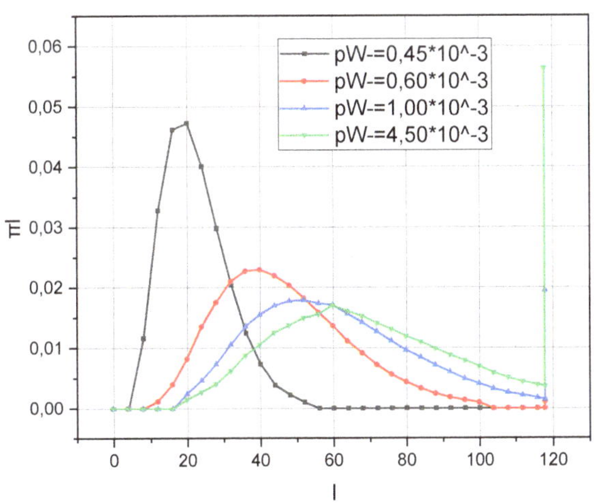

Figure 2. Visualization of the dependence of $\pi_l = f(l)$ at $p_{W^-}, l_{\max} = \text{const}$.

The transport protocol changes the size of the sliding window l according to the ratio between the maximum size of the sliding window $l_{\max}$ and the probability of losing a packet of data blocks p_{W^-}. It is possible to track three main modes in the dynamics of changing the value of the size of the sliding window. In the first mode, the size of the sliding window quickly reaches its maximum value and keeps it almost all the time during the information transaction. In the second mode, the arrival of a negative service signal W^- leads to a reduction in the size of the sliding window by half with a quick return to the maximum value $l_{\max}$. In this mode, the corresponding distribution π_l has two extremes at the points $l_{\max}$ and $l_{\max}/2$. As the probability of the appearance of a negative service signal p_{W^-} increases, the duration of the sliding window size in a stable state (third mode) decreases. The distribution π_l loses its extremum at this point $l_{\max}/2$. If the probability of the appearance of a negative service signal W^- exceeds 0.1, then the value $l_{\max} > 10$ does not have a noticeable effect on the appearance of the distribution π_l. Breaks of the smooth nature of individual curves from Figure 2 are explained by the chosen mathematical apparatus for describing the studied process which is a priori stepwise.

Figure 3 shows the mathematical expectation μ_B and the root mean square deviation σ_B of the bandwidth B as a function of the probability p_{W^-}. The curves are obtained for different values of the estimate L at a constant value of $l_{\max} = 70$.

From Figure 3a it can be seen that the curves of the mathematical expectation μ_B are monotonic decreasing functions under conditions of the increasing value of the argument p_{W^-}. Such a theoretically determined trend coincides with the real behavior of the network construct, which was investigated by the nuttcp utility. It can be seen that the corresponding value of the estimate L determines the initial value of the curve $\mu_B = f(p_{W^-})$ and determines the rate of its decreasing with increasing probability p_{W^-}. Thus, the accuracy of the estimation of the bandwidth of the communication channel is an important factor that affects the performance of information transfer in conditions of low interference (low probability of the appearance of a negative service signal W^-). This is an important conclusion because it is a factor that should determine the shift in the focus of attention of researchers from the search for more accurate methods of estimating the bandwidth of a communication channel (low probability p_{W^-}) to searching for more secure methods of encoding data packets or causes of interference (high probability p_{W^-}).

Figure 3b shows that the function $\sigma_B = f(p_{W^-})$ can be both monotonic and have an extremum on the interval $p_{W^-} \in (0,1)$. This fact allows us to predict the possibility of setting the problem of finding the optimal value of the bandwidth of the communication channel for the parametric space L, $l_{\max}$, p_{W^-}.

Figure 4 visualizes the mathematical expectation μ_B and the root mean square deviation σ_B of the bandwidth B as a function of the circular delay parameters D. The arguments are the probability values p_{W^-}.

Note, that the nature of the functional dependencies presented in Figures 3 and 4, coincides, which indicates the relationship between the parameters L D. Figure 4 shows that the curves of the mathematical expectation μ_B are monotonic decreasing functions under conditions of the increasing value of the argument p_{W^-}. Such a theoretically determined trend coincides with the real behavior of the network construct, which was investigated by the nuttcp utility. It is interesting that the curve σ_B can be both monotonic and have an extremum on the interval $p_{W^-} \in (0,1)$. This circumstance makes it possible to supplement the parametric space of the potential problem of finding the optimal value of the bandwidth of the communication channel with the parameter D.

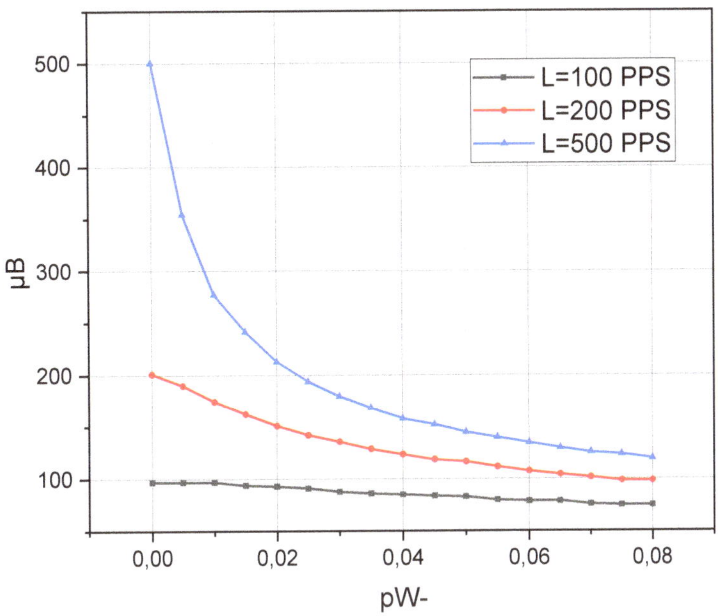

(a)

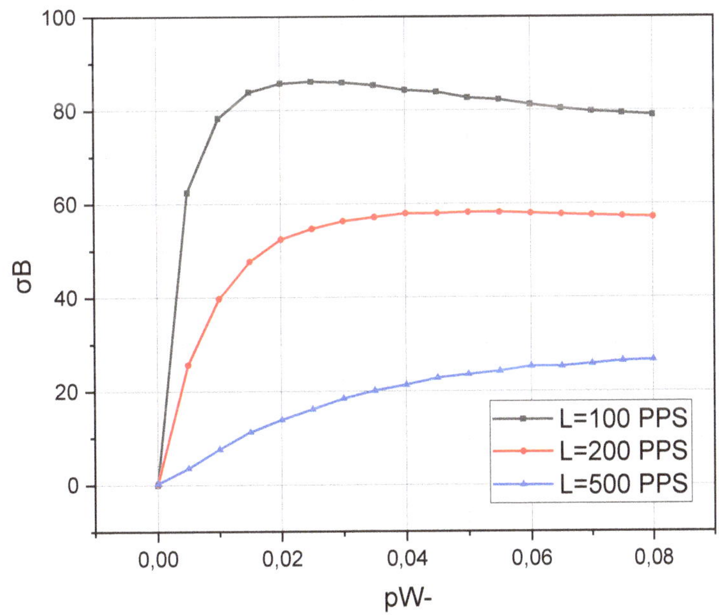

(b)

Figure 3. (**a**) Visualization of the dependence of $\mu_B = f(p_{W^-})$ at $L =$ const. (**b**) Visualization of the dependence of $\sigma_B = f(p_{W^-})$ at $L =$ const.

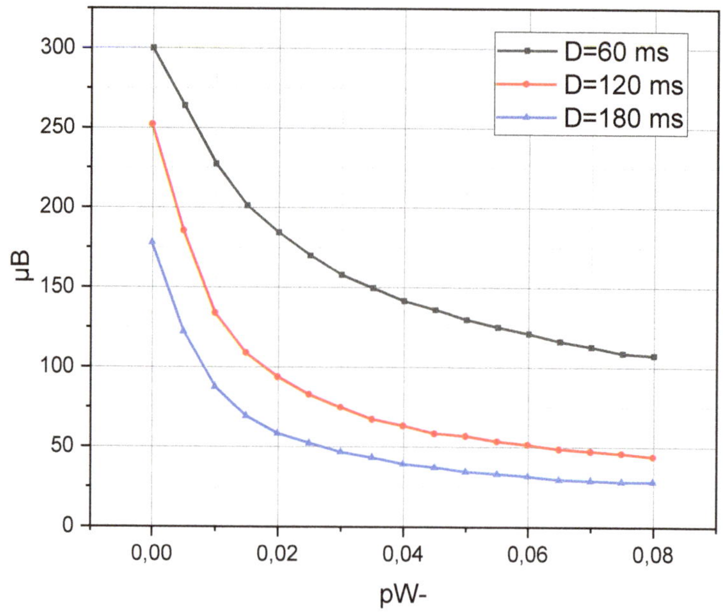

(a)

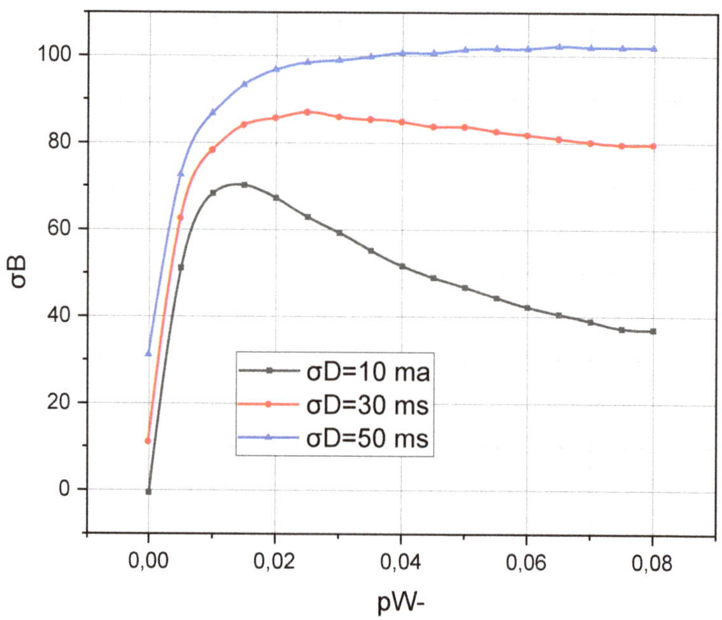

(b)

Figure 4. (**a**) Visualization of the dependence of $\mu_B = f(p_{W-})$ at $D =$ const. (**b**) Visualization of the dependence of $\sigma_B = f(p_{W-})$ at $\sigma_D =$ const.

We will conclude the experimental section by comparing the TCP Westwood+ protocol, the Window Scale parameter of which was determined based on the author's mathematical apparatus, with the BIC TCP, TCP Vegas, TCP NewReno, TCP Veno protocols without tuning (see Figure 5). The communication channel was supported by 5G technology.

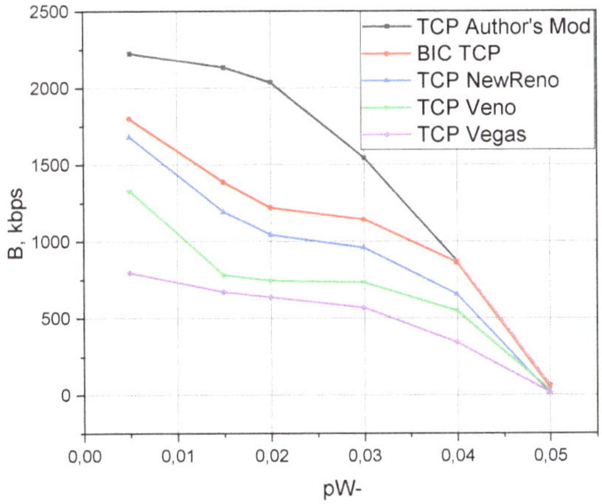

Figure 5. Empirical dependences $B = f(p_{W-})$ for a stable wireless network configuration and different transport protocols.

The results shown in Figures 3 and 4 confirm the adequacy of the mathematical apparatus proposed by the authors, and the results shown in Figure 5 testify to its applied potential. However, this potential can be fully revealed only by researching the relevant optimization problems.

5. Conclusions

The object of this research is the transport layer for managing the process of data transfer between the sensor network end IoT device and the hub using the communication capabilities of the 5G platform. In this case, the authors focus their attention on the TCP protocol. We interpreted the research object as a semi-Markov (focused on the dynamics of the size of the sliding window of the protocol) process with two nested Markov chains (the first characterizes the current size of the sliding window, and the second—the number of data blocks sent at the current value of this characteristic).

As a result, a stationary distribution of the size of the sliding window (a parameter that determines the intensity of the information flow from the addressee) was obtained both for the resulting semi-Markov process and for nested Markov chains, etc. A recursive approach to the calculation of the mentioned stationary distribution, which is characterized by linear computational complexity, is formalized. Based on the obtained stationary distribution of the size of the sliding window, a distribution function is formulated that characterizes the bandwidth of the communication channel between the entities specified in the researched process.

Future research. As we mentioned earlier, the object of our research is the transport layer for managing the process of data transfer between the sensor network end IoT device and the hub using the communication capabilities of the 5G platform. In this article, we presented the basic mathematical apparatus for studying this process. We see its further application in determining the optimal TCP parameters for exact QoS policies that will be used to manage information interaction in a 5G cluster that supports the operation of a sensor network using URLLC, mMTC, and eMBB technologies.

The results presented in the article showed the promise of using the TCP protocol in a sensor network, the end IoT devices of which generate massive traffic. At the same time, we note that the TCP protocol has some problems with data security, namely [24–26]:

- TCP is incapable of safeguarding a segment against message modification attacks due to its lack of protection for the checksum field. This field is intended to detect alterations in a segment, but it remains vulnerable to message modification attacks, allowing for the manipulation of TCP segments without detection. Additionally, there are no mechanisms for peer entities to detect message modification attacks.
- TCP does not provide data encryption capabilities, making it unable to maintain the security of segment data against message eavesdropping attacks. TCP transports unencrypted data from the application layer, leaving any valuable information exposed to potential interception.
- TCP is unable to defend connections against unauthorized access attacks because it verifies a peer entity solely based on the source IP address and port number, which can be easily modified by attackers.

We will try to remove these limitations in our future studies.

Author Contributions: Conceptualization, V.K.; methodology, V.K.; software, V.K.; validation, K.G. and K.P.; formal analysis, V.K.; investigation, V.K.; resources, K.G. and K.P.; data curation, K.G. and K.P.; writing—original draft preparation, V.K.; writing—review and editing, V.K.; visualization, V.K.; supervision, V.K.; project administration, V.K.; funding acquisition, V.K. All authors have read and agreed to the published version of the manuscript.

Funding: Project "Methodology for Increasing the Dependability of Information Systems for Critical Use with a Heterogeneous Wireless Interface", reg. no. 2022/45/P/ST7/03450, the POLONEZ BIS 2 program, implemented by the National Science Center in Krakow.

Data Availability Statement: Most data is contained within the article. All the data are available on request due to restrictions, e.g., privacy or ethics.

Acknowledgments: The authors are grateful to all colleagues and institutions that contributed to the research and made it possible to publish its results.

Conflicts of Interest: The authors declare no conflict of interest.

References

1. Kovtun, V.; Grochla, K. Investigation of the Competitive Nature of eMBB and mMTC 5G Services in Conditions of Limited Communication Resource. *Sci. Rep.* **2022**, *12*, 16050. [CrossRef] [PubMed]
2. Kovtun, V.; Izonin, I.; Gregus, M. Formalization of the Metric of Parameters for Quality Evaluation of the Subject-System Interaction Session in the 5G-IoT Ecosystem. *Alex. Eng. J.* **2022**, *61*, 7941–7952. [CrossRef]
3. Bilyk, O.; Obelovska, K. Power Consumption Analysis at MAC-Sublayer of Wireless Sensor Networks. In *Advances in Artificial Systems for Logistics Engineering*; Springer: Cham, Switzerland, 2022; pp. 27–36. [CrossRef]
4. Lin, J.; Cui, L.; Zhang, Y.; Tso, F.P.; Guan, Q. Extensive Evaluation on the Performance and Behaviour of TCP Congestion Control Protocols under Varied Network Scenarios. *Comput. Netw.* **2019**, *163*, 106872. [CrossRef]
5. Marin, A.; Rossi, S.; Zen, C. Size-Based Scheduling for TCP Flows: Implementation and Performance Evaluation. *Comput. Netw.* **2020**, *183*, 107574. [CrossRef]
6. Kovtun, V.; Altameem, T.; Al-Maitah, M.; Kempa, W. The Markov Concept of the Energy Efficiency Assessment of the Edge Computing Infrastructure Peripheral Server Functioning over Time. *Electronics* **2023**, *12*, 4320. [CrossRef]
7. Ding, L.; Tian, Y.; Liu, T.; Wei, Z.; Zhang, X. Understanding Commercial 5G and Its Implications to (Multipath) TCP. *Comput. Netw.* **2021**, *198*, 108401. [CrossRef]
8. Bruhn, P.; Kühlewind, M.; Muehleisen, M. Performance and Improvements of TCP CUBIC in Low-Delay Cellular Networks. *Comput. Netw.* **2023**, *224*, 109609. [CrossRef]
9. Li, F.; Guo, Z.; Liang, B.; Yi, X.; Wang, X.; Li, W.; Wang, Y. A Measurement Study on Device-to-Device Communication Technologies for IIoT. *Comput. Netw.* **2021**, *192*, 108072. [CrossRef]
10. Kovtun, V.; Altameem, T.; Al-Maitah, M.; Kempa, W. Information Technology for Maximizing Energy Consumption for Useful Information Traffic in a Dense Wi-Fi 6/6E Ecosystem. *Electronics* **2023**, *12*, 3847. [CrossRef]
11. Hurni, P.; Bürgi, U.; Anwander, M.; Braun, T. TCP Performance Optimizations for Wireless Sensor Networks. In Proceedings of the Wireless Sensor Networks: 9th European Conference, EWSN 2012, Trento, Italy, 15–17 February 2012; Springer: Berlin/Heidelberg, Germany, 2012; Volume 7158, pp. 17–32. [CrossRef]
12. Kim, H.-S.; Im, H.; Lee, M.-S.; Paek, J.; Bahk, S. A Measurement Study of TCP over RPL in Low-Power and Lossy Networks. *J. Commun. Netw.* **2015**, *17*, 647–655. [CrossRef]

13. Park, M.; Paek, J. TAiM: TCP Assistant-in-the-Middle for Multihop Low-Power and Lossy Networks in IoT. *J. Commun. Netw.* **2019**, *21*, 192–199. [CrossRef]
14. Gomez, C.; Arcia-Moret, A.; Crowcroft, J. TCP in the Internet of Things: From Ostracism to Prominence. *IEEE Internet Comput.* **2018**, *22*, 29–41. [CrossRef]
15. Kumar, S.; Michael, P.A.; Kim, H.-S.; Culler, D.E. Tcplp: System design and analysis of full-scale TCP in low-power networks. *arXiv* **2018**, arXiv:1811.02721. Available online: https://www.researchgate.net/profile/Hyung-Sin-Kim/publication/328800924_TCPlp_System_Design_and_Analysis_of_Full-Scale_TCP_in_Low-Power_Networks/links/5bef0defa6fdcc3a8ddbdd21/TCPlp-System-Design-and-Analysis-of-Full-Scale-TCP-in-Low-Power-Networks.pdf?origin=publicationDetail&_sg%5B0%5D=ao08rSTIp-huC9mwHDffX6nPqYWS-F-bllTl6AwYmksypELboerDfT-s8mOKnN6ukMDVtaoguaDN3cHFBjsmkw.NoAQuyppjva5rfmHUWLDodjPT7xtjCGFkUuOig7fk5BY3e-WSlFoWX-YyKmIHwc6-hqfBHzNcfGzhz0fR2kp-w&_sg%5B1%5D=9MFgJwBY3EWoM_n8znysdSBewA54Si3mSzqlmTi8SXdZ45TPwdNyFsnTsE88RUd66bFoFJJFCKp9TX9BMr6jWC0rQn-RWY8Rk3TIkPzOMqHi.NoAQuyppjva5rfmHUWLDodjPT7xtjCGFkUuOig7fk5BY3e-WSlFoWX-YyKmIHwc6-hqfBHzNcfGzhz0fR2kp-w&_iepl=&_rtd=eyJjb250ZW50SW50ZW50IjoibWFpbkl0ZW0ifQ%3D%3D&_tp=eyJjb250ZXh0Ijp7ImZpcnN0UGFnZSI6Il9kaXJlY3QiLCJwYWdlIjoiX2RpcmVjdCIsInBvc2l0aW9uIjoicGFnZUhlYWRlciJ9fQ (accessed on 8 November 2023).
16. Obelovska, K.; Snaichuk, Y.; Selecky, J.; Liskevych, R.; Valkova, T. An Approach Toward Packet Routing in the OSPF-Based Network with a Distrustful Router. *Wseas Trans. Inf. Sci. Appl.* **2023**, *20*, 432–443.
17. Auzinger, W.; Obelovska, K.; Dronyuk, I.; Pelekh, K.; Stolyarchuk, R. A Continuous Model for States in CSMA/CA-Based Wireless Local Networks Derived from State Transition Diagrams. *Proc. Int. Conf. Data Sci. Appl.* **2021**, 571–579. [CrossRef]
18. Izonin, I.; Tkachenko, R.; Krak, I.; Berezsky, O.; Shevchuk, I.; Shandilya, S.K. A Cascade Ensemble-Learning Model for the Deployment at the Edge: Case on Missing IoT Data Recovery in Environmental Monitoring Systems. *Front. Environ. Sci.* **2023**, *11*, 1295526. [CrossRef]
19. Semenov, A.; Semenova, O.; Kryvinska, N.; Tromsyuk, V.; Tsyrulnyk, S.; Rudyk, A.; Kacprzyk, J. Advanced Correlation Method for Bit Position Detection towards High Accuracy Data Processing in Industrial Computer Systems. *Inf. Sci.* **2023**, *624*, 652–673. [CrossRef]
20. Kor, A.-L.; Yanovsky, M.; Pattinson, C.; Kharchenko, V. SMART-ITEM: IoT-Enabled Smart Living. In Proceedings of the 2016 Future Technologies Conference (FTC), San Francisco, CA, USA, 6–7 December 2016. [CrossRef]
21. Zaitseva, E.; Rabcan, J.; Levashenko, V.; Kvassay, M. Importance Analysis of Decision Making Factors Based on Fuzzy Decision Trees. *Appl. Soft Comput.* **2023**, *134*, 109988. [CrossRef]
22. Durnyak, B.; Tymchenko, B.H.O.; Tymchenko, O.; Anastasiya, D. Research of Image Processing Methods in Publishing Output Systems. In Proceedings of the 2018 XIV-th International Conference on Perspective Technologies and Methods in MEMS Design (MEMSTECH), Lviv, Ukraine, 18–22 April 2018. [CrossRef]
23. Mochurad, L.I. Canny Edge Detection Analysis Based on Parallel Algorithm, Constructed Complexity Scale and CUDA. *Comput. Inform.* **2022**, *41*, 957–980. [CrossRef]
24. Kou, L.; Wu, J.; Zhang, F.; Ji, P.; Ke, W.; Wan, J.; Liu, H.; Li, Y.; Yuan, Q. Image Encryption for Offshore Wind Power Based on 2D-LCLM and Zhou Yi Eight Trigrams. *International J. Bio-Inspired Comput.* **2023**, *22*, 53–64. [CrossRef]
25. Cheng, Y.; Liu, Y.; Zhang, Z.; Li, Y. An Asymmetric Encryption-Based Key Distribution Method for Wireless Sensor Networks. *Sensors* **2023**, *23*, 6460. [CrossRef] [PubMed]
26. Hazaimeh, O.M.A.; Al Jamal, M.F.; Alomari, A.; Bawaneh, M.J.; Tahat, N. Image Encryption Using Anti-Synchronisation and Bogdanov Transformation Map. *Int. J. Comput. Sci. Math.* **2022**, *15*, 43. [CrossRef]

electronics

MDPI

Article

Information Fusion for 5G IoT: An Improved 3D Localisation Approach Using K-DNN and Multi-Layered Hybrid Radiomap

Brahim El Boudani [1,*], Tasos Dagiuklas [1], Loizos Kanaris [2], Muddesar Iqbal [1] and Christos Chrysoulas [3]

[1] School of Engineering, London South Bank University, London SE1 0AA, UK; tdagiuklas@lsbu.ac.uk (T.D.); m.iqbal@lsbu.ac.uk (M.I.)
[2] Signit Solutions Ltd., Nicosia 2311, Cyprus; l.kanaris@sigintsolutions.com
[3] School of Computing, Edinburgh Napier University, Edinburgh EH11 4BN, UK; c.chrysoulas@napier.ac.uk
* Correspondence: elboudab@lsbu.ac.uk; Tel.: +44-20-7815-7815

Abstract: Indoor positioning is a core enabler for various 5G identity and context-aware applications requiring precise and real-time simultaneous localisation and mapping (SLAM). In this work, we propose a K-nearest neighbours and deep neural network (K-DNN) algorithm to improve 3D indoor positioning. Our implementation uses a novel data-augmentation concept for the received signal strength (RSS)-based fingerprint technique to produce a 3D fused hybrid. In the offline phase, a machine learning (ML) approach is used to train a model on a radiomap dataset that is collected during the offline phase. The proposed algorithm is implemented on the constructed hybrid multi-layered radiomap to improve the 3D localisation accuracy. In our implementation, the proposed approach is based on the fusion of the prominent 5G IoT signals of Bluetooth Low Energy (BLE) and the ubiquitous WLAN. As a result, we achieved a 91% classification accuracy in 1D and a submeter accuracy in 2D.

Keywords: indoor localisation; 5G IoT; deep learning; machine learning; information fusion; tracking; Internet of Things

Citation: El Boudani, B.; Dagiuklas, T.; Kanaris, L.; Iqbal, M.; Chrysoulas, C. Information Fusion for 5G IoT: An Improved 3D Localisation Approach Using K-DNN and Multi-Layered Hybrid Radiomap. *Electronics* **2023**, *12*, 4150. https://doi.org/10.3390/electronics12194150

Academic Editor: Franco Cicirelli

Received: 27 August 2023
Revised: 25 September 2023
Accepted: 29 September 2023
Published: 5 October 2023

1. Introduction

Supported by AI (artificial intelligence) and IoT (Internet of Things), 3D positioning is a core enabler for various 5G identity and context-aware applications requiring precise and real-time simultaneous localisation and mapping (SLAM) both indoors and outdoors [1]. Typical scenarios include mobile robots (MRs) performing surgery on a patient, autonomous ground vehicles (AGVs) docking newly arrived products in a smart factory, and unmanned aerial vehicles (UAVs) monitoring crops' status [2–4]. Very recently, a Cisco report [5] predicted an increase in the number of these specialised IoT devices connected to the internet from 8.8 billion in 2018 to 13.1 billion by 2023—1.4 billion of them will be 5G-capable. In this respect, while space giants (SpaceX and Lockheed Martin) [6] are working to improve the outdoor accuracy of GPS III to below 3 m, heterogeneous 5G IoT networks (HetNets) represent themselves as an indispensable source for improving indoor positioning systems (IPSs).

A 3rd Generation Partnership Project (3GPP) release established the requirements for improved indoor/outdoor 3D localisation using a RAT-independent positioning scheme for vertical and horizontal sectors [7]. Therefore, this specification put a strong emphasis on seamless collaborations and fusion between various radio technologies, such as device-to-device communication, ultra-dense communication, millimetre wave (mm wave), sub-6 GHz, and vehicle-to-everything (V2X) [8], and protocols such as IEEE 802.15.1 (Bluetooth Low Energy), IEEE 802.11be (extremely High Throughput WLAN), and IEEE 802.11az (Next Generation Positioning) [9]. In light of this, a very good opportunity has emerged in the area of indoor localisation for both urban areas and smart cities.

To further improve positioning accuracy, researchers have focused on various hybrid approaches. For 5G IoT networks, the location of the user's equipment is estimated using a combination of signal propagation characteristics such as angle of arrival (AOA), time of arrival, time difference of arrival, received signal strength (RSS), RSS difference (RSSD), direction of arrival (DOA), and frequency difference of arrival (FDoA) [3]. These hybrid approaches were recently further surveyed in [10–12]. Among all these approaches, the RSS fingerprint-based method is the most widely used for real-time tracking. Additionally, most of the existing approaches consider the use of the RSS from specific radio technology. However, the offline phase of fingerprint collection requires a considerable amount of human resources and is also time-consuming, especially for complex buildings. For this reason, we propose a K-nearest-neighbours and deep neural network (K-DNN) algorithm to improve 3D indoor positioning. The contributions of this paper can be summarized as follows:

- A realistic information fusion scenario for 5G IoT networks was planned and deployed utilizing a 5G IoT gateway, a Bluetooth Low Energy (BLE) network, and a set of wireless IoT access points without requiring any extra information such as a magnetic-inductive sensor, acoustics, visible light, or a powerline.
- Our implementation used a novel data-augmentation concept for a received signal strength (RSS)-based fingerprint technique to produce a 3D fused hybrid fingerprint. This concept was supported by the interquartile range (IQR) method for the detection and elimination of outliers.
- To improve 3D positioning accuracy, a K-DNN cooperative algorithm was implemented on the constructed hybrid multi-layered radiomap.

The concept presented is a continuation of our previous work in [13,14] towards cooperative localisation. This paper is divided into the following parts: Section 2 covers the state of the art in fingerprint-based techniques for 3D/2D positioning and information fusion methods. The proposed system model and the underlying algorithms are presented in Section 3. The 5G IoT physical network environment is explained in Section 4. The experimental setup is covered in Section 5. Section 6 provides the performance evaluation, and Section 7 analyses the obtained results. Finally, a summary and directions for future work are presented in Section 8.

2. State of the Art

2.1. Received Signal Strength

Received signal strength (RSS) is a way to measure the signal power received by a user's equipment. This is expressed in decibel milliwatts (dBm) or milliwatts (Mw). The RSS-based method is one of the methods widely adopted by the indoor localisation research community. RSS can be used to approximate the distance between a user device (UE) and a transmitting device (Tx), as shown in Figure 1.

Using an RSS indicator (RSSI), a relative measurement of RSS, and a free-space path loss (FSPL) propagation model [15], the distance delta between a UE and Tx can be estimated via the formula below:

$$FSPL(dB) = 20log_{10}(d) + 20log_{10}(f) + \phi \tag{1}$$

where d is the distance expressed in metres; f is the frequency measured in kilohertz, megahertz, or gigahertz; and ϕ is a constant based on the frequency unit. During location determination, this formula assumes that the antennas are lossless and their polarisation is the same. However, this is not often the case in complex and unpredictable environments with continuous noise.

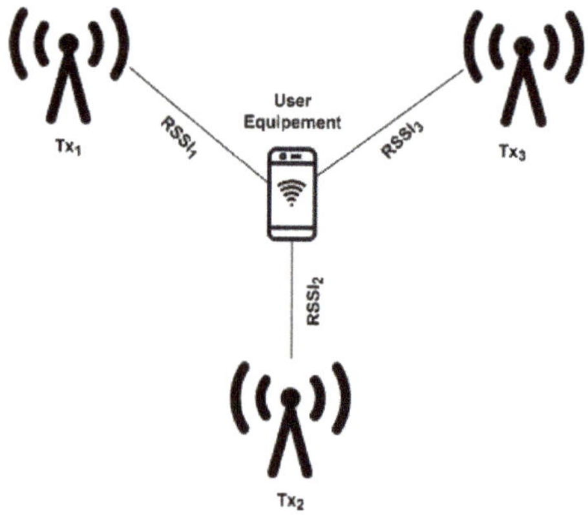

Figure 1. RSS-based positioning.

2.2. RSS Fingerprint-Based 2D and 3D Indoor Positioning

In the RSS fingerprint-based method, unlike the free-space path loss (FSPL) model, the location is estimated by matching the signal received from the user equipment with a database of preconstructed location radiomaps. The most significant advantage of this method is its ability to maintain high accuracy in a cluttered multi-path environment, according to a study conducted by [16,17]. As shown in Figure 2, this technique has two phases: offline and online. In the offline phase, a site survey or measurement campaign is conducted through which a set of RSS signals is collected and linked to its corresponding location XY in the 2D case and XYZ in the 3D case. The constructed radiomap is then used to train a localisation algorithm with a distance error loss function, such as least squares [18], weighted least means [19], maximum likelihood estimation [20], or convex optimisation [21].

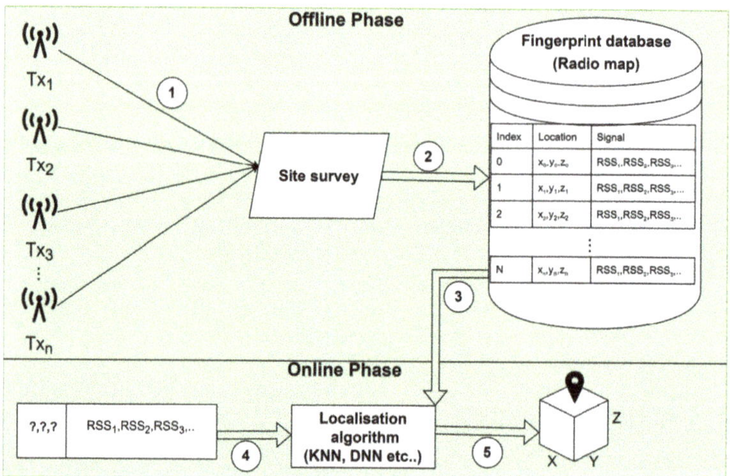

Figure 2. Fingerprint -based positioning phases.

To construct a radiomap, the most commonly used method for collecting signal fingerprints is called war-diving [22]. After identifying the indoor area of interest, the user equipment stays in each position for a specific time interval to obtain enough fingerprint information. As the monitoring device moves along the grid, the collected RSS single is stored in the database along with a reference point. Regarding 5G IoT positioning indoors, the use of this technique was investigated by Huan et al. in [23]. The authors used the Kalman filter to remove the noisy RSS values. Next, a universal kriging (UK) algorithm was used for spatial interpolation and data augmentation to reduce dependency on the fingerprinting database. Finally, the authors trained a KNN model to calculate the user equipment's location, achieving a 1.44 m positioning error. Although this approach is interesting, it was not established whether the system could perform equally in a 3D environment. Additionally, the use of a single base station might seem to save power, but it does not guarantee the same accuracy given the changes in the environment and the LOS issues in cluttered space. Similarly, Gong et al. [24] suggested a two-step KNN (2-KNN) algorithm that used reference signals from the state information of the channel (CSI). During the offline phase, a smooth rank sequence (SRS) estimated the number of received signal paths. During the online phase, a trained 2-KNN algorithm was used to determine the 2D location of the user equipment. Most studies have overlooked 3D localisation, which is essential for scenarios like robot navigation, immersive shopping, and virtual reality. This was the main motivation for us to investigate this area. Further studies on 5G and beyond (6G) can be found in the following survey papers: [25,26].

2.3. *Information Fusion for 5G IoT*

Information fusion for 5G IoT has attracted considerable attention from the research community. This technique, as shown in Figure 3, involves blending data from various sources or sensors using a data fusion system to gain better inference and improve accuracy/precision. This concept produces an effective and reliable IPS (indoor positioning system) while saving the cost of expensive infrastructure [27]. Over the last decade, researchers have attempted to merge data readings from sources such as RFID [28], GPS [29], pedometers [30], BLE (Bluetooth Low Energy) [14], VLC (visible light communication), and many other technologies, as stated in [13,31]. Very recently, Klus et al. [32] examined fusing GNSS with WLAN data in a 5G network to improve positioning. The authors implemented a neural network as their main algorithm. Based on the authors' conclusions, the proposed approach achieved an accuracy of 1 m in an open space and 3.4 m in a cluttered area. A serious limitation of this study was the GNSS's inability to penetrate walls composed of different materials, especially in complex environments. In a more recent work, Alvarez-Merino et al. [33] looked into using WiFi fine-time measurement (FTM), UWB, and cellular-based radio fusion to improve indoor location accuracy. The authors' approach showed promising results. However, unlike [32], this system did not rely on existing infrastructure but required a UWB setup that could be costly and was limited to user equipment with this capability. These limitations motivated us to propose a cost-effective setup based on BLE and WiFi. A detailed discussion of these techniques can be found in the following resources: [34–36].

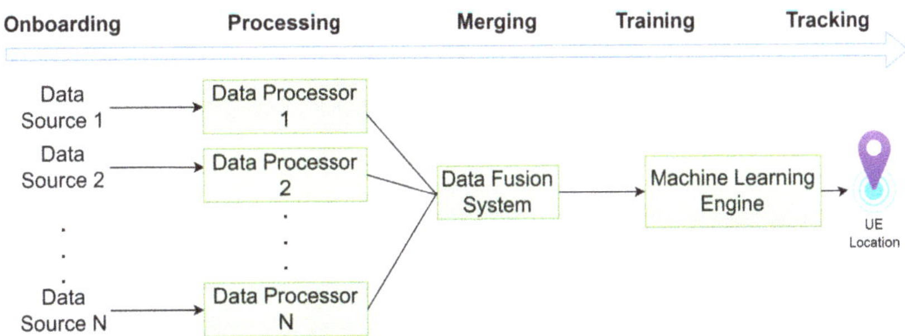

Figure 3. Information fusion localisation process.

BLE Technology

BLE has emerged as a low-cost wireless solution for localising people and assets, offering traditional Bluetooth protocol capabilities over ultra-low power consumption circuits [37]. BLE-enabled devices communicate over 2.4 GHz and use 40 channels (PHY channels) divided by a 2MHz frequency gap. Channels 37, 38, and 39 are used for advertising, while the rest are used for data transfer during a connection. This technology uses a neighbour discovery process (NDP) in which a BLE-enabled device, often referred to as a "scanner", searches for nearby BLE devices called "advertisers" [38]. Once the discovery process has finished, a list of available devices is returned based on availability and RSS value. According to the core specification, BLE 5.0 has improved drastically compared to version 4.2, offering two types of discovery process: basic and advanced [39]. Recently, BLE has become a central source of information fusion amongst the research community. Several research papers have explored the use of this technology to improve indoor localisation accuracy. Kanakareja et al. [40] investigated using the BLE protocol along with LoRa to reduce the distance error of an indoor tracker called "The Things Network" (TTN). The idea is very promising; however, it only works for environments where a low-power wide area network (LP-WAN), like a wireless sensor network, is deployed. To track the movements of elderly people indoors, Kolakowski et al. [41] used BLE and ultra-wide band technologies. While this is a very effective low-power solution, realising it requires the deployment of UWB infrastructure. Additionally, UWB suffers from clock synchronisation issues due to the time-sensitive nature of its pulses, which is not practical for real-time localisation systems [42]. Finally, to label areas like parking lots and meeting rooms with localisation information, Hu et al. [43] proposed a system called Grid-Loc that combined both active radio frequency identification (RDFID) and BLE. Similarly, this solution needed a pre-setup to start tracking and did not make use of widely existing infrastructure technologies such as wireless local-area networks (WLANs). In our work, to improve 3D localisation, we implemented a fusion of BLE and the ubiquitous WLAN. This concept is further discussed in Section 3.

2.4. *Machine Learning*

In the fingerprint-based localisation method, the application of machine learning involves training a model on a radiomap dataset that has been collected during the offline phase. Given a radiomap database, the localisation model aims to infer the state or location of the user's device from the received measurement vector σ, which includes RSS values σ_i from several access points. According to the literature, the widely used algorithms can be classified into deterministic and probabilistic algorithms. The principle behind these methodologies is based on searching a database of fingerprints and finding one or more locations whose RSS values have the highest similarity to the one currently observed.

2.4.1. Probabilistic Approach

In the probabilistic approach, the position is determined based on the likelihood that the user is in the location 'x' given vector or RSS values received during the online phase. Assuming that a set of location candidates L is $L = \{L_1, L_2, L_3, ..., L_m\}$ for any obtained RSS vector values σ, one selects L_i if

$$P(L_i|\sigma) > P(L_j|\sigma) \text{ for } j,k = 1,2,3,...n, \ \ i \neq j \tag{2}$$

where $P(L_i|\sigma)$ is the probability that a user device is at location L_i given the RSS vector σ if its likelihood is higher than that of $P(L_j|\sigma)$.

Finally, using Equation (2), the 3D location $(\hat{x}, \hat{y}, \hat{z})$ can be estimated using the weighted average probability as follows:

$$(\hat{x}, \hat{y}, \hat{z}) = \Sigma_{i=1}^{n}(P(L_i|s)(x_{L_i}, y_{L_i}, z_{L_i})) \tag{3}$$

2.4.2. Deterministic Approach

In the deterministic positioning approach, location λ is considered a non-random vector [44]. The main objective is to estimate $\hat{\lambda}$ at every step. Usually, the location estimate is treated as a linear combination of calibrated points p_i. The principle behind this approach can be summarised in the following equation:

$$\hat{\lambda} = \Sigma_{i=1}^{k} \frac{w_i}{\Sigma_{j=1}^{M} w_j} \lambda_i \quad (4)$$

Here, the set $\{\lambda_1 \ldots \lambda_i\}$ denotes the sequence of reference points associated with Δ_i, which is the distance between the respective radiomap fingerprint $\bar{r}_i$ and the measurement x taken during live positioning, i.e., $\Delta_i = ||x_i - \bar{r}||$. The norm $||.||$ in this equation can be any arbitrary formula. This can be the Mahalanobis norm [45], the Manhattan norm (1 norm) [46], or the Euclidean norm (2 norm) [44]. As this paper focusses on the latter, w_i can be written as follows:

$$\Delta_i = \sqrt{\sum_{j=1}^{N} x_{ij} - s_j)^2} \quad (5)$$

In Equation (4), w_i is a set of non-random weight coefficients assigned to each reference point based on its importance in distinguishing it from other fingerprints. Consequently, the value of w_i assigned to each fingerprint impacts the location estimation. In this case, the weight allocation expressed in Equation (4) refers to the weighted K-nearest neighbours (WKNN) algorithm [46]. A possible value for w_i can be the inverse of the RSS innovation [46], which can be expressed as follows:

$$w_i = \frac{1}{||x - \bar{r}||} \quad (6)$$

If Equation (4) is simplified, it can be assumed that all fingerprints are assigned equal weights. As a result of this assumption, w_i is eliminated, and the formula becomes the K-nearest neighbours (KNN) method. Thus, setting K = 1, the equation yields the simple nearest neighbours (NN) method [44,47]. In terms of performance, it was demonstrated in [44,46] that the KNN and WKNN methods offer a higher degree of accuracy than the NN method in the cases of K = 3 and K = 4, respectively. However, the NN method appears to perform satisfactorily and offers the same results in the presence of high-density RSS radiomaps [48].

Several researchers have addressed the question of indoor localisation in 5G networks using KNN [23,24,49–53]. Despite this, the KNN method alone failed to deal with a highly dense 3D radiomap, as studied in [13,54,55]. This motivated us to propose a combination of deep learning and KNN methods to improve localisation in complex 3D environments. Since the main focus of this paper is on the deterministic positioning approach based on deep learning and K-nearest neighbours, more complex methods such as the database correlation method (DCM), linear discriminant analysis (LDA), and the *k*-anonymity method can be found in [56–58], respectively. The following subsection deals with existing research contributions related to deep learning.

2.4.3. Deep Learning

Deep learning is a subclass of machine learning algorithms based on artificial neural networks (ANNs) and representation learning [59]. Artificial neural networks themselves were inspired by biological networks. These types of algorithms are more powerful than traditional machine learning algorithms as they use multiple connected layers to extract complex patterns from raw data [60]. The training technique used in deep learning can be supervised, semi-supervised, or unsupervised [61]. In 5G networks, the adaptation of these techniques [62,63] in indoor and outdoor localisation has shown some great results. Wafaa et al. [64] studied the use of CNNs to reduce localisation error and improve accuracy. Their approach converted a 2D fingerprint radiomap and its kurtosis values to a 3D RSS radio image. This 3D tensor was then used as an input for their proposed model. This localisation framework was tested in a 20 m $\times$ 20 area. The reported results suggest that this concept could achieve up to 94.13% accuracy in a grid size of 2 m × 2 m with 10 anchors. Although it sounds promising, this concept has not been tested in a 3D environment. Similarly, Yang et al. [65] proposed an indoor 3D localisation scheme based on a 1D CNN and BLE signal fingerprinting. This approach was tested in a 3D space of 4.0 m × 2.0 m × 3.0 m. The authors deployed eight BLE beacons and divided the 3D space into 16 grids of 1 m × 1 m × 1 m in size. Following these steps, the system was able to achieve a 0.25 m error and a precision of almost 100%. A serious limitation of this work was that the framework was tested in a small and uncluttered environment. Furthermore, to achieve the same result, according to the adopted setup, a BLE must be deployed for each 1 m^2. This is usually not cost-effective, especially for large complex buildings. To overcome these two limitations, we suggest the use of a hybrid radiomap and a combination of KNN and DNN to realise a cost-effective scalable solution. The following section covers the proposed approach in detail.

3. The Proposed Approach

Our proposed approach aimed to improve indoor positioning using several 5G IoT wireless signal data sources. This could be achieved by merging actual BLE and WiFi 3D location data with simulated BLE and WiFi location data into a multi-layered hybrid radiomap to save the tedious time spent constructing a fingerprint database. To support this data augmentation approach, K-DNN, a new cooperative positioning algorithm that combines KNN (K-nearest neighbours) and DNNs (dense neural networks) was developed to reduce the localisation error. Figure 4 provides an overview of the algorithmic flow of the proposed K-DNN system. The following subsections describe in detail the K-DNN algorithm used in this paper.

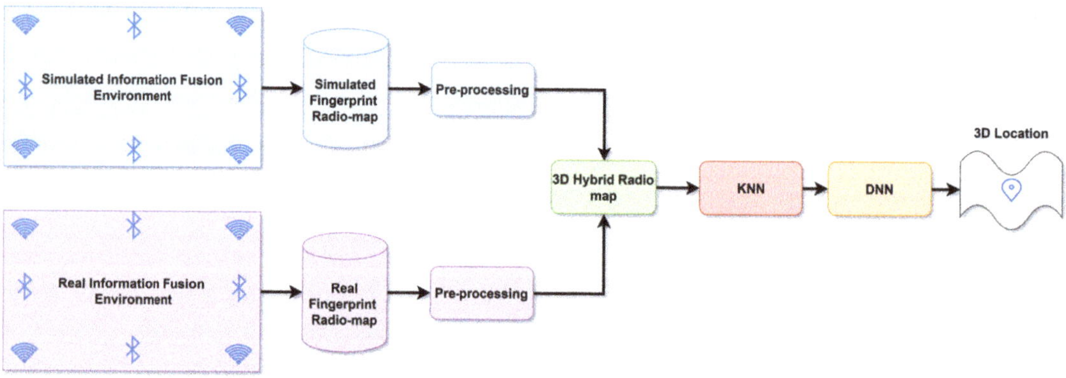

Figure 4. The flow of the proposed K-DNN system model.

3.1. *K-DNN Architecture and Hybrid 3D Localisation for 5G IoT*

K-DNN is a novel cooperative positioning algorithm. Given a set of WLAN transmitters N and a set of BLE transmitters M connected to a set of 3D locations (XYZ), two machine learning models are trained to support each other to achieve minimal distance error. During the offline phase, the algorithm receives two matrices of hybrid radiomaps. This can be mathematically expressed as

$$BLE\_RSS : \{(x_1, y_1, z_1, ble_1 \ldots ble_m), \ldots, (x_i, y_i, z_i, ble_1 \ldots ble_m)\}$$
$$WLAN\_RSS : \{(x_i, y_i, z_i, wlan_1 \ldots wlan_n), \ldots, (x_i, y_i, z_i, wlan_1 \ldots wlan_n)\}$$

K-DNN begins by eliminating outliers from the given radiomaps using the IQR (interquartile) method [66]. The cleaned fingerprint datasets are then merged into a single radiomap. Next, a min–max normalisation technique is implemented to convert the RSSI values of the BLE and WLAN into the same scale. As a final step in this phase, the KNN model is first trained to predict the 2D location (X, Y), and the DNN is trained to predict the 1D location (Z).

During the online phase, the K-DNN receives the following input:

$$BLE\_RSS_{online} : \{ble_1, \ldots, ble_m\}$$
$$WLAN\_RSS_{online} : \{wlan_1, \ldots, wlan_n\}.$$

Given this, KNN attempts to approximate the 2D (XY) locations as an output. This outcome is then fed along with the original input received by KNN into the DNN, which in turn predicts the 1D (Z) location. As a result, the 3D (XYZ) location is realised through this cooperative prediction approach. The main reason for including these two models was the nature of the 1D (classes) and 2D (continuous values) outputs.

3.2. K-DNN Model Architecture

3.2.1. KNN

The K-nearest neighbours algorithm is a non-parametric supervised machine learning algorithm used for pattern classification and regression. This means that it does not make any assumptions about the data being analysed. Since learning in KNN is supervised, the trainer has to choose the parameters to achieve the best results. This algorithm was first proposed in 1951 by Evelyn Fix, Joseph Hodges [67], and Thomas Cover [68], who later expanded on it. In the K-DNN algorithm, KNN is used to predict the 2D (XY) location. As previously highlighted in Algorithm 1, this algorithm receives a set of RSS values as input R. This can be written as $R = [RSS_1, RSS_2, \ldots, RSS_n]$

The input provided to KNN consists of a vector of seven normalised RSS values. This part of K-DNN model attempts to reduce the localisation error of the X and Y location using the Euclidean distance. The output of this model is then combined with the original input R and fed into the DNN model.

3.2.2. DNN

Deep learning is a crucial building block in the proposed K-DNN system. It allows the learning of complex patterns and data representations through multiple processing layers [61]. One of the most important architectures in deep learning is the deep neural network, also known as multiple-layer perceptron (MLP) or a deep feed-forward network [69]. The DNN considered in K-DNN is a classification model. Figure 5 shows the number of layers, neurones, and input and output parameters used in this model.

The input layer of this network receives transposed vectors of signal values and 2D locations. This can be expressed as

$$DNN_{input} = [X, Y, RSS_1, RSS_2, \ldots, RSS_n]^T \tag{7}$$

where X and Y are the 2D points predicted by KNN and RSS_i represents the signal value of the ith transmitter (BLE or WLAN).

The calculated result for this layer is then fed into the first hidden layer. Each input element from Equation (7) is multiplied by a specific weight vector $\vec{w}$. The product of this operation is then added to a bias b. The formula for this can be expressed as follows:

$$h1 = \sum_{i=1}^{n} w_i^1 I_i + b_i^1 \tag{8}$$

where I_i is the element ith of the input vector. The summation of all these inputs is then passed onto an activation function unit A. In our proposed network, this is the rectified linear unit (ReLu).

$$A_1 = max(0, h1) \tag{9}$$

Here, A_1 is the activation function of the first hidden layers. The output of this layer is 128 neurones. In the same way,

$$h2 = \sum_{i=1}^{n} w_i^2 a_i^1 + b_i^2 \tag{10}$$

The result of this hidden layer is passed onto a further activation unit A_2:

$$A_2 = max(0, h2) \tag{11}$$

Finally, the output of Equation (11) is received by hidden layer 3 to make a similar calculation for $h1$ and $h2$:

$$h3 = \sum_{i=1}^{n} w_i^3 a_i^2 + b_i^3 \tag{12}$$

The values calculated by Equation (12) are then fed into the activation function below:

$$A_3 = max(0, h3) \tag{13}$$

To predict the correct height of the mobile device, the softmax function equation below is used:

$$\theta(a_i) = \frac{\exp(a_i^3)}{\sum_j \exp(a_j^3)} \tag{14}$$

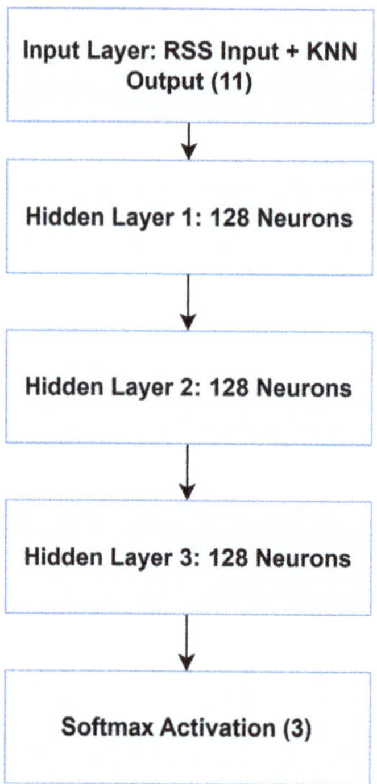

Figure 5. Layers of the DNN.

3.3. K-DNN Psuedocode

For clarity purposes, the Algorithm 1 below explains how K-DNN works.

Algorithm 1: K-DNN Algorithm for 3D Localisation

```
Input  : BLE_RSS : {(x1, y1, z1, ble1...blem)} ;              ▷ Get Hybrid BLE RSS
Input  : WLAN_RSS : {(x1, y1, z1, wlan1.....wlann)} ;         ▷ Get Hybrid WiFi RSS
Output: Λ ;                                                   ▷ Output 3D location
Require: Signal UpperThreshold μ;
Require: Signal LowerThreshold η;
Require: First quartile Q1;
Require: Third quartile Q3;
IQR← Q3-Q1 ;                                                  ▷ Calculate Interquartile
for blei in BLE_RSS and wlani in WLAN_RSS do
    if blei<Q3+(1.5*IQR) and blei>Q1-(1.5*IQR) then
        ble_r← blei ;                                         ▷ Apply IQR method to BLE
    if wlani<Q3+(1.5*IQR) and wlani>Q1-(1.5*IQR) then
        wlan_r← wlani ;                                       ▷ Apply IQR method to WLAN
    RSS← wlan_r ∪ ble_r ;                                     ▷ Fuse BLE and WLAN Radiomaps
end
for RSSi in RSS do
    R ← (RSSi−μ)/(μ−η) ;                                      ▷ Normalize signal
    X_Y ← KNN(R) ;                                            ▷ Apply first model prediction
    Z ← DNN(R, X_Y) ;                                         ▷ Apply second model prediction
    Λ ← X_Y ∪ Z ;                                             ▷ merge results output
end
return Λ
```

4. 5G IoT Physical Network Environment

In this part of the article, we explain the main components of the 5G IoT network that was used in this experiment. For clarity purposes, Figure 6 shows the logical network architecture.

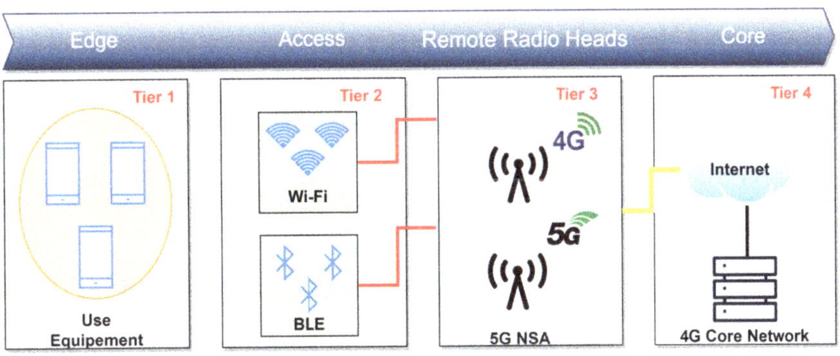

Figure 6. 5G IoT network logical architecture.

4.1. 5G Core Network

In this experimental testbed, we adopted 5G NSA (non-stand-alone) access as suggested by 3GPP release 15 [70]. This concept uses dual connectivity (eNodeB/gNodeB) to provide radio access to 5G-enabled UE (user equipment) via 4G EPC infrastructure, as demonstrated in Figure 7a. The 5G core network complied with 3GPP release 16 [71] and used an open source called Open5gs [72]. This platform implements both 5GC (5G core) and EPC (evolved packet core) using the C language. Open5G has evolved from

4G NextEPC and comes with a WebUI to manage network subscribers. The developed 5G core network was used to configure NR/LTE networks for a private cellular network infrastructure. The core network was virtualised and deployed on a 64-bit Linux machine using a VMWARE workstation. It is worth mentioning that at the time of writing this paper, other projects such as OpenAirInterface [73] and free5GC [74] have been instigated. However, these solutions are not stable yet. A detailed description of these three projects can be found in [75].

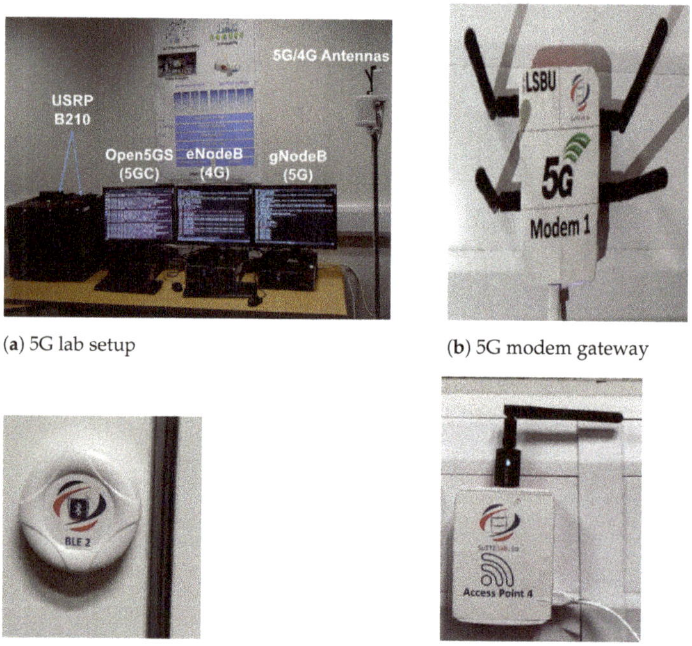

(**a**) 5G lab setup

(**b**) 5G modem gateway

(**c**) Bluetooth Low Energy 2

(**d**) 5G wireless access point 4

Figure 7. 5G IoT test environment.

4.2. eNodeB (4G)/gNodeB (5G)

The Evolved/E-UTRAN Node B is a component in the E-UTRA of 4G LTE11. This component connects subscribers to service providers through the S1-AP protocol linked to S1-MME from the mobility management entity side. The eNodeB has its own radio control functionality that manages a USRP B210 SDR (software-defined radio), as shown in Figure 7a. This component offers a radio service via the air interface. The operating frequency of this radio unit for 4G is between 800 MHz and 2600 MHz, as per the OfCom regulations. A duplexer was also used to reduce the number of antennas needed to keep the transmitter (Tx) and receiver (Rx) synchronised for both radio units. The software side of this solution was implemented on a custom-built PC powered by an i9 CPU with a total memory of 32 GB. This unit was an implementation of 3GPP release 15 [70], as previously highlighted. This meant that it used dual connectivity to offer the service to the user equipment. The 5G-capable device had to first connect to the MME through the eNodeB to attach to a gNodeB. This is why it is called the NSA mode. This unit used the X2AP protocol to communicate with the eNodeB nearby. The dedicated hardware for this base station was similar to the eNodeB. In order to reduce the clock drifting, a 5G radio was offered through a USRP B210 attached to a 5G band 7 cavity duplexer.

4.3. 5G IoT Modem

The 5G gateway implemented in this testbed consisted of a Raspberry Pi 4 model B and a Quectel 5G Quectel RM500Q-GL modem [76], as shown in Figure 7b. This gateway linked the 5G cellular network to the WLAN and BLE networks used to extract fingerprints.

4.4. Wireless Local Area Networks

During the experimental design, five IEEE 802.11ac [77] wireless access points were considered for deployment at the assigned site. Figure 7d depicts one of the access points used in this setup. In this configuration, each transmitter operated at 2.4 Ghz and a coverage range of 45 m, although dual band was possible, as this technology also supports 5 Ghz.

4.5. Bluetooth Low Energy

As a secondary source for information fusion, we considered using the IEEE 802.15.1 standard, which is BLE version 5.0 [78]. The devices used in this experiment operated at 2.4 Ghz and 350 m. Figure 7c illustrates one of the BLE units used in this setup. The following section covers the simulated environment of this architecture.

5. Test Environment

The K-DNN algorithm was tested by combining actual and simulated measurements. The experiment took place in two teaching laboratories at London South Bank University of approximately 126 m^2 (6 m wide by 21 m long by 3 m high), as shown in Figure 8.

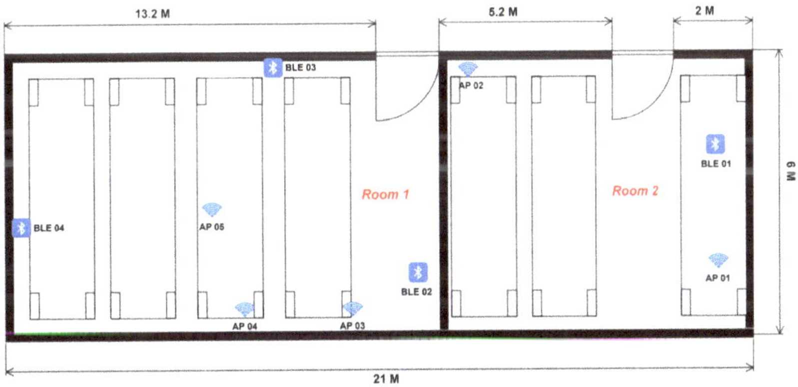

Figure 8. Floor plan with access points and BLE position.

To achieve this task, a 5G IoT network was deployed in the two laboratories. The network consisted of five IEEE 802.11 access points and four IEEE 802.15 BLE units that were randomly placed based on Table 1. Two radiomaps were constructed: the first was generated using an actual measurement campaign, and the second using TruNet wireless, a 3D ray-tracing deterministic simulator [79].

Table 1. The 5G IoT setup location and antenna orientation.

Device	X	Y	Z	Antenna Orientation
AP1	0	0	2	Vertical
AP2	6	6.5	1.5	Horizontal
AP3	0	11	1.5	Vertical
AP4	0	13	1	Horizontal
AP5	3	15	0.5	Horizontal
BLE01	4	0	1	N/A
BLE02	1	9	1.5	N/A
BLE03	6	13	2	N/A
BLE04	3	21	0.5	N/A

5.1. *Radiomap from Actual 5G IoT Measurements*

During data collection, fingerprints were collected in 2236 equally spaced locations (0.5 m spacing) at 0.5 m, 1 m, 1.5 m, and 2.5 m heights, as shown in Figure 9a. At each measurement location, 30 distinct measurements were recorded at an interval of 1 second using the iFused fingerprint data collector developed for Android-based devices, as shown in Figure 9b. The RSS values stored in the radiomap ranged from −103 dBm to −28 dBm. During the measurement campaign, the application recorded data from 5 APs and 4 BLE devices.

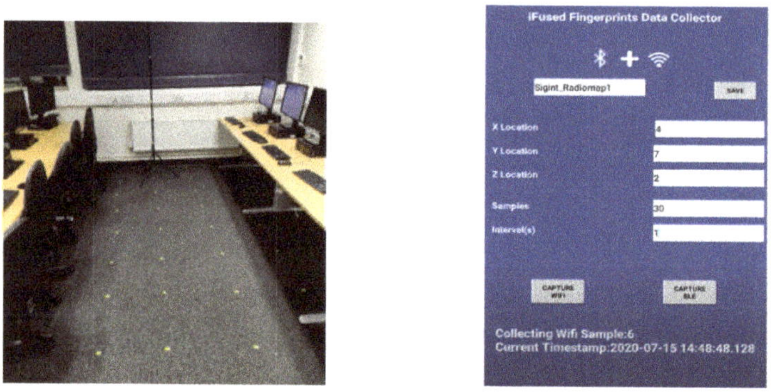

(**a**) FW-208 classroom grid setup (**b**) iFused fingerprint data collector

Figure 9. Physical environment and fingerprint data collector.

5.2. *5G IoT Simulated Radiomap*

5.2.1. TruNet Tool

As previously highlighted, we considered using a 3D ray-tracing (RT) deterministic tool called Trunet [79]. This application constructs 3D radiomaps in conjunction with calibration techniques. The main advantage is that the tool generates efficient radiomaps while saving the time and cost incurred by a measurement campaign. Figure 10a,b illustrate the building simulated using the Trunet software along with a layer of the multi-layered simulated radiomap generated for both WLAN and BLE, respectively.

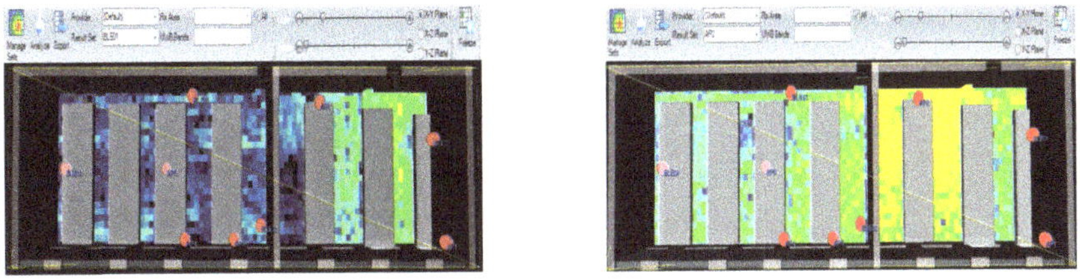

(**a**) Simulated radiomap for BLE 1 (**b**) Simulated radiomap for access point 2

Figure 10. 5G IoT simulated environment radiomap example.

5.2.2. Simulated Radiomap

The second radiomap was constructed using the TruNet simulator. RSS fingerprints were collected according to the procedure used by the authors of [13,80]. To ensure that the measurement recorded by the iFuse application matched the simulated measurements, 5 APs and 4 BLE units were configured according to the antenna radio propagation characteristics in Table 2. Furthermore, the building structure and furniture were configured based on the calibration procedure in [81]. As a result, the same 2236 measurement points were generated and defined as receiver cells. At the end of this process, two layers of fingerprints (2 m and 1.5 m high) were merged with the actual measurement radiomap.

Table 2. The BLE and WLAN radio propagation parameters.

Parameter	BLE	WLAN
Rx sensitivity (dBm)	−70	−120
Tx power (dBm)	8	12
Antenna type	Omnidirectional	Omnidirectional
Max refractions	5	12
Max reflections	5	12
Max diffractions	1	1

5.2.3. The physical Network Behaviour

It was evident that the obtained RSS signal could be affected by various types of noise from the environment. We needed to ensure that the radiomap constructed by the simulation matched with the results of the measurement campaign. Figure 11 shows a strong correlation between the real RSS values and the Trunet values measured for access points and BLE units.

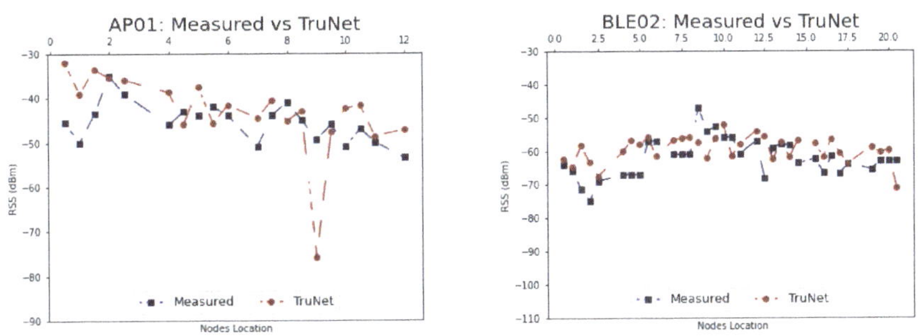

(**a**) Measured vs. simulated fingerprints for AP 01 (**b**) Measured vs. simulated fingerprints for BLE 02

Figure 11. BLE 02 and AP 01 simulated vs. real measurement comparison.

5.3. Preprocessing

5.3.1. Multi-Layered Radiomap Hybridisation

The hybridisation of a radiomap refers to the process of merging simulated and real measurements of the same environment at different height levels. This preprocessing technique merges multiple 3D layers from various available sources. In this experiment, we combined two simulated measurements (2 m and 1.5 m heights) with two layers of real measurements (0.5 m and 1m heights). This technique is novel as far as we know and has not been implemented in previous papers. It could be beneficial for scenarios involving complex buildings where extensive human resources and time are allocated. To ensure that there was a correlation between the simulated and real measurements, we compared the location IDs of the same layer belonging to the same BLE and access point. As demonstrated in Figure 12, there was a strong correlation between the measurements obtained in the simulation and the actual measurements. Furthermore, to prove the feasibility of this technique, we compared the non-fused and fused models, as presented at a later stage in this paper.

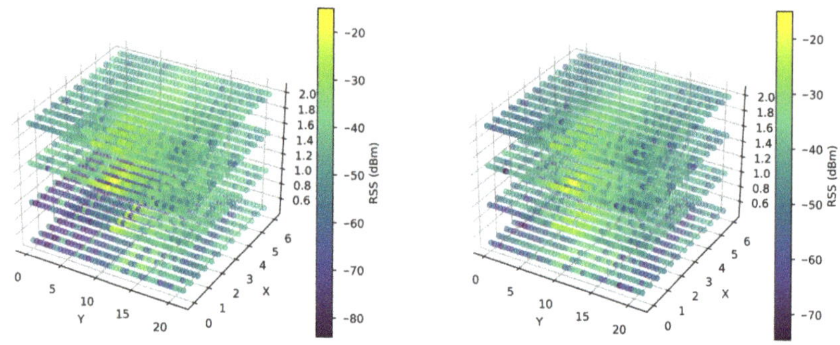

(**a**) Access point 03 hybrid radiomap (**b**) Access point 03 simulated radiomap

Figure 12. Simulated Authors. No need to move it. vs. hybrid radiomap.

5.3.2. Feature Selection

During the feature selection process, a Pearson correlation test was performed between the BLE units and APs, as this was necessary to ensure that there was no redundancy in the information provided to the K-DNN models. Figure 13 clearly shows that there was no substantially positive or negative correlation between the selected BLE unit and AP used in this experiment.

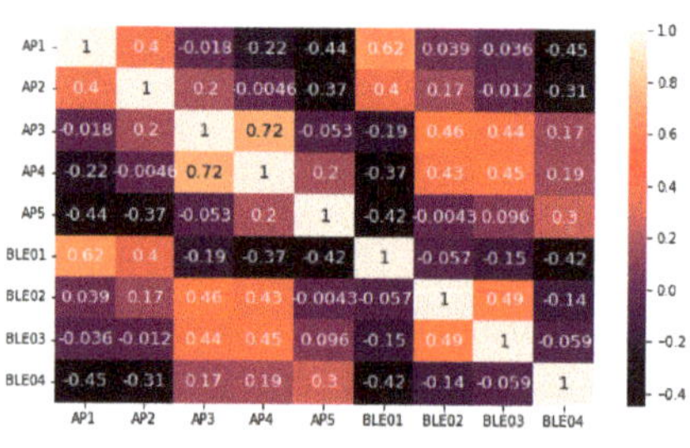

Figure 13. Pearson correlation matrix for the radiomap.

5.3.3. Outlier Elimination

Outliers are generally values that lie an abnormal distance from other values in a normal distribution. In the case of RSS-based positioning, these types of values find their way into a radiomap during the measurement campaign when a signal fluctuation occurs or when there is interference, such as human activity. To deal with this data quality problem, we applied the interquartile method introduced by Upton and Cook in [82], as shown in Figure 14.

In this work, we implemented this method to prevent K-DNN from learning extreme RSS values that were picked up by the receiver during the data collection process. After treating the outliers, 2031 observations were left to train the K-DNN model. Table 3 shows a summary of the considered features and their minimum and maximum values.

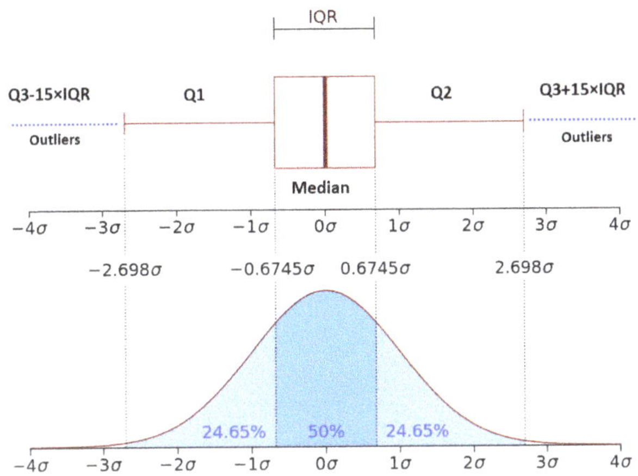

Figure 14. Outlier removal using IQR technique.

Table 3. The features used to construct the fingerprint database.

Variable	Min. Value	Max. Value	Type
X	0	6	Coordinates
Y	0	21	Coordinates
Z	0.5	2	Coordinates
AP1	−84 dBm	−28 dBm	RSS value
AP2	−86 dBm	−30 dBm	RSS value
AP3	−84 dBm	−35 dBm	RSS value
AP4	−87 dBm	−32 dBm	RSS value
AP5	−109 dBm	−37 dBm	RSS value
BLE01	−105 dBm	−32 dBm	RSS value
BLE02	−86 dBm	−32 dBm	RSS value
BLE03	−97 dBm	−35 dBm	RSS value
BLE04	−120 dBm	−42 dBm	RSS value

5.3.4. Data Normalisation

To preserve the relationship between the original data values while speeding up the learning process, a min–max normalisation technique was implemented to scale the original values between 0 and 1. Since the scaled values were negative, we extracted the absolute value. The equation used was

$$\left| \frac{RSS_i - min(RSS)}{min(RSS) - max(RSS)} \right| \quad (15)$$

where $min(RSS)$ refers to the minimum value of the threshold signal in the training signal, that is, −120 dBm, and $max(RSS)$ represents the maximum measured value, that is, −28 dBm. Each measurement of the signal that we needed to convert is denoted by RSS_i, where i is the ith row on the N BLE unit or access point transmitter. For a different scenario, it would be preferable to rely on the receiver sensitivity level as the minimum value while choosing the strongest measured signal value during the offline phase as the maximum value. This process was important for both the KNN and DNN models, as it changed the values of each access point and BLE unit to a common scale, without affecting the differences in the range of values.

5.3.5. One-Hot Encoding

One-hot encoding is the process of converting a column of continuous ordinal numeric values to binary columns based on the distinctive values [13], as shown in Algorithm 2. This process was applied in this experiment to the 1D (Z) values. Mapping the distinctive values 0.5 m, 1.0 m, 1.5 m, and 2.0 m to four binary columns was the result of this process.

Algorithm 2: One-Hot Encoding

Input :Column Z ▷ `get Z columns`
Output:Result matrix of N binary vectors unique values from Z
Dictionary D = [];
Results R = {[],[],[],[]};
for *i in Z.length* **do**
    if i ∉ D:▷ `If value not in dictionary add it`
    key = D[i]
    D[i] = Z[i]
end
return D
Map D into results R columns as binary vector $\{[Z_1], [Z_2], \ldots, [Z_n]\}$

6. Performance Evaluation

Testing the performance of K-DNN involved training the DNN using the ADAM (adaptive momentum) algorithm [83] and KNN using the elbow method [84]. The former is useful for learning highly sparse datasets, while the latter is a technique used to cross-check the model performance against the number of K chosen. Figure 15a reveals how the DNN converged in the 1000th training iteration. The KNN model achieved the lowest error rate at K = 6, as illustrated in Figure 15b.

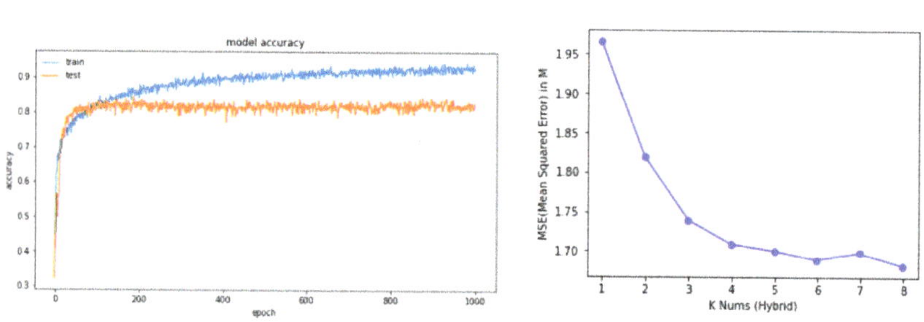

(**a**) DNN Epochs vs. model accuracy (**b**) KNN MSE vs. the number of K selected

Figure 15. 5G IoT simulated environment radiomap example.

Additionally, it is worth noting that the DNNs were trained using the hyperparameters in Table 4. In the following section, we evaluate and compare the performance of this model on different radiomaps.

Table 4. DNN hyperparameters.

Hyperparameter	Value
Learning algorithm	ADAM
Learning rate	0.001
β 1	0.9
β 2	0.999
Dropout	0.35
Momentum	0.99
Batch size	64
ϵ	1e-07
Number of hidden layers	3
Number of hidden layers in each neuron	128

7. Results Analysis

7.1. DNN Scoring

To assess the impact of the proposed approach, we trained four models using different combinations of radiomaps to draw comparisons with the concept suggested in this paper. The four models were trained as follows:

- Model 1: hybrid radiomap (proposed approach).
- Model 2: hybrid radiomap without information fusion.
- Model 3: simulated radiomap.
- Model 4: simulated radiomap without information fusion.

Using 180 random samples, as suggested by the authors in [85], we tested the misclassification performance of each DNN model at various heights: 0.5 m, 1 m, 1.5 m, and 2 m, as illustrated in Figure 16. In the graph, it is clear that the hybrid approach with information fusion achieved the lowest misclassification count out of the four models. As can be seen in Table 5, 91% of the samples—circa 164—were accurately classified. The model trained using the proposed hybrid approach without information fusion came second, with a classification rate of 87% (152 out of 180 samples). The third model was trained with information fusion and a simulated radiomap, and it performed badly compared to the two previous models. This model achieved a classification rate of 73% (132 out of 180 samples). The fourth model, which was trained using a simulated radiomap without information fusion, performed worse, with a classification score of 56 out of 180. These results demonstrate how the proposed hybrid approach outperformed the rest of the training scenarios. Given this, it could be concluded that the hybrid information fusion technique could drastically improve localisation in a 1D environment. Detailed misclassification counts for each height are provided in Table 5. The model with the poorest performance is indicated in red, while the model with the best performance is denoted in blue.

Table 5. Detailed results of DNN misclassification count.

	Hybrid	Hybrid No Fusion	Simulated	Simulated No Fusion
0.5 m	0	9	17	18
1 m	8	8	14	17
1.5 m	3	5	10	11
2 m	5	6	7	10
Total	16	28	48	56

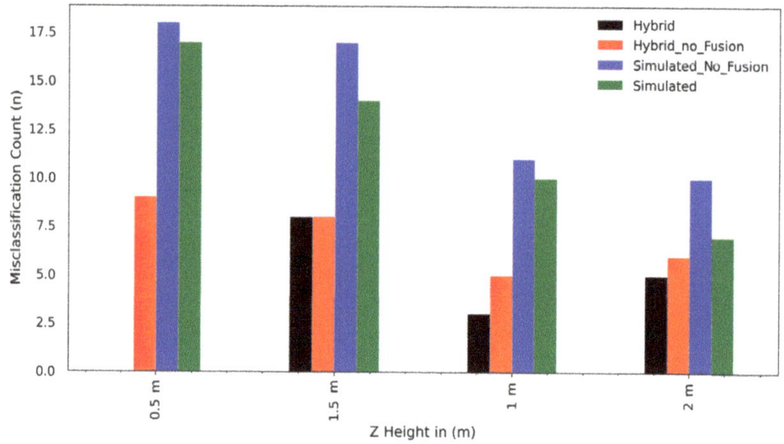

Figure 16. DNN misclassification count results by height.

7.2. KNN Scoring

As in the previous subsection, to assess the feasibility of our proposed technique in 2D localisation, four KNN models were evaluated using 180 samples. The trained models were as follows:

- KNN 1: hybrid radiomap with information fusion.
- KNN 2: hybrid radiomap without information fusion.
- KNN 3: simulated radiomap with information fusion.
- KNN 4: simulated radiomap without information fusion.

Figure 17 shows the cumulative distribution function (CDF) of the error in metres for each KNN model. For the 75th percentile, it is demonstrated that the hybrid with fusion, simulated with fusion, hybrid, and simulated models achieved 90 cm, 1 m, 1.10 m, and 1.20 m errors, respectively. Using the CDF as a metric, the proposed 3D multi-layered hybrid approach achieved a submetre accuracy, in contrast to the rest of the models. Therefore, given the results for KNN and the DNN, it can be strongly argued that the proposed K-DNN method drastically reduced the localisation error.

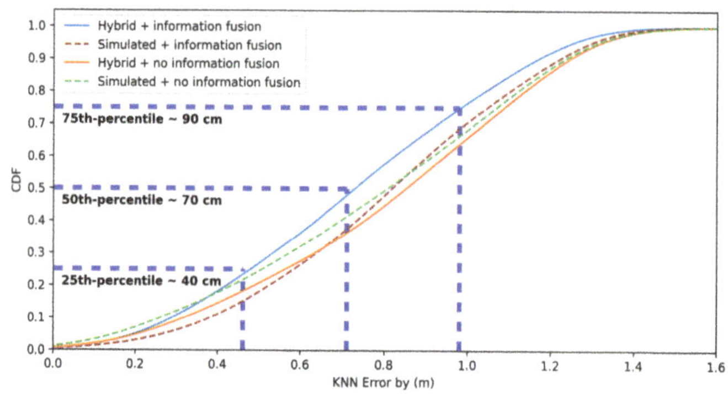

Figure 17. KNN CDF results.

8. Conclusions and Future Work

In this work, we proposed a novel algorithm for improved indoor positioning in 5G IoT networks. The proposed approach used IQR to deal with outliers and a hybrid radiomap to reduce the labour cost incurred during the data collection phase. Additionally, we demonstrated how cooperative machine learning localisation can be implemented on top of this technique. Using this approach, we showed how information fusion implemented on 3D multi-layered radiomaps can be used to reduce the localisation error to the submetre level in 2D and attain a 91% classification rate in 1D. This result could be achieved in a similar environment if the steps in Figure 4 are followed. This concept has the potential for expansion into more intricate indoor positioning scenarios, encompassing diverse radio data sources from a heterogeneous network like 5G micro-infrastructure (including microcells, femtocells, and picocells). Additionally, our proposed K-DNN model demonstrated strong performance with RSS-based IoT and wireless sensor networks. As a result, our future endeavours will focus on enhancing the model by integrating data from different azimuth angles (45°, 90°, 180°, and 360°). Another avenue of research could involve incorporating floor-level detection for buildings with multiple stories.

Author Contributions: Conceptualization, B.E.B. and L.K.; methodology, B.E.B., L.K., T.D. and M.I.; software, B.E.B.; validation, B.E.B., L.K., T.D. and C.C.; formal analysis, B.E.B. and L.K.; investigation, B.E.B.; resources, T.D. and B.E.B.; data curation, B.E.B. and L.K.; writing—original draft preparation, B.E.B. and L.K.; writing—review and editing, L.K., T.D. and C.C.; visualization, B.E.B.; supervision, L.K. and T.D.; project administration, T.D.; All authors have read and agreed to the published version of the manuscript.

Funding: This research received no external funding.

Data Availability Statement: Not applicable.

Conflicts of Interest: The authors declare no conflict of interest.

References

1. Ge, Y.; Wen, F.; Kim, H.; Zhu, M.; Jiang, F.; Kim, S.; Svensson, L.; Wymeersch, H. 5G SLAM using the clustering and assignment approach with diffuse multipath. *Sensors* **2020**, *20*, 4656. [CrossRef]
2. Walia, J.S.; Hämmäinen, H.; Kilkki, K.; Yrjölä, S. 5G network slicing strategies for a smart factory. *Comput. Ind.* **2019**, *111*, 108–120. [CrossRef]
3. Li, G.; Lian, W.; Qu, H.; Li, Z.; Zhou, Q.; Tian, J. Improving patient care through the development of a 5G-powered smart hospital. *Nat. Med.* **2021**, *27*, 936–937. [CrossRef] [PubMed]
4. Khan, S.K.; Naseem, U.; Siraj, H.; Razzak, I.; Imran, M. The role of unmanned aerial vehicles and mmWave in 5G: Recent advances and challenges. *Trans. Emerg. Telecommun. Technol.* **2021**, *32*, e4241. [CrossRef]
5. Cisco. Cisco Annual Internet Report-Cisco Annual Internet Report (2018–2023) White Paper— Cisco. Available online: https://www.cisco.com/c/en/us/solutions/collateral/executive-perspectives/annual-internet-report/white-paper-c11-741490.html. (accessed on 10 February 2022).
6. Norris, P. Satellite Programs in the USA 59. In *Handbook of Space Security*; Springer: Berlin/Heidelberg, Germany , 2020; p. 1133.
7. 3GPP. *TR-2 2.872: Study on Positioning Use Cases*; Tech. Report 16; ETSI: Sophia Antipolis, France , 2018.
8. Wymeersch, H.; Seco-Granados, G.; Destino, G.; Dardari, D.; Tufvesson, F. 5G mmWave positioning for vehicular networks. *IEEE Wirel. Commun.* **2017**, *24*, 80–86. [CrossRef]
9. Leonardo, L.; Yuhei, N.; Kurosaki, M.; Ochi, H. High Precision Localization Protocol with Diversity for 802.11 az. *IEICE Tech. Rep. IEICE Tech. Rep.* **2017**, *117*, 69–74.
10. Alsinglawi, B.; Elkhodr, M.; Nguyen, Q.V.; Gunawardana, U.; Maeder, A.; Simoff, S. RFID localisation for Internet of Things smart homes: A survey. *arXiv* **2017**, arXiv:1702.02311.
11. Deak, G.; Curran, K.; Condell, J. A survey of active and passive indoor localisation systems. *Comput. Commun.* **2012**, *35*, 1939–1954. [CrossRef]
12. Yassin, A.; Nasser, Y.; Awad, M.; Al-Dubai, A.; Liu, R.; Yuen, C.; Raulefs, R.; Aboutanios, E. Recent advances in indoor localization: A survey on theoretical approaches and applications. *IEEE Commun. Surv. Tutor.* **2016**, *19*, 1327–1346. [CrossRef]
13. El Boudani, B.; Kanaris, L.; Kokkinis, A.; Kyriacou, M.; Chrysoulas, C.; Stavrou, S.; Dagiuklas, T. Implementing deep learning techniques in 5G IoT networks for 3D indoor positioning: DELTA (DeEp Learning-Based Co-operaTive Architecture). *Sensors* **2020**, *20*, 5495. [CrossRef]
14. Kanaris, L.; Kokkinis, A.; Liotta, A.; Stavrou, S. Fusing bluetooth beacon data with Wi-Fi radiomaps for improved indoor localization. *Sensors* **2017**, *17*, 812. [CrossRef] [PubMed]
15. Guerra, A.; Guidi, F.; Dardari, D. Single-anchor localization and orientation performance limits using massive arrays: MIMO vs. beamforming. *IEEE Trans. Wirel. Commun.* **2018**, *17*, 5241–5255. [CrossRef]
16. Liu, Y.; Shi, X.; He, S.; Shi, Z. Prospective positioning architecture and technologies in 5G networks. *IEEE Netw.* **2017**, *31*, 115–121. [CrossRef]

17. Horsmanheimo, S.; Lembo, S.; Tuomimaki, L.; Huilla, S.; Honkamaa, P.; Laukkanen, M.; Kemppi, P. Indoor positioning platform to support 5G location based services. In Proceedings of the 2019 IEEE International Conference on Communications Workshops (ICC Workshops), Shanghai, China, 20–24 May 2019; pp. 1–6.
18. Wang, G.; Chen, H.; Li, Y.; Jin, M. On received-signal-strength based localization with unknown transmit power and path loss exponent. *IEEE Wirel. Commun. Lett.* **2012**, *1*, 536–539. [CrossRef]
19. Huang, J.; Liu, P.; Lin, W.; Gui, G. RSS-based method for sensor localization with unknown transmit power and uncertainty in path loss exponent. *Sensors* **2016**, *16*, 1452. [CrossRef] [PubMed]
20. Coluccia, A.; Ricciato, F. On ML estimation for automatic RSS-based indoor localization. In Proceedings of the IEEE 5th International Symposium on Wireless Pervasive Computing 2010, Modena, Italy, 5–7 May 2010; pp. 495–502.
21. Zhang, Y.; Xing, S.; Zhu, Y.; Yan, F.; Shen, L. RSS-based localization in WSNs using Gaussian mixture model via semidefinite relaxation. *IEEE Commun. Lett.* **2017**, *21*, 1329–1332. [CrossRef]
22. Tsui, A.W.T.; Lin, W.C.; Chen, W.J.; Huang, P.; Chu, H.H. Accuracy performance analysis between war driving and war walking in metropolitan Wi-Fi localization. *IEEE Trans. Mob. Comput.* **2010**, *9*, 1551–1562. [CrossRef]
23. Huang, S.; Zhao, K.; Zheng, Z.; Ji, W.; Li, T.; Liao, X. An optimized fingerprinting-based indoor positioning with Kalman filter and universal kriging for 5G internet of things. *Wirel. Commun. Mob. Comput.* **2021**, *2021*, 9936706. [CrossRef]
24. Gong, Y.; Zhang, L. Improved K-nearest neighbor algorithm for indoor positioning using 5G channel state information. In Proceedings of the 2021 IEEE 5th Information Technology, Networking, Electronic and Automation Control Conference (ITNEC), Xi'an, China, 15–17 October 2021; Volume 5, pp. 333–337.
25. Mogyorósi, F.; Revisnyei, P.; Pašić, A.; Papp, Z.; Törös, I.; Varga, P.; Pašić, A. Positioning in 5G and 6G Networks & mdash—A Survey. *Sensors* **2022**, *22*, 4757. [CrossRef]
26. Farahsari, P.S.; Farahzadi, A.; Rezazadeh, J.; Bagheri, A. A Survey on Indoor Positioning Systems for IoT-Based Applications. *IEEE Internet Things J.* **2022**, *9*, 7680–7699. [CrossRef]
27. He, S.; Shin, H.S.; Xu, S.; Tsourdos, A. Distributed estimation over a low-cost sensor network: A review of state-of-the-art. *Inf. Fusion* **2020**, *54*, 21–43. [CrossRef]
28. Aoughlis, S.; Saddaoui, R.; Achour, B.; Laghrouche, M. Dairy cows' localisation and feeding behaviour monitoring using a combination of IMU and RFID network. *Int. J. Sens. Netw.* **2021**, *37*, 23–35. [CrossRef]
29. Aikawa, S.; Yamamoto, S.; Morimoto, M. WLAN finger print localization using deep learning. In Proceedings of the 2018 IEEE Asia-Pacific Conference on Antennas and Propagation (APCAP), Auckland, New Zealand, 5–8 August 2018; pp. 541–542.
30. Hilsenbeck, S.; Bobkov, D.; Schroth, G.; Huitl, R.; Steinbach, E. Graph-based data fusion of pedometer and WiFi measurements for mobile indoor positioning. In Proceedings of the 2014 ACM International Joint Conference on Pervasive and Ubiquitous Computing, Seattle, WA, USA, 13–17 September 2014; pp. 147–158.
31. Kanaris, L.; Kokkinis, A.; Liotta, A.; Stavrou, S. Combining smart lighting and radio fingerprinting for improved indoor localization. In Proceedings of the 2017 IEEE 14th International Conference on Networking, Sensing and Control (ICNSC), Calabria, Italy, 16–18 May 2017; pp. 447–452.
32. Klus, R.; Talvitie, J.; Valkama, M. Neural network fingerprinting and GNSS data fusion for improved localization in 5G. In Proceedings of the 2021 International Conference on Localization and GNSS (ICL-GNSS), Tampere, Finland, 1–3 June 2021 ; pp. 1–6.
33. Álvarez-Merino, C.S.; Luo-Chen, H.Q.; Khatib, E.J.; Barco, R. WiFi FTM, UWB and cellular-based radio fusion for indoor positioning. *Sensors* **2021**, *21*, 7020. [CrossRef] [PubMed]
34. Abu-Mahfouz, A.M.; Hancke, G.P. Localised information fusion techniques for location discovery in wireless sensor networks. *Int. J. Sens. Netw.* **2018**, *26*, 12–25. [CrossRef]
35. Zhang, Y.; Jiang, C.; Yue, B.; Wan, J.; Guizani, M. Information fusion for edge intelligence: A survey. *Inf. Fusion* **2022**, *81*, 171–186. [CrossRef]
36. Zhuang, Y.; Sun, X.; Li, Y.; Huai, J.; Hua, L.; Yang, X.; Cao, X.; Zhang, P.; Cao, Y.; Qi, L.; et al. Multi-sensor integrated navigation/positioning systems using data fusion: From analytics-based to learning-based approaches. *Inf. Fusion* **2023**, *95*, 62–90. [CrossRef]
37. Ji, T.; Li, W.; Zhu, X.; Liu, M. Survey on indoor fingerprint localization for BLE. In Proceedings of the 2022 IEEE 6th Information Technology and Mechatronics Engineering Conference (ITOEC), Chongqing, China, 4–6 March 2022; Volume 6, pp. 129–134.
38. Shan, G.; Choi, G.; Roh, B.H.; Kang, J. An Improved Neighbor Discovery Process in BLE 5.0. In Proceedings of the 2019 IEEE 10th Annual Information Technology, Electronics and Mobile Communication Conference (IEMCON), Vancouver, BC, Canada, 17–19 October 2019; pp. 0809–0812.
39. Core Specification 5.0–Bluetooth® Technology Website. Available online: https://www.bluetooth.com/specifications/specs/core-specification-5/ (accessed on 10 May 2022).
40. Kanakaraja, P.; Kotamraju, S.K.; Nadipalli, L.S.P.S.; Aswin Kume, S.W. IoT enabled BLE and LoRa based indoor localization without GPS. *Turk. J. Comput. Math. Educ. (TURCOMAT)* **2021**, *12*, 1637–1651.
41. Kolakowski, J.; Djaja-Josko, V.; Kolakowski, M.; Broczek, K. UWB/BLE tracking system for elderly people monitoring. *Sensors* **2020**, *20*, 1574. [CrossRef]

42. Cheong, P.; Rabbachin, A.; Montillet, J.P.; Yu, K.; Oppermann, I. Synchronization, TOA and position estimation for low-complexity LDR UWB devices. In Proceedings of the 2005 IEEE International Conference on Ultra-Wideband, Zurich, Switzerland, 5–8 September 2005; pp. 480–484.
43. Hu, Q.; Yang, J.; Qin, P.; Fong, S.; Guo, J. Could or could not of Grid-Loc: Grid BLE structure for indoor localisation system using machine learning. *Serv. Oriented Comput. Appl.* **2020**, *14*, 161–174. [CrossRef]
44. Bahl, P.; Padmanabhan, V.N. RADAR: An in-building RF-based user location and tracking system. In Proceedings of the Proceedings IEEE INFOCOM 2000. Conference on Computer Communications. NINETEENTH Annual Joint Conference of the IEEE Computer and Communications Societies (Cat. No. 00CH37064), Tel Aviv, Israel , 26–30 March 2000; Volume 2, pp. 775–784.
45. Yeung, W.M.; Zhou, J.; Ng, J.K. Enhanced fingerprint-based location estimation system in wireless LAN environment. In *Emerging Directions in Embedded and Ubiquitous Computing, Proceedings of the EUC 2007 Workshops: TRUST, WSOC, NCUS, UUWSN, USN, ESO, and SECUBIQ, Taipei, Taiwan, 17–20 December 2007*; Springer: Berlin/Heidelberg, Germany , 2007; pp. 273–284.
46. Li, B. Indoor positioning techniques based on wireless LAN. In Proceedings of the 1st IEEE International Conference on Wireless Broadband & Ultra Wideband Communications,Budapest, Hungary, 10–15 July 2005.
47. Saha, S.; Chaudhuri, K.; Sanghi, D.; Bhagwat, P. Location determination of a mobile device using IEEE 802.11 b access point signals. In Proceedings of the 2003 IEEE Wireless Communications and Networking, 2003. WCNC 2003, New Orleans, LA, USA, 16-20 March 2003; Volume 3, pp. 1987–1992.
48. Honkavirta, V.; Perala, T.; Ali-Loytty, S.; Piche, R. A comparative survey of WLAN location fingerprinting methods. In Proceedings of the 2009 6th Workshop on Positioning, Navigation and Communication, Hannover, Germany, 19 March 2009; pp. 243–251. [CrossRef]
49. Yousaf, J.; Zia, H.; Alhalabi, M.; Yaghi, M.; Basmaji, T.; Shehhi, E.A.; Gad, A.; Alkhedher, M.; Ghazal, M. Drone and Controller Detection and Localization: Trends and Challenges. *Appl. Sci.* **2022**, *12*, 12612. [CrossRef]
50. Xu, L.; Yao, S.; Rao, S.; Hu, Q.; Liu, C.; Zhu, H. Indoor Positioning Based on Enhanced 5G Fingerprint Positioning Algorithm. In *Signal and Information Processing, Networking and Computers, Proceedings of the International Conference on Signal and Information Processing, Networking and Computers*; Springer: Berlin/Heidelberg, Germany, 2022; pp. 1179–1184.
51. Ruan, Y.; Chen, L.; Zhou, X.; Liu, Z.; Liu, X.; Guo, G.; Chen, R. iPos-5G: Indoor positioning via commercial 5G NR CSI. *IEEE Internet Things J.* **2022**, *10*, 8718–8733. [CrossRef]
52. Gao, X.; He, D.; Wang, P.; Zhou, Z.; Xiao, Z.; Arai, S. One-Reflection Path Assisted Fingerprint Localization Method with Single Base Station under 6G Indoor Environment. In Proceedings of the 2023 IEEE International Symposium on Circuits and Systems (ISCAS), Monterey, CA, USA, 21–25 May 2023; pp. 1–5.
53. Rathnayake, R.; Maduranga, M.W.P.; Tilwari, V.; Dissanayake, M.B. RSSI and Machine Learning-Based Indoor Localization Systems for Smart Cities. *Eng* **2023**, *4*, 1468–1494. [CrossRef]
54. Liu, W.; Chen, J. UAV-aided Radio Map Construction Exploiting Environment Semantics. *IEEE Trans. Wirel. Commun.* **2023**, *22*, 6341–6355 . [CrossRef]
55. Yang, L.; Chen, H.; Cui, Q.; Fu, X.; Zhang, Y. Probabilistic-KNN: A novel algorithm for passive indoor-localization scenario. In Proceedings of the 2015 IEEE 81st Vehicular Technology Conference (VTC Spring), Glasgow, UK, 11–14 May 2015 ; pp. 1–5.
56. Kemppi, P.; Nousiainen, S. Database correlation method for multi-system positioning. In Proceedings of the 2006 IEEE 63rd Vehicular Technology Conference, Melbourne, VIC, Australia , 7–10 May 2006; Volume 2, pp. 866–870.
57. Nuno-Barrau, G.; Páez-Borrallo, J.M. A new location estimation system for wireless networks based on linear discriminant functions and hidden Markov models. *EURASIP J. Adv. Signal Process.* **2006**, *2006*, 1–17. [CrossRef]
58. Wu, Q.; Liu, H.; Zhang, C.; Fan, Q.; Li, Z.; Wang, K. Trajectory protection schemes based on a gravity mobility model in IoT. *Electronics* **2019**, *8*, 148. [CrossRef]
59. Deng, L.; Yu, D. Deep learning: Methods and applications. *Found. Trends® Signal Process.* **2014**, *7*, 197–387. [CrossRef]
60. Burghal, D.; Ravi, A.T.; Rao, V.; Alghafis, A.A.; Molisch, A.F. A comprehensive survey of machine learning based localization with wireless signals. *arXiv* **2020**, arXiv:2012.11171.
61. LeCun, Y.; Bengio, Y.; Hinton, G. Deep learning. *Nature* **2015**, *521*, 436. [CrossRef]
62. Teo, M.I.; Seow, C.K.; Wen, K. 5G Radar and Wi-Fi Based Machine Learning on Drone Detection and Localization. In Proceedings of the 2021 IEEE 6th International Conference on Computer and Communication Systems (ICCCS), Chengdu, China, 23–26 April 2021; pp. 875–880.
63. Al-Tahmeesschi, A.; Talvitie, J.; López-Benítez, M.; Ruotsalainen, L. Deep Learning-based Fingerprinting for Outdoor UE Positioning Utilising Spatially Correlated RSSs of 5G Networks. In Proceedings of the 2022 International Conference on Localization and GNSS (ICL-GNSS), Tampere, Finland, 7–9 June 2022; pp. 1–7.
64. Njima, W.; Ahriz, I.; Zayani, R.; Terre, M.; Bouallegue, R. Deep CNN for Indoor Localization in IoT-Sensor Systems. *Sensors* **2019**, *19*, 3127. [CrossRef]
65. Yang, S.; Sun, C.; Kim, Y. Indoor 3D localization scheme based on BLE signal fingerprinting and 1D convolutional neural network. *Electronics* **2021**, *10*, 1758. [CrossRef]
66. Walfish, S. A review of statistical outlier methods. *Pharm. Technol.* **2006**, *30*, 82.
67. Fix, E.; Hodges, J.L. Discriminatory analysis. *Nonparametric Discrim. Small Sample Perform. Rep. A* **1951**, *193008*, 238–247.
68. Cover, T.; Hart, P. Nearest neighbor pattern classification. *IEEE Trans. Inf. Theory* **1967**, *13*, 21–27. [CrossRef]
69. Goodfellow, I.; Bengio, Y.; Courville, A. *Deep Learning*; MIT Press: Cambridge, MA, USA; London, UK, 2016.

70. 3GPP. Release 15. 2018. Available online: https://www.3gpp.org/release-15 (accessed on 10 August 2021).
71. 3GPP. Release 16. 2020. Available online: https://www.3gpp.org/release-16 (accessed on 10 August 2021).
72. Open5GS. Open5GS | Open Source Project of 5GC and EPC (Release-16). 2020. Available online: https://open5gs.org/ (accessed on 10 August 2021).
73. Alliance, G.S. OpenAirInterface—5G Software Alliance for Democratising Wireless Innovation. 2020. Available online: https://openairinterface.org/ (accessed on 10 August 2021).
74. free5GC. free5GC 5G Project. 2020. Available online: https://www.free5gc.org/ (accessed on 10 August 2021).
75. Kim, M.; Park, K.; Park, J.; Kim, Y.; Lee, J.; Moon, D. Analysis of Current 5G Open-Source Projects. *Electron. Telecommun. Trends* **2021**, *36*, 83–92.
76. Quectel. 5G RM500Q-GL | Quectel. 2020. Available online: https://www.tekmodul.de/wp-content/uploads/2020/05/Quectel_RM500Q-GL_5G_Specification_V1.0_Preliminary_20200313.pdf (accessed on 3 November 2022).
77. Alliance, W. Wi-Fi CERTIFIED ac | Wi-Fi Alliance—wi-fi.org. Available online: https://www.wi-fi.org/discover-wi-fi/wi-fi-certified-ac (accessed on 7 August 2023).
78. SIG, B. Bluetooth® Core Specification Version 5.0 Feature Enhancements | Bluetooth® Technology Website—bluetooth.com. Available online: https://www.bluetooth.com/bluetooth-resources/bluetooth-5-go-faster-go-further/ (accessed on 8 August 2023).
79. Fractal Networx Limited. TruNET Wireless. 2017. Available online: www.fractalnetworx.com (accessed on 10 July 2023).
80. Raspopoulos, M.; Laoudias, C.; Kanaris, L.; Kokkinis, A.; Panayiotou, C.G.; Stavrou, S. 3D Ray Tracing for device-independent fingerprint-based positioning in WLANs. In Proceedings of the 2012 9th Workshop on Positioning, Navigation and Communication, Dresden, Germany, 15–16 March 2012; pp. 109–113.
81. Jemai, J.; Piesiewicz, R.; Kurner, T. Calibration of an indoor radio propagation prediction model at 2.4 GHz by measurements of the IEEE 802.11 b preamble. In Proceedings of the 2005 IEEE 61st Vehicular Technology Conference, Stockholm, Sweden, 30 May–1 June 2005; Volume 1, pp. 111–115.
82. Upton, G.; Cook, I. *Understanding Statistics*; Oxford University Press: Oxford, UK, 1996.
83. Kingma, D.P.; Ba, J. Adam: A method for stochastic optimization. *arXiv* **2014**, arXiv:1412.6980.
84. Bholowalia, P.; Kumar, A. EBK-means: A clustering technique based on elbow method and k-means in WSN. *Int. J. Comput. Appl.* **2014**, *105*, 17–24. [CrossRef]
85. Kanaris, L.; Kokkinis, A.; Fortino, G.; Liotta, A.; Stavrou, S. Sample Size Determination Algorithm for fingerprint-based indoor localization systems. *Comput. Netw.* **2016**, *101*, 169–177. [CrossRef]

electronics

MDPI

Article

Latin-Square-Based Key Negotiation Protocol for a Group of UAVs

Guangyue Kou *, Guoheng Wei, Zhimin Yuan and Shilei Li

School of Information Security, Naval University of Engineering, Wuhan 430030, China; wgh7929@aliyun.com (G.W.); yuanzhimin@nudt.edu.cn (Z.Y.); leeshilei@aliyun.com (S.L.)
* Correspondence: m21183902@nue.edu.cn

Abstract: Unmanned aerial vehicle mobile ad hoc networks (UAVMANETs) formed by multi-UAV self-assembling networks have rapidly developed and been widely used in many industries in recent years. However, UAVMANETs suffer from the problems of complicated key negotiations and the difficult authentication of members' identities during key negotiations. To address these problems, this paper simplifies the authentication process by introducing a Latin square to improve the process of signature aggregation in the Boneh–Lynn–Shacham (BLS) signature scheme and to aggregate the keys negotiated via the elliptic-curve Diffie–Hellman (ECDH) protocol into new keys. As shown through security analysis and simulations, this scheme improves the efficiency of UAVMANET authentication and key negotiation while satisfying security requirements.

Keywords: UAVMANET; multiparty key negotiation; Latin square; BLS protocol

Citation: Kou, G.; Wei, G.; Yuan, Z.; Li, S. Latin-Square-Based Key Negotiation Protocol for a Group of UAVs. *Electronics* **2023**, *12*, 3131. https://doi.org/10.3390/electronics12143131

Academic Editors: Dionisis Kandris and Eleftherios Anastasiadis

Received: 5 June 2023
Revised: 14 July 2023
Accepted: 16 July 2023
Published: 19 July 2023

1. Introduction

Unmanned aerial vehicles (UAVs) [1] are unmanned aircraft that can be flown autonomously or remotely controlled using wireless channels for communication. The benefits of UAVs include their simple structure, flexible deployment, and low prices. In recent years, with the rapid development and large-scale application of internet-of-things (IoT) technology, the development trend of UAVs has shifted from single UAVs to the cooperative operation of multiple UAVs. UAV mobile ad hoc networks (UAVMANETs) [2] composed of multiple UAVs have become a new type of mobile self-organized networks that are widely used in commercial drone performances, joint search and rescue operations, environmental surveys, military missions, and other applications.

A UAVMANET is a special self-organizing network created by placing clusters of UAVs in open wireless channels [3], through which these UAV clusters can connect autonomously after large-scale deployment. Each node in a UAVMANET has the same status and acts as a temporary relay node while completing its flight mission [4]. The decentralized structure of UAVMANETs ensures greater self-organization, more distributed control, and more dynamic topologies than are found in traditional wireless and wired networks.

However, since the UAV clusters work in insecure open channels [5], UAVMANETs are vulnerable to malicious attackers during the self-assembly process. Such attackers can compromise UAVMANETs by eavesdropping on, jamming, and hijacking message data on the communication links [6]. Key negotiation techniques for establishing secure communication over insecure channels can be applied in UAVMANETs; however, attackers can disguise themselves as legitimate users to obtain session keys illegitimately [7]. Additionally, UAVMANET networking needs to account for the flexibility of the network members. Therefore, there is a need to establish a key agreement scheme that can guarantee the efficient generation of session keys and support any number of UAV group members to ensure the confidentiality, integrity, and availability of data communication. Such a UAVMANET key negotiation protocol should have the following features:

- Extensibility: In key negotiation, any number of UAV group members should be allowed to form a UAVMANET.
- Security: Group members should be secure during the negotiation of group keys, and the final session key information should not be able to be breached by malicious users due to group member key interactions.
- Authenticability: Participating UAVMANETs should be authenticable during key negotiations to prevent man-in-the-middle attacks.

1.1. Related Works

UAVMANETs, as a special kind of ad hoc network, have a multiparty key negotiation problem. Solutions to this problem can be divided into two categories: noninteractive key negotiation protocols and interactive key negotiation protocols. Noninteractive key negotiation protocols allow the communicating parties to negotiate the same key in a single key negotiation. After Diffie and Hellman [8] proposed the first noninteractive key negotiation protocol in 1976, many cryptographers attempted to extend this approach to multiple parties, that is, to solve the group key negotiation problem through a single key negotiation. Joux [9] first accomplished the expansion of the Diffie–Hellman (DH) protocol from two to three parties with only one round of communication but did not expand the protocol to more than three members. Garg et al. [10] proposed implementing a multilinear mapping scheme (the GGH scheme) on an ideal lattice using a hierarchical coding system as a solution to the multiparty key negotiation problem. However, this scheme was proven to be unreliable by Hu et al. [11]. Therefore, at present, it is not possible to achieve a noninteractive key negotiation protocol with more than three parties. The research on interactive key negotiation protocols is mainly based on expanding the two-party DH protocol to multiple parties [12,13] by using the DH protocol as the core scheme to form a unified key through the interaction of the protocol participants in multiple communication rounds.

Dutta et al. [14] explored the DH algorithm on a ring structure with forward and backward security but did not support the dynamic joining and leaving of members. Steiner et al. [15] improved the DH protocol by proposing a key agreement approach that can be used for multiple parties and accounts for dynamic group members. However, the number of communication rounds generated in the key update and establishment phase of the protocol is related to the number of group members; as the number of group members increases, establishing group keys becomes more time consuming. Kim [16] formulated a tree-based key management structure to improve the DH protocol and calculated the root node key by cascading the subkeys of the leaf nodes. Compared with other structures, this tree-based key management structure is better suited to the use of the DH protocol in a group environment and can more efficiently reduce the number of node keys [17–20].

Due to the unique mathematical properties of Latin square arrays, they are widely used in the field of communication [21,22]. They can also be applied in key negotiations. Because a given partial Latin square can be uniquely extended to a complete Latin square, a Latin square can be constructed for multiparty key gating. Stones et al. [23] constructed a shared key based on subsecrets using symmetric self-replication. Chum et al. [24] constructed a Latin square key-sharing scheme using hash functions. Shen et al. [25] combined a Latin square scheme with a traditional (t, n)-gated key-sharing scheme to optimize machine-to-machine communication by enhancing efficiency and security. We note that in the above applications, the Latin square is load-balancing to adjust the communication model for distributed systems. Boneh et al. [26] proposed using a Latin square to adapt a key negotiation scheme for cloud computing. This protocol supports any number of user members and incorporates key validation and fault tolerance, but its use of multiple mappings is too burdensome for computing on drones.

In the last two years of research on UAV key negotiation, Xia et al. [27] proposed an identity-based elliptic-curve key negotiation scheme to achieve authentication and key negotiation between UAVs and ground stations. However, the proposed system is only applicable to static UASs with a central node, which is less flexible. Zhang et al. [28] proposed a lightweight authentication and key negotiation protocol for UAVs. The physical unclonable function (PUF) is introduced in the protocol operation, and the authentication and key negotiation can be completed using only hash and heterodyne operations using the characteristics of the PUF, avoiding complex cryptographic operations. However, PUF-based schemes have disadvantages such as complex configuration and the need for specific PUF hardware. Tian et al. [29] proposed a UAV authentication and key negotiation protocol based on the PUF that can communicate across domains. This protocol can communicate across domains before multiple ground stations, but the scheme does not apply to UASs without a central station. Xie et al. [30] managed multiple drone tasks by building a three-tier blockchain. Therefore, this paper proposes using a Latin square to optimize the rounds and process of key negotiation in a self-organizing network of UAVs and designs a set of improved DH protocols to ensure that security and efficiency can be simultaneously addressed in the process of UAV group key negotiation. At the same time, the proposed protocol accounts for the networking characteristics of UAVMANETs and supports a flexible authentication process.

1.2. Motivation and Contributions

The main contributions of this paper are as follows:

- We propose a Latin-square-based dynamic-group key negotiation protocol with authentication. Using the strong mathematical and cryptographic properties of Latin squares, we designed the protocol to allow any number of members to form a group and negotiate the session keys through a self-organizing network of group members without the assistance of a central node for key negotiation. Compared with other key negotiation protocols, our protocol has greater decentralization and networking flexibility.
- The proposed protocol is made more efficient by combining a Latin square array with the Boneh–Lynn–Shacham (BLS) signature algorithm. By combining the signature aggregation process with the construction of a Latin square, it is ensured that each round of communication verifies and aggregates the previous round of blocks, achieving a more efficient signature scheme. The traditional protocol requires a communication cost $O(n^2)$, while the proposed protocol has only an $O(n \log n)$ communication cost. The proposed Latin-square-based signature scheme incurs only half the communication overhead of the elliptic curve digital signature algorithm (ECDSA), and this scheme uses curve hashing to manage its time overhead, unlike other schemes.
- The proposed protocol has higher efficiency and less overhead in the key negotiation phase than the traditional protocol. We optimized the broadcast scheme in the traditional key agreement protocol to communicate with specified members in the square; as a result, only an $O(n \log n)$ communication cost is required to complete key negotiation, whereas the traditional key negotiation protocol has a communication cost of $O(n^2)$. Furthermore, in the key agreement stage, we used the elliptic-curve point product algorithm, which incurs less communication overhead. Therefore, the proposed key negotiation protocol is more efficient than the traditional protocol.

1.3. Organization

This paper is organized as follows. The first section introduces the concept and main features of UAVMANETs. The second section presents the initial parameters of the protocol along with the mathematical notation used. The third section describes the model used. The fourth section presents the protocol. The fifth and sixth sections analyze the security and key properties of the protocol. The final section summarizes the full text.

2. Preliminaries

In this section, we briefly describe the key techniques to be used and clarify their connection to this paper. The symbols that appear in this paper are defined in Table 1.

Table 1. Symbolic notations used in the proposed protocol.

Notation	Description
PID_i	UAV identifier
$\mathbb{F}_p$	Domain formed by $\mathbb{G}$
$\mathbb{F}_{p^2}$	Domain formed by $\mathbb{G}_T$
$\mathbb{G}$	Additive group
$\mathbb{G}_T$	Multiplicative group
P, Q	Prime numbers
$\hat{e}$	Weil pairing on $\mathbb{G} \times \mathbb{G} \to \mathbb{G}_T$
$\mathcal{G}_1$	The base point of an elliptic curve over a finite field for authentication
$\mathcal{G}_2$	The base point of an elliptic curve over a finite field for key negotiation
Pk_i	Drone public key
Sk_i	Drone private key
p_i	The temporary public key for drones
s_i	The temporary private key for drones
$M_{t,i}$	The tth negotiated key in the ith round
$w_{i,j}$	Shared key of PID_i and PID_j
κ	Negotiated key
$H(s_i)$	Hash of s_i
$Sign_i$	Signature

2.1. BLS Signature Protocol

Building BLS signatures requires the utilization of curve hashing and the Weil pairing technique.

Curve hashing means that the result of hashing a message corresponds to a point on an elliptic curve, and the construction method is to determine the corresponding points on the elliptic curve for various points whose hash values are plotted on the X coordinate axis.

A Weil pairing is the mapping of two points on a curve to a single number using a special function. Let E be the elliptic curve defined by the equation $y^2 = x^3 + 1$ over $\mathbb{F}_{p^2}$, let $P \in \mathbb{F}_p$ be a point of order Q, and let $\mathbb{G}$ be the subgroup of points generated by P, where $\mathbb{G}_T$ is a subgroup of $\mathbb{F}_{p^2}$. Then, the map $\varphi(Q)$ is an automorphism of the group of points on the curve E. To obtain a nondegenerate map, we define the modified Weil pairing $\hat{e} : \mathbb{G} \times \mathbb{G} \to \mathbb{G}_T$ as follows:

$$\hat{e}(P, Q) = \hat{e}(P, \varphi(Q)) \quad (1)$$

- Bilinearity: $\hat{e}(aP, bQ) = \hat{e}(P, Q)^{ab}$ for all $P, Q \in \mathbb{G}, a, b \in \mathbb{F}_p$.
- Nondegeneracy: If $P \in \mathbb{G}$, then $\hat{e}(P, P)$ is a generator of $\mathbb{G}_T$.
- Computability: There exists an efficient algorithm to compute $\hat{e}(P, Q)$ for all $P, Q \in \mathbb{G}$.

When generating a BLS signature, we first hash the curve of the message and then multiply the coordinate points on the curve obtained from the corresponding curve hash by the private key to obtain the signature. The result is the points on the curve. The signature generation process is shown in Figure 1.

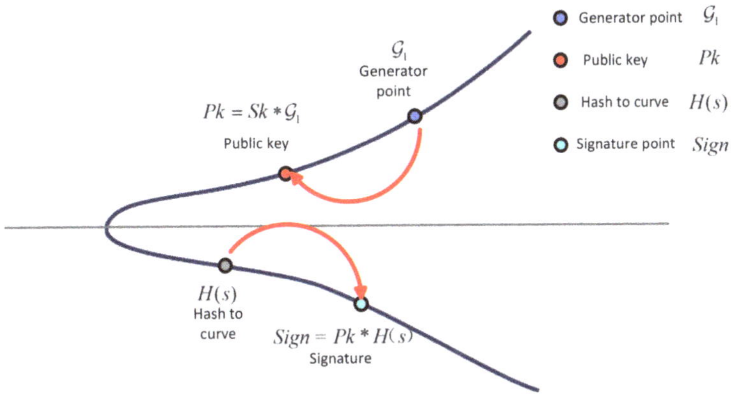

Figure 1. BLS signature generation process.

It is necessary to verify that $\hat{e}(Pk, H(s)) = \hat{e}(\mathcal{G}_1, Sign)$ when verifying a signature.

2.2. Latin Square

A Latin square is an $n \times n$ square matrix with exactly n different elements in each of the n rows of elements. The pseudocode for the process of Latin square construction is shown in Algorithm 1.

Algorithm 1 Construction of a Latin square

```
for x = 1; x ≤ k; x + + do
for y = 1; y ≤ k; y + + do
    a_{x,y} = (x + y − 1);
  end for
end for
```

In this paper, using the mathematical properties of a Latin square array, the row elements are used as the communication directions for pairing, and the pairing process forms a new Latin square array to finally aggregate the signature information and key negotiation information. Taking a 4×4 Latin square array as an example, the specific process is shown in Figure 2.

$$\begin{pmatrix} \mathbf{a}_0 & \mathbf{a}_1 & \mathbf{a}_2 & \mathbf{a}_3 \\ \mathbf{a}_1 & \mathbf{a}_2 & \mathbf{a}_3 & \mathbf{a}_0 \\ \mathbf{a}_2 & \mathbf{a}_3 & \mathbf{a}_0 & \mathbf{a}_1 \\ \mathbf{a}_3 & \mathbf{a}_0 & \mathbf{a}_1 & \mathbf{a}_2 \end{pmatrix} \rightarrow \begin{pmatrix} \mathbf{a}_0 \times \mathbf{a}_1 & \mathbf{a}_1 \times \mathbf{a}_2 & \mathbf{a}_2 \times \mathbf{a}_3 & \mathbf{a}_3 \times \mathbf{a}_0 \\ \mathbf{a}_1 \times \mathbf{a}_2 & \mathbf{a}_2 \times \mathbf{a}_3 & \mathbf{a}_3 \times \mathbf{a}_0 & \mathbf{a}_0 \times \mathbf{a}_1 \\ \mathbf{a}_2 \times \mathbf{a}_3 & \mathbf{a}_3 \times \mathbf{a}_0 & \mathbf{a}_0 \times \mathbf{a}_1 & \mathbf{a}_1 \times \mathbf{a}_2 \\ \mathbf{a}_3 \times \mathbf{a}_0 & \mathbf{a}_0 \times \mathbf{a}_1 & \mathbf{a}_1 \times \mathbf{a}_2 & \mathbf{a}_2 \times \mathbf{a}_3 \end{pmatrix} \rightarrow \begin{pmatrix} \mathbf{a}_0 \times \mathbf{a}_1 \times \mathbf{a}_2 \times \mathbf{a}_3 & \bullet & \bullet & \bullet \\ \bullet & \bullet & & \\ \bullet & & \bullet & \\ \bullet & & & \bullet \end{pmatrix}$$

Figure 2. Latin square member aggregation process.

2.3. Elliptic-Curve Diffie–Hellman Key Exchange

The elliptic-curve Diffie–Hellman (ECDH) key exchange algorithm is a DH algorithm built on elliptic curves, which uses the dot product operation $(w_i * \mathcal{G}_2) * w_j = (w_j * \mathcal{G}_2) * w_i$ on elliptic curves to negotiate keys. In this paper, the ECDH algorithm is used for UAV key negotiation. The basic units for generating public and private keys and the basic elements for conducting key negotiation are constructed as follows:

- PID_i uses a self-generated random number s_i as a temporary private key, constructs an elliptic curve using the $\mathcal{G}_2$ generated by a ground station (GS), and calculates the public key p_i.

- PID_j uses a self-generated random number s_j as a temporary private key, constructs an elliptic curve using the $\mathcal{G}_2$ generated by the GS, and calculates the public key p_j.
- PID_i and PID_j exchange their public keys p_i and p_j on an open channel.
- PID_i computes the negotiated key $\kappa = p_j * s_i$.
- PID_j computes the negotiated key $\kappa = p_i * s_j$.
- PID_i and PID_j have the same $\kappa = s_j * \mathcal{G}_2 * s_i$.

In this paper, we complete the key negotiation problem in a group by applying the ECDH algorithm several times in multiple rounds of communication to aggregate the keys, finally ensuring that all members of the group negotiate the same key.

3. The Models

3.1. System Model

Figure 3 illustrates the communication model of the UAVMANET system. In this system, there are two kinds of entities: a ground station (GS) and UAV nodes [31]. The GS, as a trusted third party in this system, does not participate in key negotiations and is only responsible for providing registration services for members in their first communication. Only members who complete registration can participate in the dynamic activities of the group. The group system consists of several UAVs registered by the GS, communicating through a self-organizing network. When new members need to join, they need to register their unique identifiers (IDs) through the GS. Then, after obtaining the identity information and relevant system parameters provided by the GS, they can interact with other group members to form new session keys. Only UAVs that have registered with the GS, and thus have unique IDs and initial parameters, participate in the authentication and key negotiation process. Therefore, this system is flexible and decentralized.

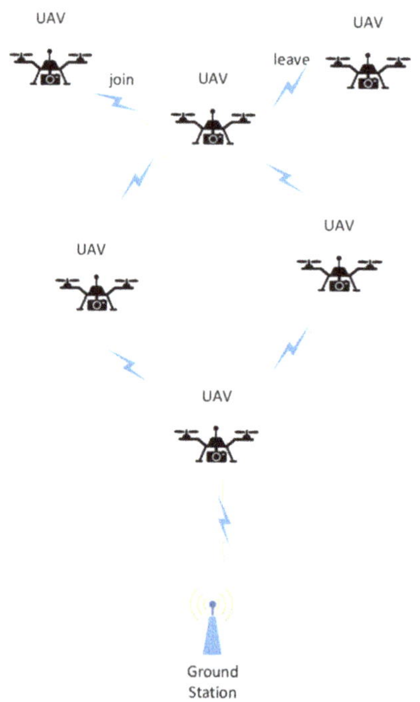

Figure 3. The communication model of the UAVMANET system.

3.2. Security Model

For this paper, two games, $Game_0$ and $Game_1$, were defined to prove the security of the authentication process and the key agreement process, respectively, of the protocol. The operational model is shown in Figure 4.

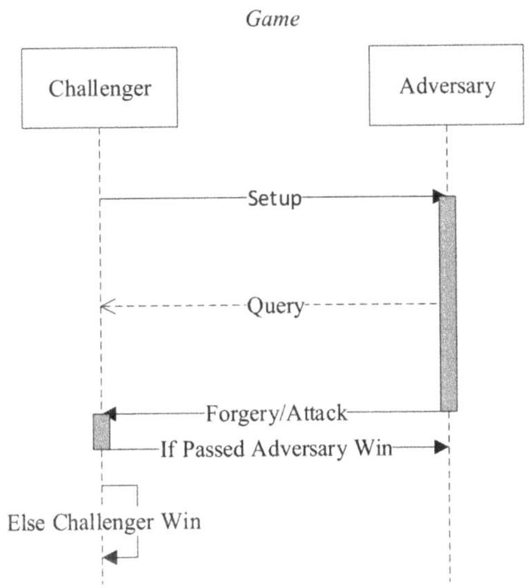

Figure 4. Operational flow chart of the security model.

$Game_0$ proves the security of the protocol authentication process. It is a game between an adversary and a challenger under the model of existential unforgeability against chosen-message attacks (EU-CMA) and is designed as follows.

Setup: The challenger $\mathcal{C}$ generates and publishes the initial parameters $\mathbb{PG} = (\mathbb{G}, \mathbb{G}_T, g, p, e)$ by executing the initial phase, which generates Pk_i.

Query: The adversary $\mathcal{A}$ selects a drone set $\{PID_1,\ PID_2 \cdots PID_{2^k}\}$ and can repeatedly ask the challenger $\mathcal{C}$ for a public key Pk_i and signature $Sign_i$.

Forgery: When $\mathcal{A}$ finishes querying $\mathcal{C}$, $\mathcal{A}$ forges a signature from the information obtained. If $\mathcal{A}$ forges a correct signature based on the information already queried, then $\mathcal{A}$ wins the game.

$Game_1$ proves the security of the protocol's key negotiation process. It is a game between an adversary $\mathcal{B}$ and a challenger $\mathcal{D}$. The game is designed as follows.

Setup: The challenger $\mathcal{D}$ generates and publishes the initial parameters by executing the initial phase, which generates Pk_i.

Query: The adversary $\mathcal{B}$ chooses a drone set $\{PID_1,\ PID_2 \cdots PID_n\}$ and can repeatedly ask the challenger $\mathcal{D}$ for a short-term key p_i. The challenger $\mathcal{D}$ replies with the short-term key p_i. ($\{PID_1,\ PID_2 \cdots PID_n\} \subset \{PID_1,\ PID_2 \cdots PID_{2^k}\}$, meaning that the adversary $\mathcal{B}$ does not have access to all keys.)

Attack: When $\mathcal{B}$ finishes querying $\mathcal{D}$, the protocol is attacked to recover the negotiated key; if $\mathcal{B}$ can compute the correct key κ, $\mathcal{B}$ wins the game.

4. The Proposed Protocol

This section describes the specific process of a multi-round DH cipher negotiation protocol based on the construction of a Latin square (Figure 5). The protocol is divided into three phases. In the first phase, a Latin square array is constructed for the cluster members for system initialization. Based on the constructed Latin square, the cluster members will select the nodes to perform key negotiations in each round. In the second phase, the cluster members authenticate their identity information. In the third phase, corresponding cluster members perform key negotiations in accordance with the rules of the Latin square.

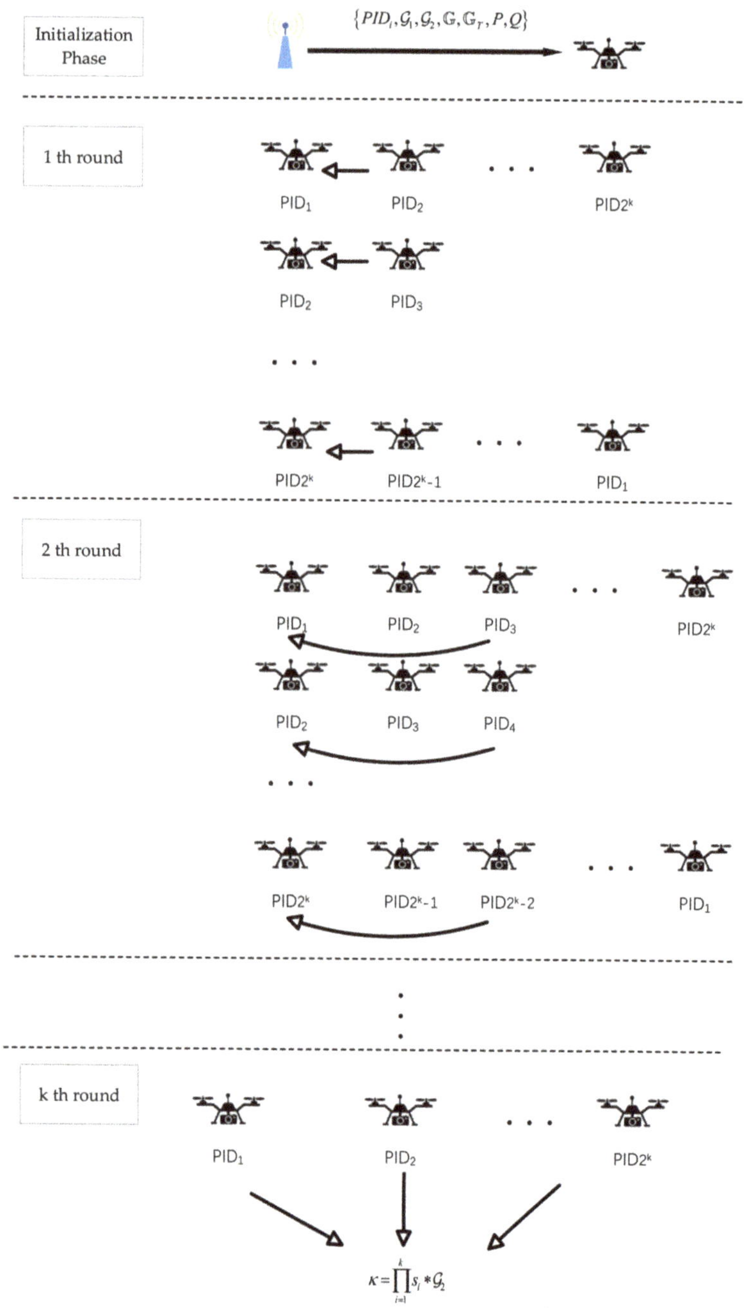

Figure 5. Protocol phase diagram.

4.1. System Initialization

Before the protocol starts, each UAV obtains a unique identity PID_i by registering with the GS. The base point $\mathcal{G}_1$ of an elliptic curve over a finite field is used to generate the authentication keys Pk_i and Sk_i, and the base point $\mathcal{G}_2$ of an elliptic curve over a finite field is used to generate the temporary keys p_i and s_i for negotiation. The GS generates the parameters $\mathbb{PG} = (\mathbb{G}, \mathbb{G}_T, P, Q, \hat{e})$, which are necessary for bilinear mapping, and the GS sends $\{PID_i, \mathcal{G}_1, \mathcal{G}_2, \mathbb{G}, \mathbb{G}_T, P, Q\}$ to each UAV member when it registers with the network.

After a UAV has joined the network, it performs the initialization operation by using the $\{PID_i, \mathcal{G}_1, \mathcal{G}_2, \mathbb{G}, \mathbb{G}_T, P, Q\}$ sent by the GS to generate its long-term key $Pk_i = Sk_i \times \mathcal{G}_1$ and its temporary key $p_i = s_i \times \mathcal{G}_2$, and it calculates $H(s_i)$. The signature $\hat{e}(Pk_i, H(s_i)) = \hat{e}(\mathcal{G}_1, Sign_i)$ is constructed based on the parameters $\mathbb{PG} = (\mathbb{G}, \mathbb{G}_T, P, Q, \hat{e})$.

4.2. Latin Square Construction

Suppose that there are three members in a group, denoted by $\mathbf{a_0}$, $\mathbf{a_1}$, and $\mathbf{a_2}$. For this three-member group, the following standard-type Latin square (Latin square in standard form) can be built:

$$\begin{pmatrix} \mathbf{a_0} & \mathbf{a_1} & \mathbf{a_2} \\ \mathbf{a_1} & \mathbf{a_2} & \mathbf{a_0} \\ \mathbf{a_2} & \mathbf{a_0} & \mathbf{a_1} \end{pmatrix}$$

To generalize this Latin square to a generic k-order standard-type Latin square model, in the proposed protocol, the total number of Latin square members n is first used to calculate $k = \log_2 n$. If k is not an integer, then to maintain the structure of the protocol, virtual members $2^k - n$ to 2^k are added to maintain the structure of the protocol and facilitate the construction of the Latin square.

The generated k-order standard Latin square matrix is shown below. For each member of the matrix, in the xth row and yth column, the element of the matrix is $\mathbf{a_{xy}} = (x + y - 1)$, corresponding to the UAV node $PID_{(x+y) \bmod 2^k}$ in the UAV swarm. By placing the IDs of the UAVs into the elements one by one, the constructed square communication matrix model for the UAV swarm can be obtained as shown below.

$$\begin{pmatrix} \mathbf{a_{11}} & \mathbf{a_{12}} & \mathbf{a_{13}} & \cdots & \mathbf{a_{12^k}} \\ \mathbf{a_{21}} & \mathbf{a_{22}} & & & \\ \mathbf{a_{31}} & & \mathbf{a_{33}} & & \cdots \\ \vdots & & & \ddots & \\ \mathbf{a_{2^k1}} & & \cdots & & \mathbf{a_{11}} \end{pmatrix} \Rightarrow \begin{pmatrix} \mathbf{PID_1} & \mathbf{PID_2} & \mathbf{PID_3} & \cdots & \mathbf{PID_{2^k}} \\ \mathbf{PID_2} & \mathbf{PID_3} & & & \\ \mathbf{PID_3} & & \mathbf{PID_5} & & \vdots \\ \vdots & & & \ddots & \\ \mathbf{PID_{2^k}} & & \cdots & & \mathbf{PID_1} \end{pmatrix}$$

Taking a member PID_1 as an example, in the first round of communication, PID_1 receives a message $Msg_{1,1}$ from PID_2 to negotiate the key $M_{1,1}$ after authentication. In the second round of communication, PID_1 negotiates the key $M_{2,1}$ with PID_3 after authentication, and in the nth round, PID_1 negotiates the key $M_{n,1}$ with PID_{2^n} after authentication. When $n = k$, indicating the last round of communication, PID_1 and $PID_{2^{k-1}+1}$ obtain the final group key κ. (Note: In this protocol, the default key for virtual members is 1).

After construction through the above process, the UAVs communicate in each round in accordance with the rules of the constructed Latin square, and the two UAVs corresponding to each round interact with each other to aggregate their authentication information and keys and form a new Latin square. Finally, a consistent key is obtained through this aggregation process. The process of signature aggregation confirms the legitimacy of the aggregated key; each member can verify the legality of the whole process, and any illegitimate user will cause errors in the final aggregated signature. Thus, the Latin square construction process ensures efficient authentication and key negotiation. In accordance with the nature of a Latin square, the aggregated information exchanged between the two communicating parties for each round of authentication and key negotiation does not contain duplicate elements, as shown in Figure 6.

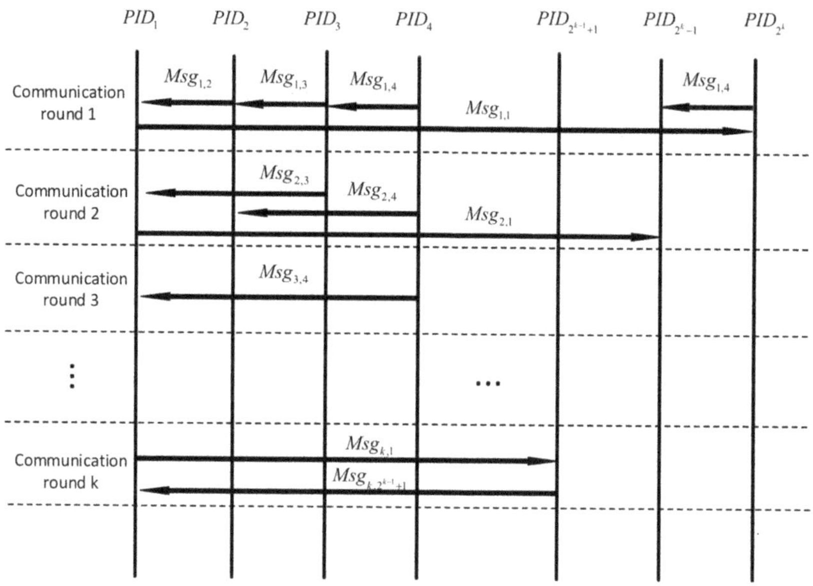

Figure 6. Operational process of the proposed protocol.

4.3. *The Proposed Protocol*

The operation of the protocol proposed in this paper is divided into two phases: the authentication phase and the group key negotiation phase.

4.3.1. Authentication Phase

In this phase, the relevant parameters and signatures for authentication are first generated by a single drone. Subsequently, aggregated signatures are formed through interactions with the relevant drones in the corresponding Latin square, and finally, authentication is completed. All participating drones obtain an aggregated signature in this way and can authenticate the identity of any member of the group in any communication round.

In the process of signature aggregation, not only is the information of the participating members authenticated, but the key negotiation process is also recorded, and untrusted individuals can be backtracked by tracing the aggregated blocks. Thus, the aggregated signature results can be used as proof of legitimate participation in the key negotiation process.

Step 1 Generation of public and private keys with individual signatures:

In this phase, each drone PID_i that has registered with the GS generates its own public key Pk_i and private key Sk_i for authentication using the generator $\mathcal{G}_1$ sent by the GS:

$$Pk_i = Sk_i * \mathcal{G}_1 \tag{2}$$

The key s_i of the drone PID_i is signed as follows. First, the hash calculation $H(s_i)$ is performed on s_i, and the result is then multiplied by Sk_i to obtain the signature result $Sign_i = Sk_i * H(s_i)$, which is transformed into a point on the elliptic hash curve. The drone sends $Msg_i = \{Pk_i||Sign_i||H(s_i)||\cdots\}$ (the data represented by the ellipses are the second-stage key agreement data) to the corresponding node for authentication.

Step 2 Aggregation of signatures on Latin squares:

A single member generates a signature message by the member communication rules specified by the Latin square constructed as described in the previous section and then starts the first round of communication. During the communication process, the members participating in each communication round aggregate the signatures from the previous communication round. Eventually, each member can generate a uniform aggregated signature and can verify the signatures from the previous rounds. Based on the difference in the aggregated signatures, the communication round in which an object was sent can be located.

After the drone PID_i receives $Msg_{1,i+1} = \{Pk_{i+1}||Sign_{i+1}||H(s_{i+1})||\cdots\}$ from PID_{i+1} in round 1, it can calculate the aggregated signature using the mathematical properties of an elliptic curve, $Sign = Sign_i + Sign_{i+1}$, while aggregating the key $Pk = Pk_i + Pk_{i+1}$.

Drone PID_i has the aggregated signature $Sign = Sign_{i,1} + Sign_{i,2} + \cdots + Sign_{i,2^n}$ and the aggregated key $Pk = Pk_i + Pk_{i+1} + \cdots Pk_{i+2^n}$ in round n $(1 < n < k)$; it receives the following PID_{i+2^n}:

$$Msg_{n,i} = \{Pk||Sign||H(s)\} \tag{3}$$

where

$$Pk = Pk_{i+2^n+1} + Pk_{i+2^n+2} + \cdots Pk_{i+2^{n+1}} \tag{4}$$

$$Sign = Sign_{i+2^n+1} + Sign_{i+2^n+2} + \cdots Sign_{i+2^{n+1}} \tag{5}$$

$$H(s) = H(s_{i+2^n+1}) + H(s_{i+2^n+2}) + \cdots H(s_{i+2^{n+1}}) \tag{6}$$

$Msg_{n,i}$ is obtained by using the mathematical properties of elliptic curves to calculate the aggregated signature $Sign = Sign_i + Sign_{i+1} + \cdots + Sign_{i+2^{n+1}}$ while aggregating the key $Pk = Pk_i + Pk_{i+1} + \cdots Pk_{i+2^{n+1}}$.

After the kth round of aggregation, PID_i can obtain the aggregated signature $Sign = \sum_{i=1}^{2^k} Sign_i$ and verify signatures with the aggregated public key $Pk = \sum_{i=1}^{2^k} Pk_i$. Similarly, each drone in the cluster can obtain the aggregated signature $Sign$ during the Latin square construction process.

Step 3 Identity verification:

Since an improved BLS signature scheme is used in the signature aggregation process, each round can be considered as a separate block, and performing authentication requires only verifying each block. That is, the following equation should be satisfied: $\hat{e}(\mathcal{G}_1, Sign_i) = \hat{e}(Pk_0, H(s_0)) \times \hat{e}(Pk_1, H(s_1)) \times \ldots \times \hat{e}(Pk_i, H(s_i))$.

After receiving Msg_{i+1} from PID_{i+1} in round one, the drone PID_i verifies the signature using the public key Pk_{i+1} of PID_{i+1}:

$$\hat{e}(Pk_{i+1}, H(s_{i+1})) = \hat{e}(Sk_i \times \mathcal{G}_1, H(s_{i+1})) = \hat{e}(\mathcal{G}_1, Sk_i \times H(s_{i+1})) = \hat{e}(\mathcal{G}_1, Sign_i) \tag{7}$$

Drone PID_i uses the aggregated public key $Pk_{i+2^n+1} + Pk_{i+2^n+2} + \cdots Pk_{i+2^{n+1}}$ received from $PID_{(i+2^n)\mathrm{mod}2^k}$ to verify the signature after the first n $(n < k)$ rounds when the message Msg_n is received.

Specifically, after receiving Msg_n from $PID_{(i+2^n)\mathrm{mod}2^k}$ in the nth $(n < k)$ round, the drone PID_i uses the received aggregated public key $Pk_{i+2^n+1} + Pk_{i+2^n+2} + \cdots Pk_{i+2^{n+1}}$ to verify the signature as follows:

$$\begin{aligned} \hat{e}(Pk, H(s)) &= \hat{e}\left(Pk_{i+2^n+1} + Pk_{i+2^n+2} + \cdots Pk_{i+2^{n+1}}, H(s)\right) \\ &= \hat{e}\left(\mathcal{G}_1 \times \left(Sk_{i+2^n+1} + Sk_{i+2^n+2} + \cdots Sk_{i+2^{n+1}}\right), H(s)\right) \\ &= \hat{e}\left(\mathcal{G}_1, Sign_{i+2^n+1} + Sign_{i+2^n+2} + \cdots Sign_{i+2^{n+1}}\right) \\ &= \hat{e}\left(\mathcal{G}_1, Sign_{i+2^n+1}\right) \times \hat{e}\left(\mathcal{G}_1, Sign_{i+2^n+2}\right) \times \cdots \hat{e}\left(\mathcal{G}_1, Sign_{i+2^{n+1}}\right) \end{aligned} \tag{8}$$

Here, the signature block of $PID_{(i+2^n)\mathrm{mod}2^k}$ can be verified only if each signature $Sign_{i+2^n+1}, Sign_{i+2^n+2}, \cdots, Sign_{i+2^{n+1}}$ in the signature block of $PID_{(i+2^n)\mathrm{mod}2^k}$ is valid. In the previous rounds of verification, signature aggregation was performed by other drones, thus saving considerable work.

Finally, after the kth round of verification, PID_i verifies the signature $Sign = \sum_{i=1}^{2^k} Sign_i$ using the aggregated public key $Pk = \sum_{i=1}^{2^k} Pk_i$. If the signature is verified, all drones in the entire cluster are legitimate users. Every drone in the cluster can be verified via this method.

4.3.2. Key Negotiation Phase

In this phase, individual drones first generate their public and private keys for negotiation. An aggregated key is then formed by the drones in the corresponding Latin square via the $ECDH$ key negotiation protocol. In the next round, the aggregated key is passed in the same way to form a new aggregated key. Finally, all cluster members can negotiate a common key κ without pass-through in the following process.

In the process of key aggregation, the keys are aggregated on an elliptic curve so that members of a group of arbitrary size can negotiate a common key without the participation of the GS in a distributed manner. Thus, the difficult problem of negotiating keys over wireless channels is solved.

Step 1 Generation of public and private keys:

A single drone PID_i generates its own public key p_i and private key s_i for key negotiation using $\mathcal{G}_2$ obtained from the GS:

$$p_i = s_i * \mathcal{G}_2 \tag{9}$$

Step 2 Calculation of negotiated and aggregated keys on the Latin square:

In the first round of communication, UAV PID_i receives p_{i+1} from PID_{i+1} and calculates the negotiated key:

$$M_{1,i} = p_{i+1} * s_i = \mathcal{G}_2 * s_{i+1} * s_i \tag{10}$$

In the second round of communication, UAV PID_i receives $M_{1,i+2}$ from PID_{i+2} and calculates the negotiated key:

$$M_{2,i} = M_{1,i} * \mathcal{G}_2 * M_{1,i+2} \tag{11}$$

Drone PID_i forms the aggregated key before the nth round ($2 < n < k$):

$$M_{n-1,i} = M_{n-2,i} * \mathcal{G}_2 * M_{n-2,(i+2^{n-1})\text{mod}2^k} \tag{12}$$

The following aggregated key is received from $PID_{(i+2^n)\text{mod}2^k}$:

$$M_{n-1,(i+2^n)\text{mod}2^k} = M_{n-2,(i+2^n)\text{mod}2^k} * \mathcal{G}_2 * M_{n-2,(i+2^{n-1}+2^n)\text{mod}2^k} \tag{13}$$

The key for this round is calculated as follows:

$$M_{n,i} = M_{n-1,i} * \mathcal{G}_2 * M_{n-1,(i+2^n)\text{mod}2^k} \tag{14}$$

After $k-1$ rounds of negotiation, PID_i obtains the key $M_{k-1,i}$, and $PID_{(i+2^{k-1})\text{mod}2^k}$ obtains the key $M_{n-1,(i+2^{k-1})\text{mod}2^k}$. Therefore, the negotiated shared key κ is obtained as follows in the κth round:

$$\kappa = M_{k-1,i} * \mathcal{G}_2 * M_{k-1,(i+2^{k-1})\text{mod}2^k} \tag{15}$$

By recursively expanding $M_{k-1,i}$ and $M_{n-1,(i+2^{k-1})\text{mod}2^k}$ as described above, we can obtain

$$\kappa = \prod_{i=1}^{k} s_i * \mathcal{G}_2 \tag{16}$$

Similarly, all UAVs in the cluster can obtain the shared key by this method.

5. Security Analysis

5.1. Informal Security Proof

Theorem 1. *Each member of the cluster can verify that the negotiated key* $\kappa = \prod_{i=1}^{k} s_i * \mathcal{G}_2$ *is correct and confidential.*

Proof. The negotiated key of UAV cluster member PID_i is $\kappa = M_{k-1,i} * \mathcal{G}_2 * M_{n-1,(i+2^{k-1})\mathrm{mod}2^k}$, where $M_{k-1,i} = M_{k-2,i} * \mathcal{G}_2 * M_{k-2,(i+2^{k-1})\mathrm{mod}2^k}$, $M_{k-1,(i+2^k)\mathrm{mod}2^k} = M_{k-2,(i+2^k)\mathrm{mod}2^k} * \mathcal{G}_2 * M_{k-2,(i+2^{k-1}+2^k)\mathrm{mod}2^k}$, and so on are calculated recursively downward to obtain $\kappa = \prod_{i=1}^{k} s_i * \mathcal{G}_2$. $\kappa = \prod_{i=1}^{k} s_i * \mathcal{G}_2$ can be transformed into $\kappa = \prod_{i=1}^{k} M_{1,i}$, where computing the private key in each $M_{1,i}$ can be considered equivalent to solving the elliptic curve discrete logarithm problem (ECDLP) puzzle. Therefore, in upward recursion, the aggregated key for each round is also secure. □

Theorem 2. *In the protocol authentication phase, each UAV member*PID_i*in the cluster can form an aggregated public key* $Pk = \sum_{i=1}^{2^k} Pk_i$*for the verification of the aggregated signature* $Sign = \sum_{i=1}^{2^k} Sign_i$ *and can verify that the signature is valid.*

Proof. UAV member PID_i in the cluster has formed the following aggregated public key in round $k-1$:

$$Pk_{i\mathrm{mod}2^k} + Pk_{(i+2)\mathrm{mod}2^k} + \cdots Pk_{(i+2^{k-1})\mathrm{mod}2^k} \tag{17}$$

PID_i receives the following aggregated public key from $PID_{(i+2^{k-1})\mathrm{mod}2^k}$:

$$Pk_{(i+2^{k-1}+1)\mathrm{mod}2^k} + Pk_{(i+2^{k-1}+2)\mathrm{mod}2^k} + \cdots Pk_{(i+2^{k-2}+2^{k-1})\mathrm{mod}2^k} \tag{18}$$

The above two aggregated public keys can be summed to obtain $Pk = \sum_{i=1}^{2^k} Pk_i$, and similarly, $Sign = \sum_{i=1}^{2^k} Sign_i$. The signature is verified as follows:

$$\begin{aligned} \hat{e}(Pk, H(s)) &= \hat{e}\left(\sum_{i=1}^{2^k} Pk_i, H(s)\right) \\ &= \hat{e}\left(\mathcal{G}_1 \times \sum_{i-1}^{2^k} Sk_i, H(s)\right) \\ &= \hat{e}\left(\mathcal{G}_1, \sum_{i=1}^{2^k} Sign_i\right) \\ &= \sum_{i=1}^{2^k} \hat{e}(\mathcal{G}_1, Sign_i) \end{aligned} \tag{19}$$

This proves the theorem. □

5.2. Formal Security Proofs

The formal security proofs are now performed for $Game_0, Game_1$ to prove the unforgeability of the protocol with key negotiation against eavesdropping attacks.

$Game_0$:

Definition 1. *The Computational Diffie–Hellman (CDH) Problem.*

On the already determined cyclic group $\mathbb{G}$, let $g^a, g^b \subset \mathbb{G}$. Calculating $e(g,g)^{ab}$ is difficult.

Let $Adv^{CDH}(\mathcal{A})$ denote the advantage that A has in trying to break the proposed protocol, defined as follows:

$$Adv^{CDH}(\mathcal{A}) = \Pr[win_A] \tag{20}$$

Let the adversary $\mathcal{A}$ be attempting to forge a signature with a nonnegligible advantage σ in solving the CDH problem, expressed as

$$Adv^{CDH}(\mathcal{A}) \geq \sigma \tag{21}$$

Forgery by the adversary $\mathcal{A}$ is considered successful when the following condition is met:

$$\Pr[win_A] \geq \mu \tag{22}$$

According to the security model introduced above, the adversary $\mathcal{A}$ and the challenger $\mathcal{C}$ run $Game_0$ as follows.

First, the challenger $\mathcal{C}$ runs the **Setup** phase to generate the cyclic group $\mathbb{G}$, $g^a, g^b \subset \mathbb{G}$, and its public key Pk_i, private key Sk_i, and signature $Sign_i$.

Then, the adversary $\mathcal{A}$ performs the **Query** operation, and the challenger $\mathcal{C}$ provides the public key Pk_i of any UAV PID_i in the UAV set $\{PID_1, PID_2 \cdots PID_{2^k}\}$ and the short-term private key hash $H(s_i)$.

When the Query operation has been executed x times, the adversary $\mathcal{A}$ performs the Forgery operation. $\mathcal{A}$ forges a signature based on the obtained data, and the forged aggregated key is $Pk = \sum_{i=1}^{x} Pk_i$ according to the algorithm in the protocol. The decryption algorithm can be used to verify the aggregated signature $Sign = \sum_{i=1}^{x} Sign_i$ with advantage $Adv^{CDH}(\mathcal{A}) \geq \sigma$ on the basis of solving the CDH problem.

$$e\left(\sum_{i=1}^{x} Pk_i, \sum_{i=1}^{x} H(s_i)\right) = e\left(\mathcal{G}_1, \sum_{i=1}^{x} Sign_i\right) \tag{23}$$

To achieve successful forgery, the adversary $\mathcal{A}$ must solve the CDH problem, that is, given $\mathbb{G}, g, g^{Pk_i}, g^{H(s_i)}$, verify $g^{Sign_i \cdot \mathcal{G}_1} = g^{Pk_i \cdot H(s_i)}$. Since there are 2^k members in the whole UAV cluster, once the adversary $\mathcal{A}$ has made x queries to obtain x keys, $\mathcal{A}$ still needs to guess $2^k - x$ keys. Let the key length be d; then, the probability that the adversary $\mathcal{A}$ wins $Game_0$ is

$$\Pr[win_A] = \frac{1}{2^{(2^k - x)\cdot d}} \cdot Adv^{CDH}(\mathcal{A}) \geq \frac{1}{2^{(2^k - x)\cdot d}} \cdot \sigma \geq \mu \tag{24}$$

If the authentication protocol can be forged, then the advantage in $\Pr[win_A] \geq \mu$ cannot be ignored. If $2^{(2^k - x)\cdot d}$ is also nonnegligible, then the CDH problem has been solved, contradicting Definition 1. Therefore, the authentication part of the protocol is not forgeable.

$Game_1$:

Definition 2. *Elliptic Curve Discrete Logarithm Problem (ECDLP).*

Consider the discrete logarithm problem on an elliptic curve with elements p_i on the elliptic curve and base point $\mathcal{G}_2$. Finding s_i under the condition that $p_i = s_i \cdot \mathcal{G}_2$ holds is difficult.

Let $Adv^{ECDLP}(\mathcal{B})$ denote the advantage that $\mathcal{B}$ has in trying to break the proposed protocol, defined as follows:

$$Adv^{ECDLP}(\mathcal{B}) = \Pr[win_{\mathcal{B}}] \tag{25}$$

Let the adversary $\mathcal{B}$ be attempting to forge a signature with a nonnegligible advantage σ in solving the ECDLP, to break the ECDLP. This is expressed as

$$Adv^{ECDLP}(\mathcal{B}) \geq \sigma \tag{26}$$

An attack by the adversary $\mathcal{B}$ is considered successful when the following condition is met:

$$\Pr[win_B] \geq \mu \tag{27}$$

According to the security model introduced earlier, the adversary $\mathcal{B}$ and the challenger $\mathcal{D}$ run $Game_1$ as follows.

First, the challenger $\mathcal{D}$ runs the Setup phase, generating the base point $\mathcal{G}_2$, the temporary public key p_i, and the temporary private key s_i.

Then, the adversary $\mathcal{B}$ performs the Query operation, and the challenger $\mathcal{D}$ provides the temporary public key p_i of any UAV PID_i in the set $\{PID_1,\ PID_2 \cdots PID_{2^k}\}$.

When the Query operation has been executed x times, the adversary $\mathcal{B}$ performs the Attack operation, attempting to compute the key based on the obtained data. The decryption algorithm is used to solve the ECDLP on the basis of the advantage $Adv^{ECDLP}(\mathcal{B}) \geq \sigma$ in calculating s_i. The final negotiated key is obtained as follows by the algorithm in the protocol:

$$\kappa_x = \prod_{i=1}^{x} s_i * \mathcal{G}_2 \tag{28}$$

To achieve successful forgery, the adversary $\mathcal{B}$ must solve the ECDLP, that is, the element p_i and the base point $\mathcal{G}_2$ on the given elliptic curve should identify s_i under the condition that $p_i = s_i \cdot \mathcal{G}_2$. Since the whole UAV cluster has 2^k members, once the adversary $\mathcal{B}$ has made x queries to obtain x keys,$\mathcal{B}$ still needs to guess $2^k - x$ keys. Let the key length be d; then, the probability of the adversary $\mathcal{B}$ winning $Game_0$ is

$$\Pr[win_B] = \frac{1}{2^{(2^k - x)\cdot d}} \cdot Adv^{ECDLP}(\mathcal{B}) \geq \frac{1}{2^{(2^k - x)\cdot d}} \cdot \sigma \geq \mu \tag{29}$$

If the authentication protocol can be forged, then the advantage in $\Pr[win_B] \geq \mu$ cannot be ignored. If $2^{(2^k - x)\cdot d}$ is also not negligible, this means that the ECDLP has been solved, contradicting Definition 2. Therefore, this protocol can resist eavesdropping attacks.

6. Comparative Analysis

This section compares the computational complexity and time overhead, among other characteristics, of the proposed protocol with those of related protocols presented in previous studies [32–34]. The experimental simulations were implemented on a laptop computer with the following specifications: 11th Gen Intel(R) Core(TM) i7-11800H @ 2.30 GHz (16 CPUs). The simulations were implemented using the Python programming language with the PyCryptodome and pypbc libraries, and we chose the class A curve in pypbc to implement bilinear pairing. Table 2 lists the execution times of some operations for comparison with those listed in the literature. For the calculation of the results, the average of 1000 operations was taken.

Table 2. The execution times of operations used in the protocol.

Operation	Symbol	Execution Time (ms)
Elliptic curve key generation	t_{ecc}	3.999
Exponentiation	t_{mi}	3.887
Elliptic curve point addition	$t_{ecc-add}$	0.001
Elliptic curve point multiplication	$t_{ecc-mul}$	0.431
Bilinear pairing operation	t_{bp}	4.232
Map-to-point hash operation	t_{mtp}	4.549
Point addition related to bilinear pairing	t_{bp-add}	0.094
Multiplication of a scalarwith a point based on bilinear pairing	t_{bp-mul}	1.812

In [32], Wei et al. proposed the CL-AAGKA protocol based on group key agreement (GKA). Through identity-based authentication, the key negotiation protocol can be authenticated without certificates. The computational overhead for a single node is $3(n+1)t_{bp} + (2n+1)t_{bp-mul} + 2nt_{ecc-mul}$. With the participation of n nodes, the computational complexity of the system is $O(n^2)$.

In [33], Zhang et al. proposed the IBAAGKA protocol, which is a communication protocol without key escrow based on asymmetric group key agreement (AGKA). Strong unforgeable stateful identity-based batch multi-signatures (IBBMS) were used to ensure that the computational overhead of a single node would be $(n+5)t_{mi} + (5n+1)t_{bp-mul} + 4t_{bp}$; accordingly, the computational complexity of the system is $O(n^2)$ with the participation of n nodes.

In [34], Shen et al. proposed a protocol whose communication model has a reduced computational complexity of $O(n \log n)$ compared to the above two protocols. However, it uses many bilinear pair-based operations for authentication and key negotiation, and its overhead for a single node is $2t_{bp} + 2t_{bp-mul} + (6\log_2 n - 1)t_{mi}$.

The protocol proposed in this paper uses the concept of Latin squares to optimize the communication model, enabling authentication and key negotiation without broadcasting and requiring multicast communication only between nodes. Compared with the above three protocols, the computational complexity is reduced to $O(n \log n)$. In the authentication phase of the protocol, the short BLS-based signature scheme is improved to enable signature aggregation on the Latin square. The mainstream DSA and ECDSA require 320 bits, whereas the BLS short signature algorithm requires only 160 bits. In the key negotiation phase, keys are aggregated using the dot product operation on an elliptic curve, which has a smaller computational overhead than the bilinear pair operation. The overhead for a single node in this scheme is $t_{bp} + t_{mtp} + t_{ecc} + (4\log_2 n - 1)t_{bp-mul} + (2\log_2 n - 1)t_{ecc-mul}$. Table 3 shows a performance comparison of the four protocols.

Table 3. Performance comparison of four protocols.

Protocol	Type of Message Distribution	System Communication Cost	The Computational Cost for Each Node
CL-AAGKA	broadcast	$O(n^2)$	$3(n+1)t_{bp} + (2n+1)t_{bp-mul} + 2nt_{ecc-mul}$
IBAAGKA	broadcast	$O(n^2)$	$(n+5)t_{mi} + (5n+1)t_{bp-mul} + 4t_{bp}$
Shen et al.'s protocol	multicast	$O(n \log n)$	$2t_{bp} + 2t_{bp-mul} + (6\log_2 n - 1)t_{mi}$
Proposed protocol	multicast	$O(n \log n)$	$t_{bp} + t_{mtp} + t_{ecc} + (4\log_2 n - 1)t_{bp-mul} + (2\log_2 n - 1)t_{ecc-mul}$

In this comparison, we used the class A elliptic curve in the pypbc library in Python to calculate the time overhead of each protocol for the cases of 16, 32, 64, and 128 group members. The calculated run times of the four protocols are compared in the form of line graphs in Figure 7. With 16 members, the protocol proposed in this paper is 2.5 times faster than the Shen et al. protocol, 5.7 times faster than the IBAAGKA protocol, and 12.8 times faster than the CL-AAGKA protocol. In the case of 128 members, the protocol proposed in this paper is 3.9 times faster than the Shen et al. protocol, 15.5 times faster than the IBAAGKA protocol, and 61.5 times faster than the CL-AAGKA protocol.

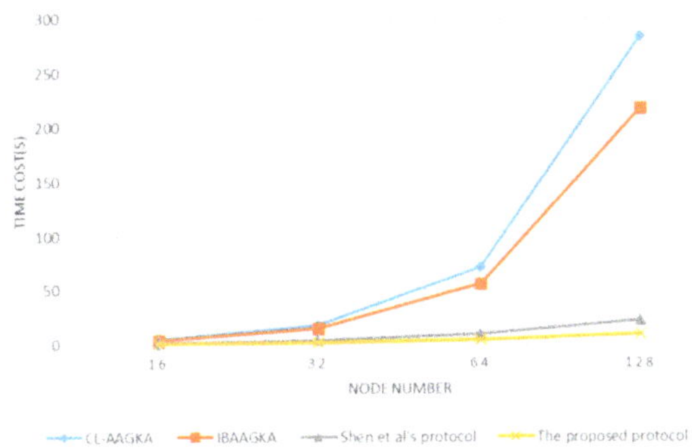

Figure 7. Time-cost comparison of the four protocols.

As the number of simulated UAV nodes increases, the run times of the CL-AAGKA and IBAAGKA protocols show exponential growth trends. In comparison, the execution times of the Shen et al. protocol and the protocol proposed in this paper grow more slowly, showing a clear time-overhead advantage. Compared to the Shen et al. protocol, the proposed protocol achieves a lower time overhead by aggregating BLS signatures over a Latin square array with the use of the elliptic-curve dot product operation, which incurs less communication overhead.

7. Conclusions

In this paper, we focused on the problems of authentication and key negotiation for a group of UAVs in the context of networking and proposed an aggregated signature-based UAV key negotiation protocol based on the concept of Latin squares. The proposed protocol is well adapted to the characteristics of UAVs communicating via wireless channels and enables the computation of a common key without the participation of a central node in the negotiation process. This paper combined the BLS signature algorithm with the Latin square approach for the first time and proposed a method for completing key negotiation through the aggregation of keys on a Latin square. The proposed protocol is highly flexible and has greater operational efficiency than existing protocols, making it more valuable in UAV environments with limited computing resources.

However, the groups formed by the protocol proposed in this paper need to be studied in more detail when the members join dynamically, and the protocol proposed in this paper needs to be improved and enhanced for situations where the group members change frequently. In the future, we will work on this basis to design a more flexible group key negotiation protocol, focusing on scenarios with frequent changes of group members.

Author Contributions: Conceptualization, G.W.; methodology, G.W.; validation, G.K.; formal analysis, G.K. and Z.Y.; investigation, G.K.; writing—original draft preparation, G.K.; writing—review and editing, G.K. and G.W.; supervision, G.W.; project administration, G.W.; and funding acquisition, S.L. All authors have read and agreed to the published version of the manuscript.

Funding: This research was funded by National Defense Science and Technology Foundation Enhancement (No. 2019-JCJQ-JJ-042), China.

Data Availability Statement: Not applicable.

Conflicts of Interest: The authors declare no conflict of interest.

Abbreviations

UAVs	Unmanned aerial vehicles
IoT	Internet of things
UAVMANET	Unmanned aerial vehicle mobile ad hoc network
BLS	Boneh—Lynn—Shacham signature algorithm
DH	Diffie—Hellman key negotiation protocol
GGH	Goldreich, Goldwasser, and Halevi mapping scheme
GS	Ground station
EU-CMA	Existential unforgeability against chosen-message attacks model
GKA	Group key agreement
AGKA	Asymmetric group key agreement
IBBMS	Identity-based batch multi-signatures

References

1. Bouachir, O.; Abrassart, A.; Garcia, F.; Larrieu, N. A mobility model for uav ad hoc network. In Proceedings of the 2014 International Conference on Unmanned Aircraft Systems (ICUAS), Orlando, FL, USA, 27–30 May 2014.
2. Sahingoz, K.O. Networking models in flying ad-hoc networks (fanets): Concepts and challenges. *J. Intell. Robot. Syst.* **2014**, *74*, 513–527. [CrossRef]
3. Iqbal, S. A study on uav operating system security and future research challenges. In Proceedings of the 2021 IEEE 11th Annual Computing and Communication Workshop and Conference (CCWC), Electr Network, Las Vegas, NV, USA, 27–30 January 2021.
4. Samanth, S.; KV, P.; Balachandra, M. Security in internet of drones: A comprehensive review. *Cogent Eng.* **2022**, *9*, 2029080. [CrossRef]
5. Zhang, L.; Wang, S.; Zhou, H.; Chen, Y.; Gui, S. Secure communication scheme of unmanned aerial vehicle system based on mavlink protocol. *J. Comput. Appl.* **2020**, *40*, 2286.
6. Wei, L.; Bing-Wen, F.; Jian, W. Survey on research of mini-drones security. *Chin. J. Netw. Inf. Secur.* **2016**, *2*, 39–45.
7. Zhi, Y.; Fu, Z.; Sun, X.; Yu, J. Security and privacy issues of uav: A survey. *Mob. Netw. Appl.* **2020**, *25*, 95–101. [CrossRef]
8. Diffie, W.; Hellman, M.E. New directions in cryptography. *IEEE Trans. Inf. Theory* **1976**, *22*, 644–654. [CrossRef]
9. Joux, A. A one round protocol for tripartite diffie–hellman. *J. Cryptol.* **2004**, *17*, 263–276. [CrossRef]
10. Garg, S.; Gentry, C.; Shai, I.; Ibm, H. Candidate multilinear maps from ideal lattices. In Proceedings of the Annual International Conference on the Theory and Applications of Cryptographic Techniques, Athens, Greece, 26–30 May 2013.
11. Hu, Y.; Jia, H. Cryptanalysis of ggh map. In Proceedings of the Advances in Cryptology–EUROCRYPT 2016: 35th Annual International Conference on the Theory and Applications of Cryptographic Techniques, Vienna, Austria, 8–12 May 2016.
12. Ingemarsson, I.; Tang, D.; Wong, C. A conference key distribution system. *IEEE Trans. Inf. Theory* **1982**, *28*, 714–720. [CrossRef]
13. Steiner, M.; Tsudik, G.; Waidner, M. In Cliques: A new approach to group key agreement. In Proceedings of the 18th of International Conference on Distributed Computing Systems, Amsterdam, The Netherlands, 26–29 May 1998.
14. Dutta, R.; Barua, R.; Sarkar, P. Pairing-based cryptographic protocols: A survey. *Cryptol. Eprint Arch.* **2004**.
15. Steiner, M.; Tsudik, G.; Waidner, M. Key agreement in dynamic peer groups. *IEEE Trans. Parallel Distrib. Syst.* **2000**, *11*, 769–780. [CrossRef]
16. Kim, Y.; Perrig, A.; Tsudik, G. Tree-based group key agreement. *ACM Trans. Inf. Syst. Secur. (TISSEC)* **2004**, *7*, 60–96. [CrossRef]
17. Lee, S.; Kim, Y.; Kim, K.; Ryu, D.-H. An efficient tree-based group key agreement using bilinear map. In Proceedings of the Applied Cryptography and Network Security: First International Conference, ACNS 2003, Kunming, China, 16–19 October 2003.
18. Kumar, A.; Tripathi, S. Ternary tree based group key agreement protocol over elliptic curve for dynamic group. *Int. J. Comput. Appl.* **2014**, *86*, 17–25. [CrossRef]
19. Barua, R.; Dutta, R.; Sarkar, P. Extending joux's protocol to multi party key agreement. In Proceedings of the Progress in Cryptology-INDOCRYPT 2003: 4th International Conference on Cryptology in India, New Delhi, India, 8–10 December 2003.
20. Dutta, R.; Barua, R. Provably secure constant round contributory group key agreement in dynamic setting. *IEEE Trans. Inf. Theory* **2008**, *54*, 2007–2025. [CrossRef]
21. Chao, C.-M.; Fu, H.-Y. Supporting fast rendezvous guarantee by randomized quorum and latin square for cognitive radio networks. *IEEE Trans. Veh. Technol.* **2015**, *65*, 8388–8399. [CrossRef]
22. Bao, L.C.; Yang, S.H.; Ieee. Latin square based channel access scheduling in large wlan systems. In Proceedings of the IEEE Wireless Communications and Networking Conference (WCNC), Cancun, Mexico, 28–31 March 2011.
23. Stones, R.J.; Su, M.; Liu, X.G.; Wang, G.; Lin, S. A latin square autotopism secret sharing scheme. *Des. Codes Cryptogr.* **2016**, *80*, 635–650. [CrossRef]
24. Chum, C.S.; Zhang, X. Applying hash functions in the latin square based secret sharing schemes. In Proceedings of the International Conference on Security & Management, Las Vegas, NV, USA, 25–28 July 2010.

25. Shen, J.; Zhou, T.Q.; Liu, X.G.; Chang, Y.C. A novel latin-square-based secre sharing for m2m communications. *Ieee Trans. Ind. Inform.* **2018**, *14*, 3659–3668. [CrossRef]
26. Boneh, D.; Lynn, B.; Shacham, H. Short signatures from the weil pairing. *J. Cryptol.* **2004**, *17*, 297–319. [CrossRef]
27. Xia, T.; Wang, M.; He, J.; Lin, S.; Shi, Y.; Guo, L. Research on identity authentication scheme for uav communication network. *Electronics* **2023**, *12*, 2917. [CrossRef]
28. Zhang, L.; Xu, J.; Obaidat, M.S.; Li, X.; Vijayakumar, P. A puf-based lightweight authentication and key agreement protocol for smart uav networks. *IET Commun.* **2022**, *16*, 1142–1159. [CrossRef]
29. Tian, C.; Jiang, Q.; Li, T.; Zhang, J.; Xi, N.; Ma, J. Reliable puf-based mutual authentication protocol for uavs towards multi-domain environment. *Comput. Netw.* **2022**, *218*, 109421. [CrossRef]
30. Xie, H.; Zheng, J.; He, T.; Wei, S.; Shan, C.; Hu, C. B-uavm: A blockchain-supported secure multi uav task management scheme. *IEEE Internet Things J.* **2023**. [CrossRef]
31. Mu, J.; Zhang, R.; Cui, Y.; Gao, N.; Jing, X. Uav meets integrated sensing and communication: Challenges and future directions. *IEEE Commun. Mag.* **2023**, *61*, 62–67. [CrossRef]
32. Wei, G.Y.; Yang, X.B.; Shao, J. Efficient certificateless authenticated asymmetric group key agreement protocol. *KSII Trans. Internet Inf. Syst.* **2012**, *6*, 3352–3365. [CrossRef]
33. Zhang, L.; Wu, Q.H.; Domingo-Ferrer, J.; Qin, B.; Dong, Z.M. Round-efficient and sender-unrestricted dynamic group key agreement protocol for secure group communications. *IEEE Trans. Inf. Forensic Secur.* **2015**, *10*, 2352–2364. [CrossRef]
34. Shen, J.; Zhang, T.; Jiang, Y.; Zhou, T.; Miao, T. A novel key agreement protocol applying latin square for cloud data sharing. *IEEE Trans. Sustain. Comput.* **2022**. [CrossRef]

electronics

MDPI

Article

Air Pollution Monitoring via Wireless Sensor Networks: The Investigation and Correction of the Aging Behavior of Electrochemical Gaseous Pollutant Sensors

Ioannis Christakis , Odysseas Tsakiridis, Dionisis Kandris * and Ilias Stavrakas

Department of Electrical and Electronic Engineering, Faculty of Engineering, University of West Attica, Thivon Avenue 250, GR-12241 Athens, Greece
* Correspondence: dkandris@uniwa.gr

Abstract: The continuously growing human activity in large and densely populated cities pollutes air and consequently puts public health in danger. This is why air quality monitoring is necessary in all urban environments. However, the creation of dense air monitoring networks is extremely costly because it requires the usage of a great number of air monitoring stations that are quite expensive. Instead, the usage of wireless sensor networks (WSNs) that incorporate low-cost electrochemical gas sensors provides an excellent alternative. Actually, sensors of this kind that are recommended for low-cost air quality monitoring applications may provide relatively precise measurements. However, the reliability of such sensors during their operational life is questionable. The research work presented in this article not only experimentally examined the correlation that exists between the validity of the measurements obtained from low-cost gas sensors and their aging, but also proposes novel corrective formulae for gas sensors of two different types (i.e., NO_2, O_3), which are aimed at alleviating the impact of aging on the accuracy of measurements. The following steps were conducted in order to both study and lessen the aging of electrochemical sensors: (i) a sensor network was developed to measure air quality at a place near official instruments that perform corresponding measurements; (ii) the collected data were compared to the corresponding recordings of the official instruments; (iii) calibration and compensation were performed using the electrochemical sensor vendor instructions; (iv) the divergence between the datasets was studied for various periods of time and the impact of aging was studied; (v) the compensation process was re-evaluated and new compensation coefficients were produced for all periods; (vi) the new compensation coefficients were used to shape formulae that automatically calculate the new coefficients with respect to the sensors' aging; and (vii) the performance of the overall procedure was evaluated through the comparison of the final outcomes with real data.

Keywords: wireless sensor networks; IoT; smart cities; environmental monitoring; air pollution; air quality monitoring; sensor aging; electrochemical gas sensors; NO_2; O_3

Citation: Christakis, I.; Tsakiridis, O.; Kandris, D.; Stavrakas, I. Air Pollution Monitoring via Wireless Sensor Networks: The Investigation and Correction of the Aging Behavior of Electrochemical Gaseous Pollutant Sensors. *Electronics* **2023**, *12*, 1842. https://doi.org/10.3390/electronics12081842

Academic Editor: Juan-Carlos Cano

Received: 17 March 2023
Revised: 7 April 2023
Accepted: 11 April 2023
Published: 13 April 2023

1. Introduction

Continuous advances achieved in the technology of microelectromechanical systems (MEMS) have made it feasible to mass-produce inexpensive devices that, despite their small dimensions, have enhanced capabilities of sensing, processing, and communicating [1]. The wireless interconnection of several devices of this type, referred to as sensor nodes, gave birth to wireless sensor networks (WSNs) [2]. The architecture of a WSN is illustrated in Figure 1. A typical sensor node comprises a processing unit, one or more sensors, a transceiver, and a power unit, which in most cases is a battery. WSNs, by taking advantage of the collaborative use of their sensor nodes, are able to not only monitor the ambient conditions over wide areas of interest but also process sensed data and wirelessly transmit them over long distances, via gateways referred to as base stations [3]. For this reason, although their operation is hampered by problems of various kinds such as connectivity loss, congestion, vulnerable security, inadequate coverage, and most of all by the extremely restricted energy adequacy of their sensor nodes [4–12], WSNs have the potential to support almost any field of human activity and therefore have an ever-increasing range of applications [13–17]. The monitoring of air quality is not only one of the numerous applications of WSNs but also an absolutely fundamental operation performed by the use of the Internet of Things (IoT) [18] in so-called "smart cities", where urban operations are performed efficiently with minimum human intervention [19,20].

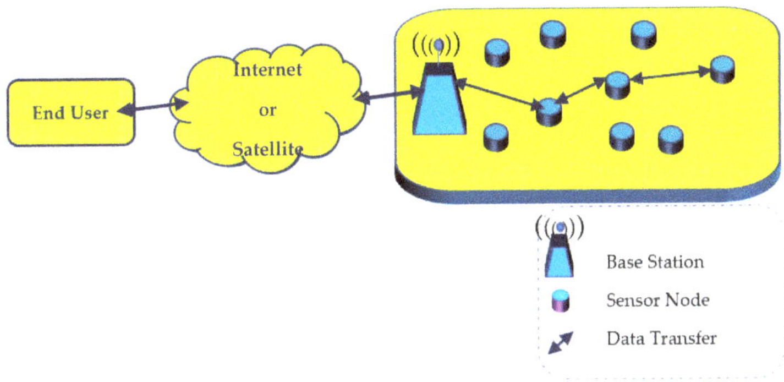

Figure 1. Typical architecture of a WSN.

Indeed, there is no doubt that air pollution is among the greatest threats against public health. This is why air quality monitoring is necessary in all populated areas. However, in wide urban areas this operation is extremely costly because it requires the creation of dense monitoring networks comprising expensive scientific stations [21,22]. On the other hand, a grid network of densely installed low-cost air quality monitoring devices can also provide detailed information on the ambient air quality. Actually, the production of low-cost devices for gas pollutant sensing is feasible thanks to recent technological advances. At the same time, various research works have tried to optimize the effectiveness of such sensors by applying techniques such as machine learning and other artificial intelligence tools [23–28]. In this way, the development and installation of low-cost air quality sensors has become increasingly important [29] as a partial solution to the problem of supplementary coverage of an area. Actually, devices of this kind do not aim at replacing corresponding scientific instruments of high accuracy, but rather at being an additional source of air quality information, provided that their results are reliable [30]. This is why the development of such sensing devices has been investigated in several research works [31–38].

The research work that is presented in this article experimentally evaluated the reduction in measurement accuracy of low-cost gas pollutant electrochemical sensors because of their aging and proposes corresponding corrective formulas. In what follows, in Section 2, the relative theoretical background is established. In Section 3, both the infrastructure is established and the procedures followed in the experiments performed are described. In Section 4, the results of the experiments carried out are both presented and discussed and the formulae proposed for corrective use are described and experimentally evaluated. Finally, in Section 5, concluding remarks are drawn.

2. Theoretical Background

Nowadays, there are many different types of low-cost gaseous pollutant sensors that are commercially available [39]. The most commonly used types are the electrochemical sensors because they are able to detect and measure the concentrations of various gases with a relatively high sensitivity and short response time [40].

Typically, the monitoring of pollutant gases in urban areas by using electrochemical sensors requires the ability to sense lower concentrations of gases than those existing in industrial areas [41]. The detection and analysis of low concentrations necessitates the usage of models that take into consideration the impact of exogenous factors such as temperature, humidity, and pressure, which affect the operation of the electrodes of electrochemical sensors [42], so that the measured values are accurate. Practically, the output of an electrochemical sensor is an electric current of the order of μA, which is then converted to voltage that is analogous to the concentration of the gas detected.

However, various research studies have demonstrated that the measurements made by low-cost air quality sensors have, in many cases, low reliability compared to the measurements made by using reference monitoring instruments [40,43–47]. Specifically, these studies showed that the accuracy of such sensors is influenced by not only exogenous factors related to environmental conditions, such as air temperature, relative humidity, and interferences with other gases (cross-sensitivity), but also endogenous factors of the sensors such as warming-up time, gain, and initial manufacturer calibration [48,49].

Also, many research studies have been conducted regarding the evaluation of low-cost air quality sensors. Actually, there are two methods with which to study and evaluate the performance of low-cost gas sensors. The first method is to evaluate the gas sensors in a laboratory environment, i.e., under controlled conditions [45,50–55]. The second method proposes the evaluation of gas sensors in the field with nearby reference measuring instruments, so that low-cost sensors are calibrated by comparing the measured data that they produce with the data obtained from reference instruments [56–60]. The second method seems to lead to more accurate and reliable results, because the environmental parameters such as temperature and humidity are taken into consideration, contrary to what happens in the corresponding procedure that is conducted in laboratory environments. Linear regression (LR), multiple linear regression (MLR) and machine learning models are the most widely used methods to calibrate low-cost sensors [61,62]. In addition, a learning systems-based generative adversarial network (GAN) research team [63] proposed a GAN-based automatic property generation (GAPG) approach to generate verification properties supporting model checking. Conversely, the time-series feature of the IoT makes the data density and the data dimension higher; as such, anomaly detection is important to ensure hardware and software security, and research work [64] has proposed a memory-augmented encoder approach to detect anomalies in IoT data, which aims to use reconstruction errors to determine data anomalies. While the aging of electrochemical sensors is a given problem, the treatment of measurements during aging remains on the table. Based on this, the investigation of equations that include correction factors to compensate and improve the measurements during their lifetime was the subject of research of this work.

Regarding the electrochemical gas sensors, their operation is based on the chemical reactions performed between environmental air and electrodes that are within a liquid vessel that is incorporated inside the sensor units [65]. The creation of dense air monitoring networks is both more affordable and easier to be deployed by using low-cost electrochemical sensors rather than high-cost monitoring instruments. On the other hand, when using low-cost electrochemical sensors, not only is the calibration of the sensors affected by ambient conditions in external environments but also sensors must be replaced at 1–2-year intervals due to the rapid wear of their chemical elements [66].

Actually, the aging of the sensors and the accuracy of their response during their lifetime have not been sufficiently studied. Research works have shown that aging biases the voltage recording at certain environmental O_3 concentrations (approximately 20% after 9 months of continuous operation), thus necessitating frequent calibration of the oxidizing gas sensor [67]. It has also been demonstrated that the deterioration of sensors due to aging is a non-reversible process. Specifically, it has been found that over long deployments (>2 years), the sensor likely becomes insensitive to NO_2 and O_3. Therefore, the prompt identification and replacement of nonfunctional sensors is essential in order to ensure reliable data acquisition in long-term field deployments [68]. The investigation of aging correction factors of low-cost electrochemical sensors were part of this work, as well as how well the use of aging correction models can realistically reproduce measurements during their lifetime.

3. Analysis of the Experimental System and Procedure

As aforementioned, the research work presented in this article not only studies the deterioration that is caused due to aging in the performance of low-cost sensors that are used for air monitoring in smart cities, but also introduces a method that aims at maintaining high levels of accuracy despite the aging of sensors. For this reason, a WSN comprising low-cost electrochemical sensors [69] was developed by the authors of this article to monitor the concentrations of nitrogen dioxide (NO_2) and ozone (O_3) in the ambient air, Then, in the long run (i.e., 3 months, 6 months, and 8 months) the performance of the low-cost sensors was compared with the reference instruments and was re-evaluated. In what follows in Section 3, the experimental system developed and the initial procedures followed for the calibration of the sensors are described.

3.1. System Overview

For first time in April 2021, the air quality monitoring system developed was installed at the center area of Athens, Greece at a location which is denoted as point A in Figure 2. The values of the measurements performed were compared to the corresponding data, which were obtained from the website of the Ministry of Environment and Energy of Greece (PERPA) [70]. These reference data were derived from the measurements performed by the official pollution measuring stations of PERPA, which are placed also in the center area of Athens at a location denoted as point B in Figure 2. The distance between these two locations is 900 m. Both locations share the same urban conditions. Actually, the spatial coverage of a monitoring site represents the quantification of the variability of concentrations of a specific pollutant around the site [71,72], while the assessment of representativeness aims at the delimitation of areas of the concentration field with similar characteristics at specific locations, as well the spatial surrogate data (similar emission sources and land-cover characteristics) [73–75].

Figure 2. Map of the locations of sensor network (A) and the PERPA station (B).

A synoptic overview of the overall system developed is illustrated in Figure 3.

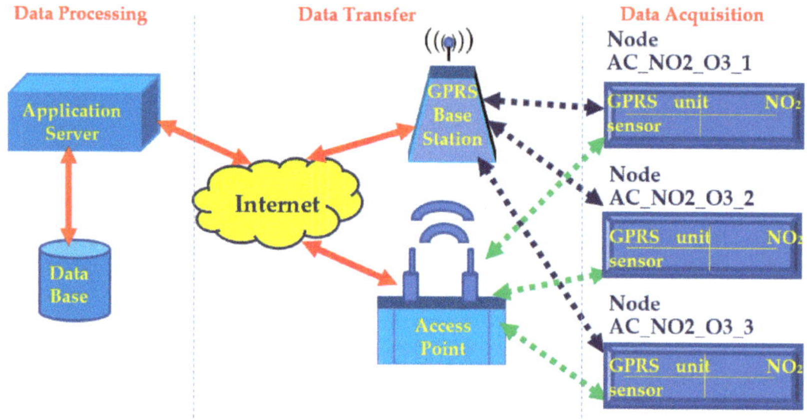

Figure 3. Schematic overview of the air monitoring system developed.

As illustrated in Figure 3, the data acquisition was carried out by three sensor nodes, which namely are: AC_NO2_O3_1, AC_NO2_O3_2, and AC_NO2_O3_3. Their names represent the location (i.e., Athens Center), the gases they detect (i.e., NO_2 and O_3) and the node's identification number. Each one of them incorporated both a NO_2 sensor and an O_3 sensor couple in order to measure the concentrations of these two chemical substances in ambient air. Each sensor node also incorporates a General Packet Radio Service (GPRS) unit and a Wi-Fi unit. By alternately using these modules, each node could correspondingly send sensed data to the Internet either across cellular communication networks via a GPRS base station or across Wi-Fi via an access point. Next, the data transmitted via the Internet reached an application server that runs under the Linux operating system. In this server, data processing and data visualization to the end user took place via an application that had been appropriately developed by using the Grafana open-source interactive visualization platform. Finally, the storage of the sensed data was performed in a database that was developed by using the influxDB open-source time-series database platform.

3.2. Sensor Nodes

As aforementioned, the system developed used three sensor nodes. They are displayed in Figure 4.

Figure 4. Picture of the three sensor nodes of the air quality monitoring system developed.

Each one of the three sensor nodes consisted of a microcontroller with high processing power, low power consumption and a sufficient number of ports for peripherals, i.e., sensors, Wi-Fi and GPS. The interior of one of the three nodes is displayed in Figure 5, while more details on their design and implementation can be found in [76].

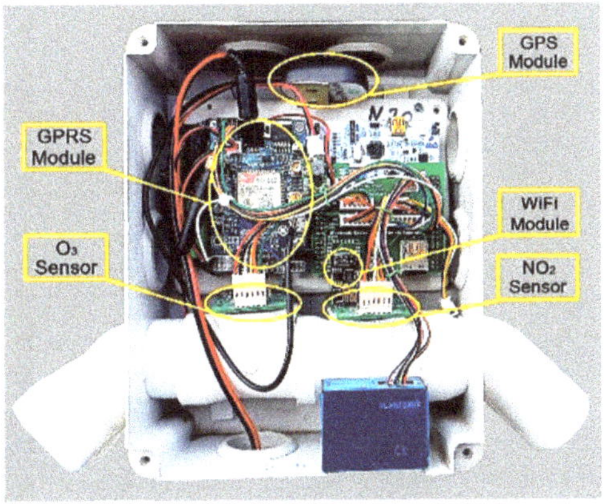

Figure 5. Photograph of the interior of the nodes of the air quality monitoring system.

3.3. Sensor Calibration

At the beginning of the experimental process, a couple of brand-new electrochemical sensors were incorporated in each one of the sensor nodes. The specific sensors were namely: Alphasense OX-B431 [77] for O_3 and Alphasense NO2-B43F for NO_2 [78]. Both of these sensors generally consisted of four electrodes supported by an individual sensor board (ISB).

These sensors provided the output measurements in mV range, corresponding to the concentration of the measured gas and any potential cross-sensitivity. Cross-sensitivity is defined as the chemical reaction of the measuring element by another gas other than the target gas, with the result that the measurements of an electrochemical sensor are affected. The cross-sensitivity from other gases can be seen in the technical specifications of the manufacturer datasheet of the ozone sensor (OX-B431) [75] and of the nitrogen dioxide sensor (NO2-B43F) [76].

Specifically, the NO_2 sensor (NO2-B43F) presented cross-sensitivity (% measured gas @5 ppm) for the gases $H_2S < -80$, $NO < 5$, $Cl_2 < 100$, $SO_2 < -3$, and $CO < -3$; (% measured gas @100 ppm) for the gases $H_2 < 0.1$, $C_2H_4 < 0.1$, $NH_3 < 0.5$, and halothane (not detected). Similarly, the O_3 sensor (OX-B431) presented cross-sensitivity (% measured gas @5 ppm) for the gases $H_2S < -80$, $NO < 5$, $Cl_2 < 100$, $SO_2 < -3$, and $CO < -3$; (% measured gas @100 ppm) for the gases $H_2 < 0.1$, $C_2H_4 < 0.1$, $NH_3 < 0.5$, and halothane <0.1.

In order to be able to read the gas concentration in each low-cost sensor, the manufacturer provides a set of steps that must be followed in order to perform an initial calibration. These steps take into consideration two factors for each sensor—the interface board and the electrodes—as well as environmental conditions such as the temperature. After this initial calibration is completed, further data elaboration is required to correct the calibrated data using known environmental measurements, and thus improve the accuracy of the extracted results.

Regarding the initial calibration of its sensors, Alphasense proposes several potential functions in its Application Note AAN-803-01 [79]. In the system developed in this specific research work, Equation (1) was selected to perform the initial calibration of the sensors used. Specifically, Equation (1) provides the calibrated voltage output (i.e., WE_c: working electrode corrected) that finally represents the concentration of the gas detected, using the provided sensitivity of each individual sensor. This step incorporates the measured working electrode reading (WE_u), the auxiliary electrode reading (AE_u), and a temperature parameter n_T according to Application Note AAN 803-01. To complete the calibration for each individual sensor and ISB, Alphasense provides a set of background electrode noise values, corresponding to working electrode electronic zero (WE_e) and auxiliary electrode electronic zero (AE_e), which are also indicated in Equation (1):

$$WE_c = (WE_u - WE_e) - n_T \times (AE_u - AE_e) \quad (1)$$

The gas pollutant concentration measurement of x gas is given by dividing the calibrated voltage output (i.e., WE_c: working electrode corrected) by the $Sensor_{\_Sensitivity}$, as shown in Equation (2):

$$GASx_m = WE_c / Sensor_{\_Sensitivity} \quad (2)$$

where $GASx_m$ is the corrected measurement concentration, WE_c is the calibrated value of x gas (see Equation (1)), and the sensor sensitivity is given by the manufacturer for each sensor.

Following the above procedure and according to the manufacturer, a methodology must be designed and followed by the end user in order to improve calibration by taking into consideration the impact of temperature, aging, etc. In this research work, after various attempts, the corrective formula that was found to fit best the corresponding values of the official instrumentation installed (i.e., the measuring stations of PERPA, Point B) [70] is the one described by Equation (3):

$$GASx_c = (GASx_m + C1)/C2 \quad (3)$$

where $GASx_c$ is the corrected value of x gas and $GASx_m$ is the value of the measured concentration of x gas after applying Equation (2), while C1 and C2 are coefficients calculated for each station individually after a period of operation in the field, side by side with the reference equipment.

Using Equation (3), the values of $GASx_c$ are expressed in ppb. The conversion from ppb to $\mu g/m^3$ is achieved by multiplying the gas concentration in ppb by the conversion factors, for an ambient pressure of 1 atmosphere and a temperature of 20 degrees Celsius. The conversion factors are, for ozone (O_3) 1.995, and for nitrogen dioxide (NO_2) 1.912. It must be noted that regarding NO_2 the correction coefficient C2 obtains three distinct values depending on the level of NO_2 concentration. Following the procedure of correction, the three coefficients are calculated as C2a, when $NO_2 < 3$, C2b, when $3 \leq NO_2 \leq 30$ and C2c, when $NO_2 > 30$.

4. Experimental Procedure Results and Discussion

Based on the aforementioned calibration and correction procedure in Section 3, several experiments were conducted in order to verify the impact of aging on the corrective formula expressed by Equation (3).

Specifically, during the period of calibration and initial correction (i.e., 14 April 2021 to 10 May 2021) the temporal variation of the corrected NO_2 and O_3 values of the three nodes for the corresponding values of the reference instruments are illustrated in Figures 6 and 7, respectively. Henceforth, the corrected values calculated for the three nodes are plotted in scatter correlation plots to evaluate the conversion performance.

Observing Figures 6 and 7, it is obvious that the corrected values of the measurements from low-cost sensors tend to follow the corresponding values of the reference instruments. Observing Figure 7, it becomes evident that the low-cost sensors failed to reach the minimum values when these were obtained from the reference instruments. This can be attributed to the cross sensitivity of the sensor that was activated from other existing environmental oxides. In Figure 8a–c, the cross-correlation performance of the calibration and correction regarding the NO_2 measurements of the three nodes is depicted. The corresponding behavior for the O_3 measurements is shown in Figure 9a–c. Both Figures 8 and 9 evince the consistency of the sensors. It is observed that C1 coefficient shows linear correlation. The expected temporal limitation of the C1 trend is significantly longer than the 2-year lifetime of the sensors, as provided by the manufacturers.

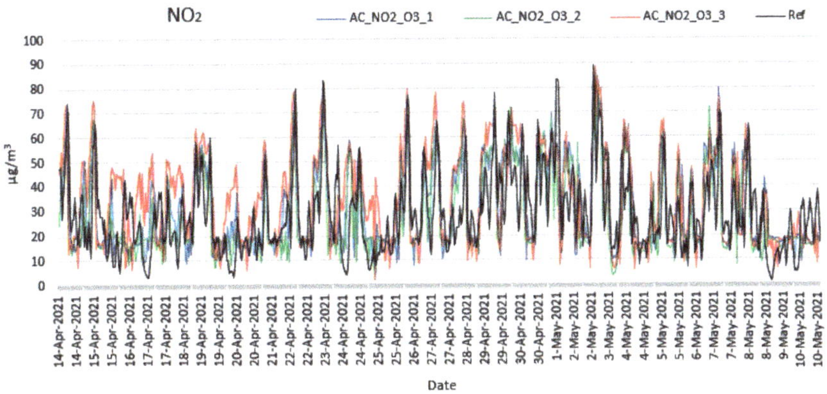

Figure 6. Measurements of NO_2 concentration from low-cost and reference sensors (14 April 2021 to 10 May 2021).

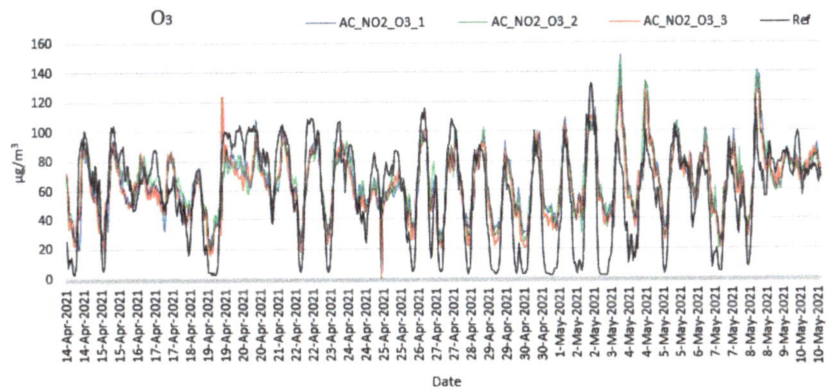

Figure 7. Measurements of O_3 concentration from low-cost and reference sensors (14 April 2021 to 10 May 2021).

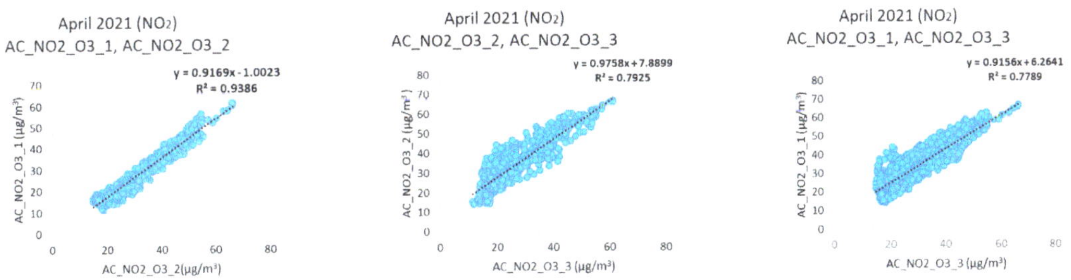

(a) (b) (c)

Figure 8. Nitrogen dioxide correlations among the three low−cost sensors used: (**a**) NO_2 correlation between AC_NO2_O3_1, and AC_NO2_O3_2; (**b**) NO_2 correlation between AC_NO2_O3_2, and AC_NO2_O3_3; (**c**) NO_2 correlation between AC_NO2_O3_1, and AC_NO2_O3_3.

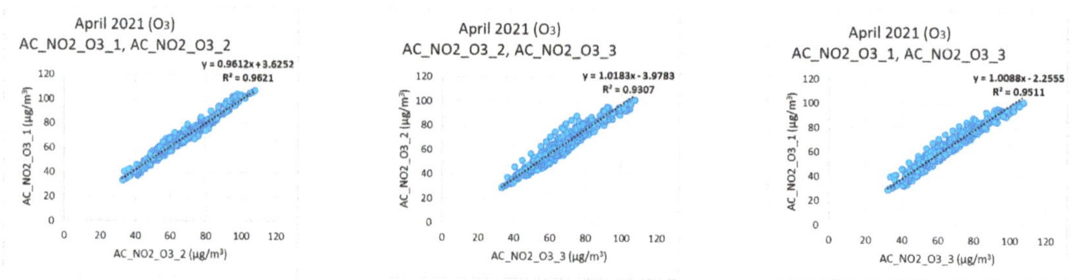

(a) (b) (c)

Figure 9. Ozone correlations among the three low−cost sensors. (**a**) O_3 correlation between AC_NO2_O3_1, and AC_NO2_O3_2; (**b**) O_3 correlation between AC_NO2_O3_2, and AC_NO2_O3_3; (**c**) O_3 correlation between AC_NO2_O3_1, and AC_NO2_O3_3.

After the evaluation of the low-cost sensors and the validation of the high degree of cross-correlation, the measurements of the low-cost sensors were compared against the measurements of reference instruments as depicted in Figure 10a,b. For simplicity reasons henceforth the data of only one of the low-cost sensors are presented in the plots to provide the degree of cross-correlation. Figure 10a,b show the correlation for the NO_2 and O_3 concentration between the low-cost node AC_NO2_O3_2 and the reference instruments for the period of training (i.e., April 2021). In addition, Table 1 summarizes the values of coefficients C1 and C2. Specifically for NO, the coefficient C2 is divided into three sub-coefficients C2a, C2b, C2c according to the measurement of the concentration of nitrogen dioxide) for all the low-cost measurement stations for all the tests that were conducted to evaluate the correlation, correction, and aging involution functions.

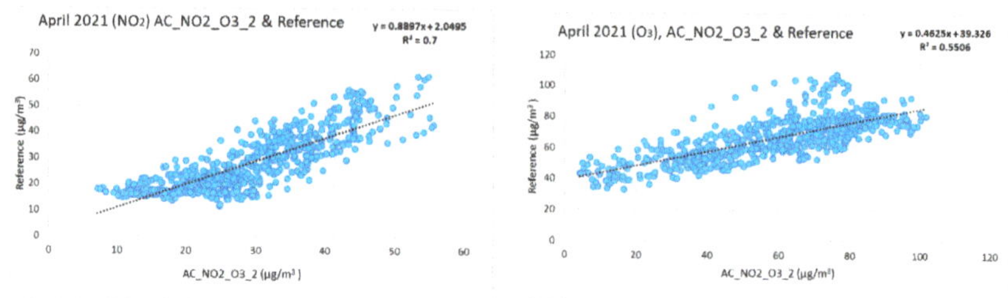

(a) (b)

Figure 10. Correlation for the NO_2 and O_3 concentration between the low-cost sensor AC_NO2_O3_2 and the reference instruments. (**a**) Correlation of NO_2 low-cost sensor (April coefficients C1, C2) and reference; (**b**) Correlation of O_3 low-cost sensor (April coefficients C1, C2) and reference.

Next, the time dependent deviation from accuracy was studied using the protocol:

- The values of C1 and C2 were maintained as constant and the correlation degree R^2 was studied between the reference instruments and the low-cost stations during July 2021, October 2021 and December 2021. The results in this step are presented in Table 2, manifesting the gradual deterioration of R^2.

Additionally, in Table 2 the reference/low-cost correlation fitting is presented as equation, $y = \alpha x + b$, where y is the dependent variable (reference instrument $\mu g/m^3$ value), α is the regression coefficient, x (low-cost station $\mu g/m^3$ value) is the independent variable and b is a constant. Observing the above findings, it becomes evident that the performance of the sensing stations gradually deteriorates as the R^2 is decreased over time. For this reason, it was decided to make a temporal change on the C1 and C2 and re-examine the performance of the sensing device.

Table 1. Coefficients C1 and C2 for the low-cost NO_2 and O_3 gas sensors of the three nodes during all the period studied.

Gas	Node	April 2021				July 2021				October 2021				December 2021			
		C1	C2			C1	C2			C1	C2			C1	C2		
			a	b	c		a	b	c		a	b	c		a	b	c
NO_2	AC_NO2_O3_1	50	15	2.2	1	47	15	3.5	1.2	43	15	3.5	1	41	15	2	1
	AC_NO2_O3_2	50	15	2	1	47	15	2.5	1	44	15	3.3	2	42	15	2.5	1
	AC_NO2_O3_3	74	15	2.4	1.5	72	15	3.4	1	69	15	3.5	1	67	15	3	1
		C1	C2			C1	C2			C1	C2			C1	C2		
O_3	AC_NO2_O3_1	43	1.8			45	2.8			48	2.4			50	2		
	AC_NO2_O3_2	45	1.8			47	2.5			49	2.1			51	2		
	AC_NO2_O3_3	40	1.7			43	2.4			45	3			47	1.6		

Table 2. Correlation results between the reference instruments and the low-cost sensors while C1 and C2 are kept constant and equal to those calculated during the installation of nodes.

Nodes	Coeff.	April 2021		July 2021		October 2021		December 2021	
		NO_2	O_3	NO_2	O_3	NO_2	O_3	NO_2	O_3
AC_NO2_O3_1	A	0.8151	0.4625	−0.3521	0.826	0.7902	0.5071	0.4829	0.531
	B	11.408	39.326	50.281	63.485	13.657	39.892	39.602	25.26
	R^2	0.4889	0.5506	0.0529	0.1963	0.3864	0.6494	0.4511	0.679
AC_NO2_O3_2	A	0.9373	0.5156	−0.2162	0.7511	0.7648	0.4751	0.3962	0.466
	B	4.2728	33.426	31.447	62.033	39.329	30.472	27.553	13.46
	R^2	0.6959	0.6141	0.0375	0.1717	0.1619	0.5399	0.3326	0.217
AC_NO2_O3_3	A	0.8897	0.4658	−0.1519	0.788	0.3447	0.3995	0.3266	0.488
	B	2.0495	38.05	32.619	57.386	9.5425	75.144	30.408	21.56
	R^2	0.7	0.5364	0.0161	0.2617	0.2388	0.4625	0.2448	0.706

It becomes evident from Table 2 that NO_2 values during July show low cross-correlation. As will be discussed later on, the impact of the aging compensation equations on the NO_2 value is positive, as it improves the corresponding R^2 despite the fact that it remains at low values. Furthermore, in Greece, during the summertime, the environmental temperature pushes the sensors to their functional limits. Specifically, according to the manufacturer, these sensors are operational at the temperature range between −20 °C and 50 °C.

- New individual values for C1 and C2 were calculated (different ones for each period of study, i.e., July 2021, October 2021 and December 2021) and the new R^2 was calculated, indicating the temporal variation of C1 and C2 for both gases during their aging. The results extracted are displayed in Table 3. In addition, in Table 3 the reference/low-cost correlation fitting is presented as equation, $y = \alpha x + b$, where y is the dependent variable (reference instrument $\mu g/m^3$ value), α is the regression coefficient, x (low-cost station $\mu g/m^3$ value) is the independent variable, and b is a constant.

The extracted results from Tables 2 and 3 are shown at Figures 11 and 12. Figure 11 shows the correlations (R^2) of NO_2 and O_3 low-cost sensor measurements and reference, for the months July, October, and December 2021 with the calculated coefficients C1 and C2 of April, as well as the correlations of NO_2 and O_3 measurements, for the months July, October, and December 2021 with the calculated coefficients C1 and C2 of each month.

Figure 12 shows coefficients A and B of equation $Y = AX + B$ for the NO_2 and O_3 measurements for July, October, and December 2021, by applying calculated coefficients C1 and C2 for April as well as calculated coefficients C1 and C2 of each month.

Table 3. Summary of the correlation results among the reference instruments and the low-cost sensors while the coefficients C1 and C2 vary and are recalculated according to the aging functions (indicated by Equations (4) and (5)).

Nodes	Coeff.	April 2021		July 2021		October 2021		December 2021	
		NO_2	O_3	NO_2	O_3	NO_2	O_3	NO_2	O_3
AC_NO2_O3_1	A	0.8151	0.4625	−0.2377	0.5311	0.6429	0.3718	0.4322	0.477
	B	11.408	39.326	37.704	42.24	8.7848	34.555	27.746	29.72
	R^2	0.4889	0.5506	0.0565	0.1961	0.4058	0.6325	0.4067	0.678
AC_NO2_O3_2	A	0.9373	0.5156	−0.2236	0.5409	0.5061	0.4081	0.4293	0.481
	B	4.2728	33.426	29.386	46.249	23.494	29.889	16.356	15.89
	R^2	0.6959	0.6141	0.057	0.1718	0.1551	0.5402	0.4733	0.468
AC_NO2_O3_3	A	0.8897	0.4658	−0.1616	0.5583	0.3382	0.2284	0.5357	0.517
	B	2.0495	38.05	32.57	43.127	10.227	45.797	12.01	31.65
	R^2	0.7	0.5364	0.0282	0.2617	0.2496	0.4642	0.4954	0.705

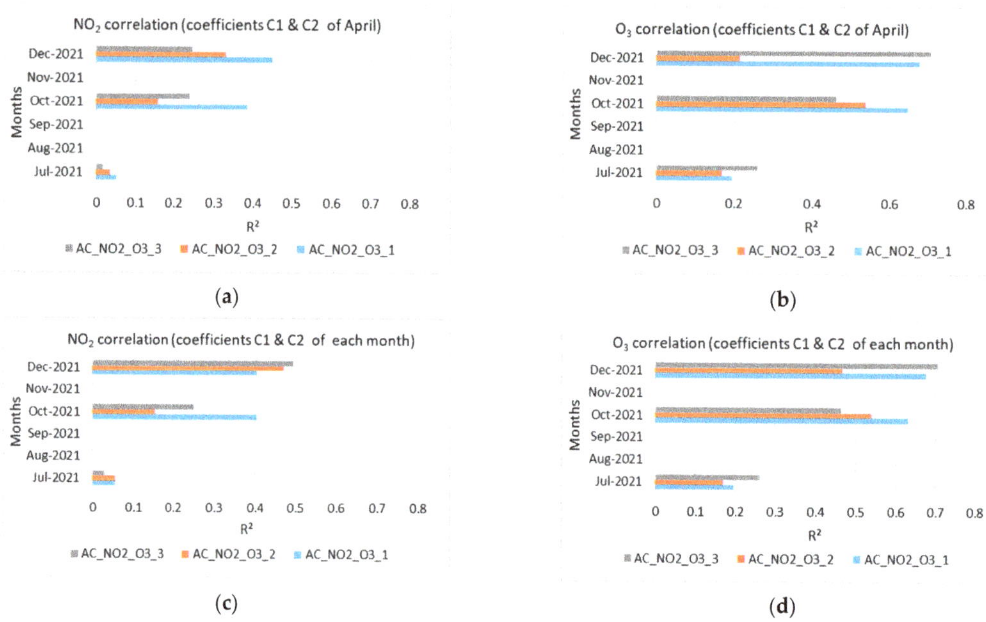

Figure 11. Correlation for the NO_2 and O_3 measurements for July, October, and December 2021 of (**a**) NO_2 sensor (April calculated coefficients C1, C2) and reference; (**b**) O_3 sensor (April calculated coefficients C1, C2) and reference; (**c**) NO_2 sensor (calculated coefficients C1, C2 per month) and reference; (**d**) O_3 low-cost sensor (calculated coefficients C1, C2 per month) and reference.

- Observation of the variation of C1 and C2 fitting was made to obtain the temporal variation of C1 and C2 and extrapolate all intermediate values. While performing this last step in order to obtain the aging formula, it was observed that coefficient C1 for NO_2 during the operation of the sensors showed a decrease by one unit per month of the initial value from the beginning of the operation of the sensor. Next, coefficient C1 concerning the aging of the sensor was introduced as $C1_{Age}$. In this way, the coefficient $C1_{Age}$ for NO_2 sensor is described in Equation (4) and the corresponding behavior for the O_3 sensor is described in Equation (5)

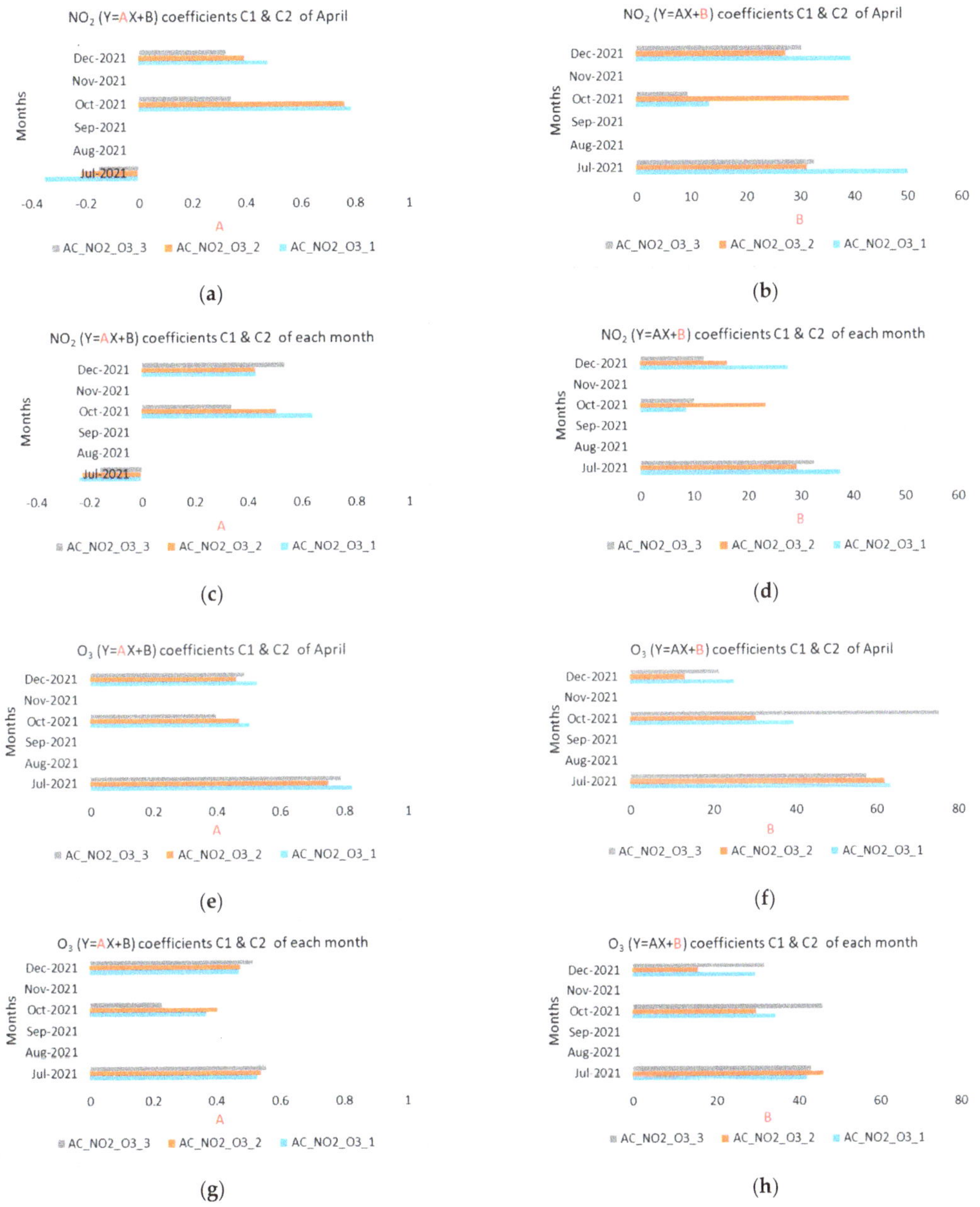

Figure 12. Coefficients A and B of the NO_2 and O_3 measurements for the months of July, October, and December 2021 (**a**) NO_2 coefficient A using April coefficients C1, C2; (**b**) NO_2 coefficient B using April coefficients C1, C2; (**c**) NO_2 coefficient A using coefficients C1, C2 of each month; (**d**) NO_2 coefficient B using coefficients C1, C2 of each month; (**e**) O_3 coefficient A using April coefficients C1, C2; (**f**) O_3 coefficient B using April coefficients C1, C2; (**g**) O_3 coefficient A using coefficients C1, C2 of each month; (**h**) O_3 coefficient B using coefficients C1, C2 of each month.

$$C1_{Age}\,(NO_2) = C1_{init}\,(NO_2) - n \tag{4}$$

$$C1_{Age}\,(O_3) = C1_{init}\,(O_3) + n \tag{5}$$

where $C1_{init}$ represents the initial value of the coefficient from the beginning of the sensor's operation, and n expresses the sum of the months of in-time service of the sensor.

Next, the temporal variation of all coefficients (i.e., C1, C2x) was plotted to evaluate the aging functions' performance. The extracted results are depicted in Figure 13a–d for NO_2 sensors and Figure 14a,b for O_3 sensors. These figures prove the variability of C1 and C2 during the operation time (8 months) of the sensors.

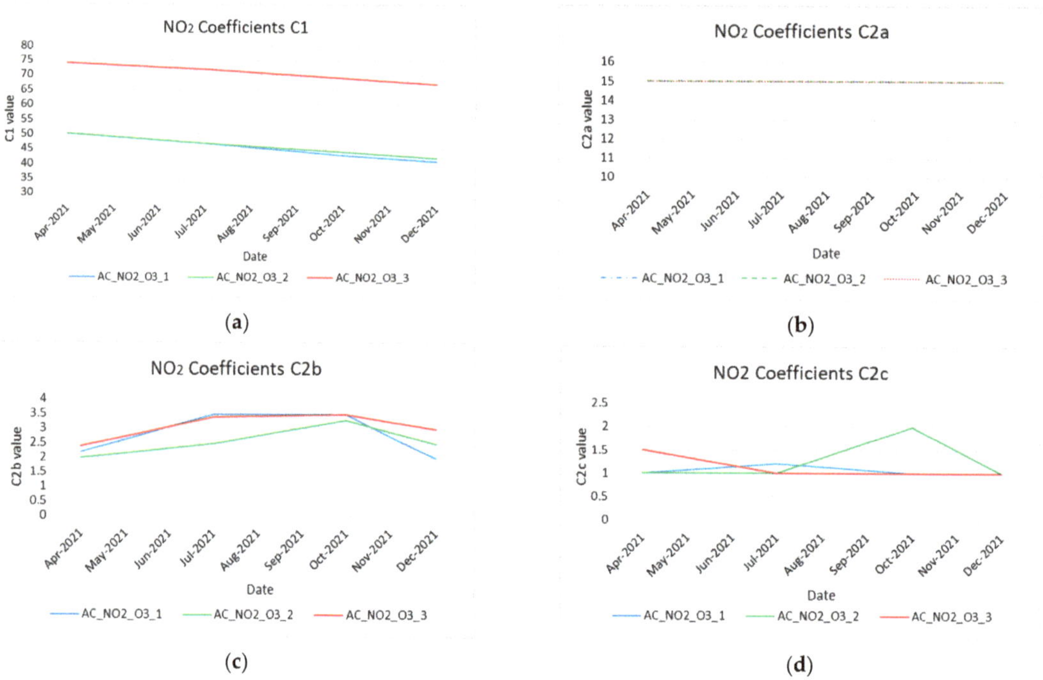

Figure 13. Coefficient variation of NO_2 low-cost sensors during their operational time. (**a**) NO_2 low-cost sensors' coefficients C1 variation; (**b**) NO_2 low-cost sensors' coefficients C2a variation; (**c**) NO_2 low-cost sensors' coefficients C2b variation; (**d**) NO_2 low-cost sensors' coefficients C2c variation.

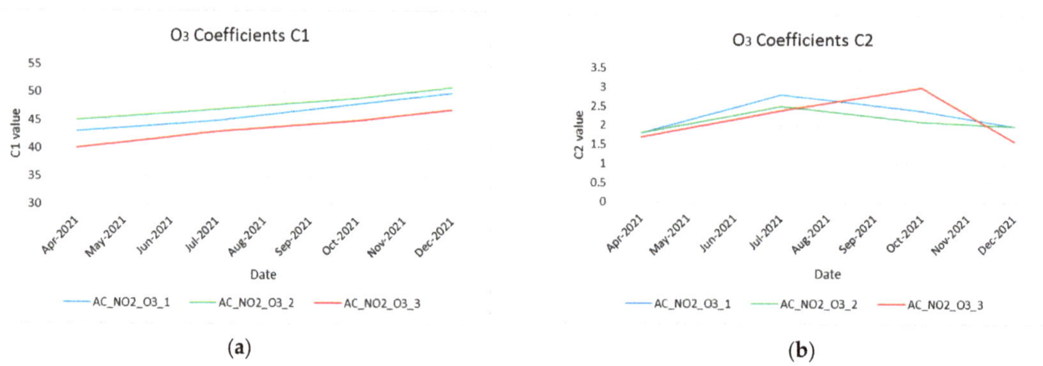

Figure 14. Coefficient variation of O_3 low−cost sensors during their operational time. (**a**) O_3 low-cost sensors' coefficients C1 variation; (**b**) O_3 low−cost sensors' coefficients C2 variation.

Both in the nitrogen dioxide sensors and in the ozone sensors, it is observed that the C1 coefficient shows a linearity, which is expected according to the manufacturer; the sensor during its lifetime (about two years) operates in a linear range.

Coefficient C2 concerns the scale of the measured values in relation to aging of the sensor. According to the datasheets of O_3 [77] and NO_2 [78] sensors, their sensitivity depends on the temperature [79]. The mean temperature values per month in 2021 in the center of Athens are presented in Table 4, where the average temperatures for the months that the experiment took place are highlighted [80].

Table 4. Annual climatological summary for 2021 [80].

Annual Climatological Summary				
Name: athens984 City: Athens State: Attica, Greece Elev: 60m Lat: 37°58′42″ N Long: 23°42′56″ E Temperature (°C), Hheat Base 18.3, Cool Base 18.3				
Year	Month	Mean MAX	Mean MIN	Mean
2021	1	15.0	7.9	11.5
2021	2	15.8	7.0	11.3
2021	3	16.2	8.3	12.1
2021	4	20.0	11.3	15.7
2021	5	27.3	17.8	22.4
2021	6	30.1	21.1	25.4
2021	7	34.1	25.9	29.9
2021	8	34.3	25.5	29.7
2021	9	28.5	20.5	24.2
2021	10	21.7	15.0	18.0
2021	11	19.1	12.6	15.7
2021	12	14.9	8.3	11.7

Coefficient C2 (referred as $C2_{Scale}$) contrary to $C1_{Age}$ is not so closely related to the sensor lifetime. Instead, ambient temperature severely impacts the recordings of sensors. This is not only observed but is also stated by the sensor manufacturer. Thus, $C2_{Scale}$ is the corrected value after taking into consideration the temperature according to [75,77]. Coefficient $C2_{Scale(O_3)}$ for O_3 measurements is given by Equation (6):

$$C2_{Scale(O_3)} = C2_{init(O_3)} + \left(S_{init} - S_{current}\right)/10 \quad (6)$$

where $C2_{init(O_3)}$ is the initial value of C2 from the beginning of the O_3 sensor operation, S_{init} is the sensitivity of the sensor during the first day of its lifetime at a specific temperature and $S_{current}$ is the corresponding sensitivity at the temperature of the running month [77]. As described in Table 1 the coefficients $C2a_{Scale(NO_2)}$ ($NO_2 < 3$) and $C2c_{Scale(NO_2)}$ ($3 < NO_2 < 30$) are maintained practically constant where $C2a_{Scale(NO_2)} = 15$ and $C2c_{Scale(NO_2)} = 1$. Contrary to the above, the coefficient $C2b_{Scale(NO_2)}$ ($NO_2 > 30$) is strongly affected by the temperature. The coefficient $C2a_{Scale(NO_2)}$ for NO_2 measurements is given by Equation (7):

$$C2_{Scale(NO_2)} = C2_{init(NO_2)} + \left(S_{init} - S_{current}\right)/10 \quad (7)$$

where $C2_{init(NO_2)}$ is the initial value of the coefficient from the beginning of the NO_2 sensor operation, S_{init} is the sensitivity of the sensor during the first day of its lifetime at a specific temperature, and $S_{current}$ is the corresponding sensitivity at the temperature of the running month [78]. For each sensor, the corresponding datasheets provided by the manufacturer [75,76] include a graph which describes the relation of sensor sensitivity according to the temperature of the environment. The calculation of the corresponding sensitivity at a specific temperature is achieved by extracting the slope of the graph at each temperature case.

After having performed the above process, it was concluded that Equations (4)–(7) had an obvious impact on the performance of the sensors. Specifically, it was observed that the performance of the sensors was kept practically stable during their lifetime with respect to the corresponding results extracted when the C1 and C2 factors were maintained as initially set during their calibration. It could be seen that the adoption of a continuous correction process was required in order to maintain the validity of pollutant gas measurements.

To proceed in further evaluation of the initial calibration and the correction process due to aging, a boxplot presentation method was adopted. Actually, a boxplot, also known as a box and whisker plot, is a graphical representation of statistical data that displays the distribution of a dataset by showing the median, quartiles, and range of the data. Boxplots are useful for both comparing the distribution of different datasets and visualizing the distribution of a single dataset. They can also help to identify any potential outliers in the data.

Specifically, the boxplots depicted in Figure 15 show the variation of the measured quantities, between the inexpensive and reference sensors. Figure 15a illustrates the O_3 variation for October 2021 using the coefficients C1 and C2 that were calculated for April 2021, while Figure 15b shows the corresponding variation using the coefficients calculated using the data from October 2021. Figures 15c and 15d, respectively, show the correction degree using the coefficients C1 and C2 for NO_2.

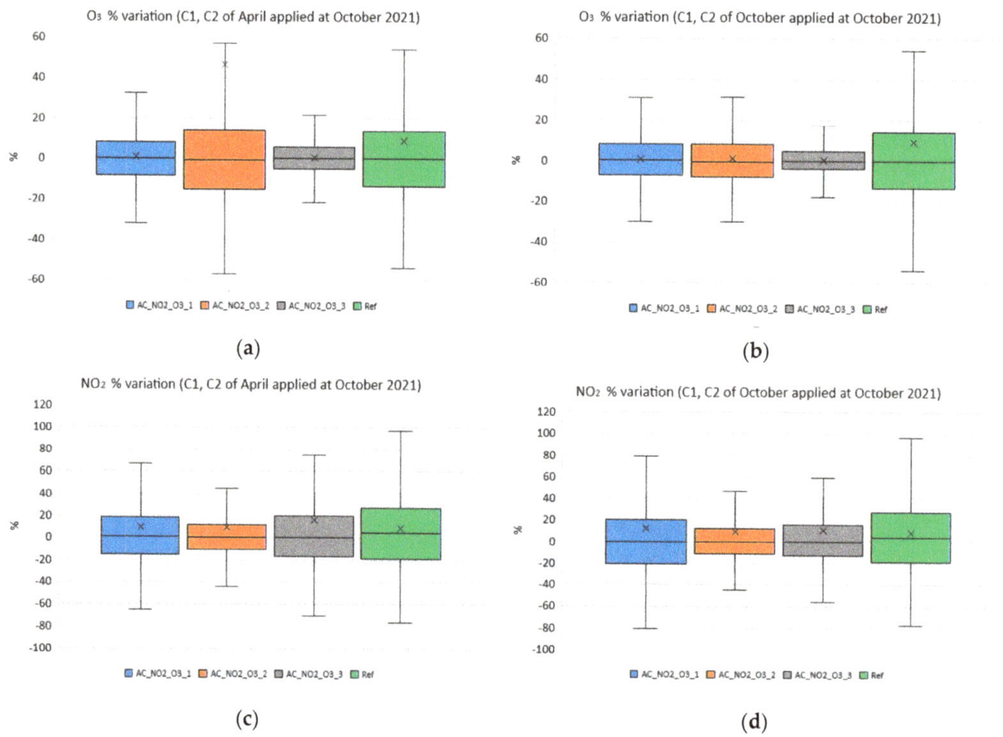

Figure 15. Variation of low−cost sensors NO_2 and O_3. (**a**) Variation of O_3 sensors using the coefficients calculated for April 2021 and applied in October 2021; (**b**) variation of O_3 sensors using the coefficients calculated for October 2021 and applied in October 2021; (**c**) variation of NO_2 sensors using the coefficients calculated for April 2021 and applied in October 2021; (**d**) variation of NO_2 sensors using the coefficients calculated for October 2021 and applied in October 2021.

Observing the results in the boxplots illustrated in Figure 15, it can be straight forwardly concluded that the deviation of the recorded values with respect to the reference instruments were significantly lower when changing the C1 and C2 coefficients (see Figure 15b,d) when compared to the corresponding deviation when using the same C1 and C2 values during the sensors' lifetime (see Figure 15a,c). This fact supports the need for adopting changes for the correction process during the lifetime of a low-cost electrochemical sensor. Furthermore, it becomes obvious that the adopted Equations (4)–(7) can improve the performance of the sensors during their lifetime instead of keeping their values constant during the lifetime of the sensors. It is still remaining to study whether these equations can be further improved.

Finally, in order to demonstrate the impact of the change on the coefficients on the performance, the corresponding boxplots were plotted and are shown in Figure 16. Figure 16a–h show the variation of NO_2 and O_3 for each month for the coefficients C1 and C2 recalculated according to the aging formulas for each one of the studied months. Specifically, Figure 16a,b correspond to the statistical variation of the reference/low-cost using the aging-corrected C1 and C2 for NO_2 and O_3 during April 2021 (period A). The corresponding box plots for July 2021 (period B) are depicted in Figure 16c,d. The October 2021 (period C) corresponding results are presented in Figure 16e,f for NO_2 and O_3, respectively. Lastly, the extracted statistical boxplot results regarding December of 2021 (period D) are shown in Figure 16g,h.

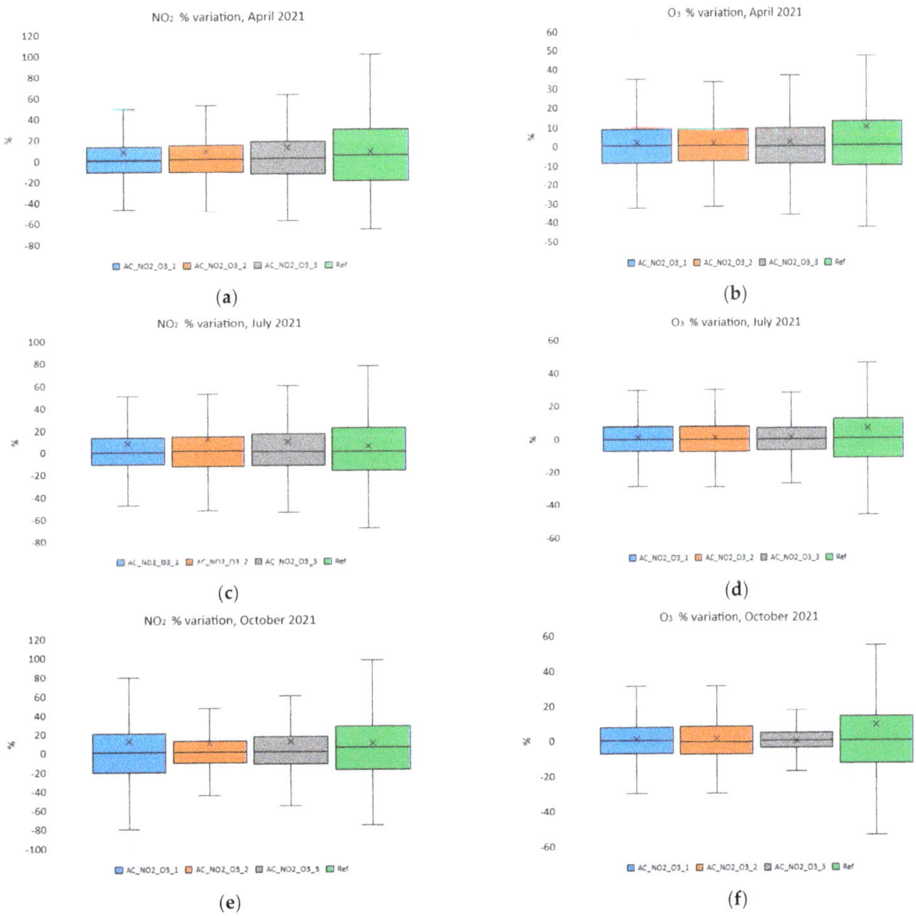

Figure 16. *Cont.*

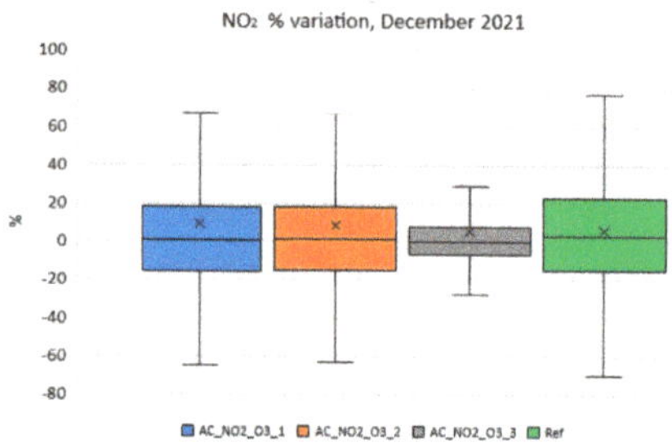

(g)

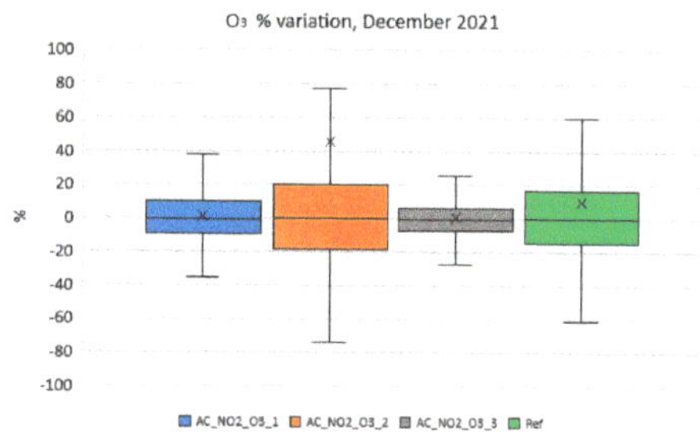

(h)

Figure 16. Variation of NO_2 and O_3 using the coefficients calculated monthly. (**a**) Variation of NO_2 using the coefficients calculated in period A; (**b**) variation of O_3 using the coefficients calculated in period A; (**c**) Variation of NO_2 using the coefficients calculated in period B; (**d**) variation of O_3 using the coefficients calculated in period B; (**e**) variation of NO_2 using the coefficients calculated in period C; (**f**) variation of O_3 using the coefficients calculated in period C; (**g**) variation of NO_2 using the coefficients calculated in period D; (**h**) variation of O_3 using the coefficients calculated in period D.

Observing Figure 16, it becomes obvious that the use of the methodology proposed in this article is indeed not only essential but also very effective in keeping the performance of inexpensive electrochemical sensors constant during their lifetime.

5. Conclusions

There is no doubt that air quality monitoring in urban environments is one of the most important applications of WSNs in smart cities. It is also true that there are many commercially available instruments that are able to perform air quality monitoring with high accuracy. However, they are very expensive to use in large-scale installations. At the same time, technological advances have made it feasible to mass-manufacture inexpensive electrochemical sensors that are able to measure the concentrations of pollutant gases in air with relatively good accuracy. On the other hand, aging caused by a variety of factors, including exposure to high levels of gases, temperature fluctuations, and moisture, deteriorates the sensitivity of such sensors, leading to inaccurate readings. One common way to address sensor aging is to periodically perform calibration of the sensors in order to check the existing accuracy and properly improve it if needed. Actually, most of the published works that deal with the aging of low-cost sensors involve machine learning, neural networks, and other similar processing-demanding methods. Such approaches require significant CPU and memory resources, having a direct impact on the processing capability specifications and the cost of such a measuring system. Additionally, incorporating such methodologies on the measuring unit significantly increases energy consumption. Even for the case when the processing is conducted at a central point, the need for high processing power remains, since each networked measuring system must be treated separately. The herein proposed solution incorporates a simple compensation algorithm of good performance that can be easily executed at any sensing node. In addition, due to the simplicity of the required actions, power consumption is practically unaffected.

Specifically, in this research work, the impact of aging on the accuracy of low-cost gaseous pollutant sensors used in WSNs was studied. Specifically, an air quality monitoring WSN containing three sensor nodes was established in the center of Athens, Greece. Each one of the specific sensor nodes encompassed a couple of sensors that monitored the concentration of ozone and nitrogen dioxide in the ambient air. For a period of eight months the values of the measurements of these sensors were compared with those made by the air monitoring instruments that are officially used by the State.

The first conclusion drawn from this research work is that the sensor's aging—which may be caused by a variety of factors, including exposure to high levels of gases, temperature fluctuations, and moisture—indeed impacts its sensitivity, leading to inaccurate readings. One common way to address sensor aging is to periodically calibrate the sensors to ensure their accuracy.

However, this research work evinced that there is a very effective methodology to keep the sensors' performance stable during their lifecycle. Actually, coefficients C1 and C2 used in the methodology proposed express the performance of a sensor during its operational life. Specifically, coefficient C1 is directly related to aging and its value changes for each month of the operational time of the sensor according to a formula which is differentiated according to the pollutant gas detected. At the same time, coefficient C2 aims at the micrometric correction of the sensor values according to the average temperature of the month of operation under study.

The suitable use of these two coefficients in the formulae proposed showed excellent results for both NO_2 and O_3 low-cost air quality sensors, in the sense that not only their aging was treatable but also high reliability of the measurements can be achieved for the entire lifetime of the sensors. In this way, air quality monitoring can be performed via low-cost sensors with no need for recalibration with official reference instruments at regular intervals. So, it is feasible to create dense air quality monitoring networks in urban areas without high acquisition costs. This is greatly beneficial to the attainment of not only inexpensive but also accurate air monitoring via WSNs in smart cities.

Author Contributions: Conceptualization, I.S. and I.C.; methodology, I.S., O.T., D.K. and I.C.; software, I.C.; validation, I.S., O.T., D.K. and I.C.; formal analysis, D.K. and I.C.; investigation, I.C.; resources, D.K. and I.C.; data curation, O.T. and I.C.; writing—original draft preparation, I.C. and D.K.; writing—review and editing, O.T., D.K. and I.S.; visualization, I.C.; supervision, D.K. and I.S.; project administration, I.S.; All authors have read and agreed to the published version of the manuscript.

Funding: This research received no external funding.

Institutional Review Board Statement: Not applicable.

Informed Consent Statement: Not applicable.

Data Availability Statement: All of the data created in this study are presented in the context of this article.

Conflicts of Interest: The authors declare no conflict of interest.

References

1. Judy, J.W. Microelectromechanical Systems (MEMS): Fabrication, Design and Applications. *Smart Mater. Struct.* **2001**, *10*, 1115–1134. [CrossRef]
2. Akyildiz, I.F.; Su, W.; Sankarasubramaniam, Y.; Cayirci, E. A Survey on Sensor Networks. *IEEE Commun. Mag.* **2002**, *40*, 104–112. [CrossRef]
3. Yick, J.; Mukherjee, B.; Ghosal, D. Wireless Sensor Network Survey. *Comput. Netw.* **2008**, *52*, 2292–2330. [CrossRef]
4. Evangelakos, E.A.; Kandris, D.; Rountos, D.; Tselikis, G.; Anastasiadis, E. Energy Sustainability in Wireless Sensor Networks: An Analytical Survey. *J. Low Power Electron. Appl.* **2022**, *12*, 65. [CrossRef]
5. Nakas, C.; Kandris, D.; Visvardis, G. Energy Efficient Routing in Wireless Sensor Networks: A Comprehensive Survey. *Algorithms* **2020**, *13*, 72. [CrossRef]
6. Farsi, M.; Elhosseini, M.A.; Badawy, M.; Arafat Ali, H.; Zain Eldin, H. Deployment Techniques in Wireless Sensor Networks, Coverage and Connectivity: A Survey. *IEEE Access* **2019**, *7*, 28940–28954. [CrossRef]
7. Kandris, D.; Vergados, D.J.; Vergados, D.D.; Tzes, A. A Routing Scheme for Congestion Avoidance in Wireless Sensor Networks. In Proceedings of the 6th Annual IEEE Conference on Automation Science and Engineering (CASE 2010), Toronto, ON, Canada, 21–24 August 2010; pp. 21–24.
8. Ploumis, S.E.; Sgora, A.; Kandris, D.; Vergados, D.D. Congestion Avoidance in Wireless Sensor Networks: A Survey. In Proceedings of the 2012 16th Panhellenic Conference on Informatics, PCI 2012, Piraeus, Greece, 5–7 October 2012; pp. 234–239. [CrossRef]
9. Kandris, D.; Alexandridis, A.; Dagiuklas, T.; Panaousis, E.; Vergados, D.D. Multiobjective Optimization Algorithms for Wireless Sensor Networks. *Wirel. Commun. Mob. Comput.* **2020**, *2020*, 1–5. [CrossRef]
10. Sharma, N.; Singh, B.M.; Singh, K. QoS-Based Energy-Efficient Protocols for Wireless Sensor Network. *Sustain. Comput. Inform. Syst.* **2021**, *30*, 100425. [CrossRef]

11. Tarnaris, K.; Preka, I.; Kandris, D.; Alexandridis, A. Coverage and K-Coverage Optimization in Wireless Sensor Networks Using Computational Intelligence Methods: A Comparsative Study. *Electronics* **2020**, *9*, 675. [CrossRef]
12. Yu, J.Y.; Lee, E.; Oh, S.R.; Seo, Y.D.; Kim, Y.G. A Survey on Security Requirements for WSNs: Focusing on the Characteristics Related to Security. *IEEE Access* **2020**, *8*, 45304–45324. [CrossRef]
13. Kandris, D.; Nakas, C.; Vomvas, D.; Koulouras, G. Applications of Wireless Sensor Networks: An Up-to-Date Survey. *Appl. Syst. Innov.* **2020**, *3*, 14. [CrossRef]
14. Pantazis, N.A.; Nikolidakis, S.A.; Kandris, D.; Vergados, D.D. An Automated System for Integrated Service Management in Emergency Situations. In Proceedings of the 2011 Panhellenic Conference on Informatics, PCI 2011, Kastoria, Greece, 30 September–2 October 2011; pp. 154–157. [CrossRef]
15. Papadakis, N.; Koukoulas, N.; Christakis, I.; Stavrakas, I.; Kandris, D. An Iot-Based Participatory Antitheft System for Public Safety Enhancement in Smart Cities. *Smart Cities* **2021**, *4*, 919–937. [CrossRef]
16. Khedo, K.K.; Bissessur, Y.; Goolaub, D.S. An Inland Wireless Sensor Network System for Monitoring Seismic Activity. *Future Gener. Comput. Syst.* **2020**, *105*, 520–532. [CrossRef]
17. Nikolidakis, S.A.; Kandris, D.; Vergados, D.D.; Douligeris, C. Energy Efficient Automated Control of Irrigation in Agriculture by Using Wireless Sensor Networks. *Comput. Electron. Agric.* **2015**, *113*, 154–163. [CrossRef]
18. Farooq, M.U.; Waseem, M.; Mazhar, S.; Khairi, A.; Kamal, T. A Review on Internet of Things (IoT). *Int. J. Comput. Appl.* **2015**, *113*, 135–144. [CrossRef]
19. Ali, A. A Framework for Air Pollution Monitoring in Smart Cities by Using IoT and Smart Sensors. *Informatica* **2022**, *46*. [CrossRef]
20. Thangammal, C.B.; Ilamathi, K.; Poonkuzhali, P.; Aarthi, R. An IoT Enabled Air Quality Monitoring Mobile App for Smart Cities. In Proceedings of the 2nd International Conference on Recent Trends in Machine Learning, IoT, Smart Cities and Applications, Hyderabad, Telangana, India, 28–29 March 2021; pp. 275–285. [CrossRef]
21. Snyder, E.G.; Watkins, T.H.; Solomon, P.A.; Thoma, E.D.; Williams, R.W.; Hagler, G.S.W.; Shelow, D.; Hindin, D.A.; Kilaru, V.J.; Preuss, P.W. The Changing Paradigm of Air Pollution Monitoring. *Environ. Sci. Technol.* **2013**, *47*, 11369–11377. [CrossRef]
22. Kumar, P.; Morawska, L.; Martani, C.; Biskos, G.; Neophytou, M.; Di Sabatino, S.; Bell, M.; Norford, L.; Britter, R. The Rise of Low-Cost Sensing for Managing Air Pollution in Cities. *Environ. Int.* **2015**, *75*, 199–205. [CrossRef]
23. Balogun, A.-L.; Tella, A.; Baloo, L.; Adebisi, N. A Review of the Inter-Correlation of Climate Change, Air Pollution and Urban Sustainability Using Novel Machine Learning Algorithms and Spatial Information Science. *Urban Clim.* **2021**, *40*, 100989. [CrossRef]
24. Zhao, B.; Yu, L.; Wang, C.; Shuai, C.; Zhu, J.; Qu, S.; Taiebat, M.; Xu, M. Urban Air Pollution Mapping Using Fleet Vehicles as Mobile Monitors and Machine Learning. *Environ. Sci. Technol.* **2021**, *55*, 5579–5588. [CrossRef]
25. Lautenschlager, F.; Becker, M.; Kobs, K.; Steininger, M.; Davidson, P.; Krause, A.; Hotho, A. OpenLUR: Off-The-Shelf Air Pollution Modeling with Open Features and Machine Learning. *Atmos. Environ.* **2020**, *233*, 117535. [CrossRef]
26. Wang, A.; Xu, J.; Tu, R.; Saleh, M.; Hatzopoulou, M. Potential of Machine Learning for Prediction of Traffic Related Air Pollution. *Transp. Res. Part D Transp. Environ.* **2020**, *88*, 102599. [CrossRef]
27. Aditya, C.R.; Deshmukh, C.R.; Nayana, D.K.; Vidyavastu, P.G. Detection and prediction of air pollution using machine learning models. *Int. J. Eng. Trends Technol.* **2018**, *59*, 204–207.
28. Xi, X.; Zhao, W.; Rui, X.; Wang, Y.; Bai, X.; Yin, W.; Don, J. A comprehensive evaluation of air pollution prediction improvement by a machine learning method. In Proceedings of the 2015 IEEE International Conference on Service Operations and Logistics, and Informatics (SOLI), Yasmine Hammamet, Tunisia, 15–17 November 2015; pp. 176–181.
29. McKercher, G.R.; Salmond, J.A.; Vanos, J.K. Characteristics and Applications of Small, Portable Gaseous Air Pollution Monitors. *Environ. Pollut.* **2017**, *223*, 102–110. [CrossRef] [PubMed]
30. Lewis, A.; Peltier, W.R.; von Schneidemesser, E. *Low-Cost Sensors for the Measurement of Atmospheric Composition: Overview of Topic and Future Applications*; Research report; World Meteorological Organization: Geneva, Switzerland, 2018.
31. Cross, E.S.; Williams, L.R.; Lewis, D.K.; Magoon, G.R.; Onasch, T.B.; Kaminsky, M.L.; Worsnop, D.R.; Jayne, J.T. Use of Electrochemical Sensors for Measurement of Air Pollution: Correcting Interference Response and Validating Measurements. *Atmos. Meas. Tech.* **2017**, *10*, 3575–3588. [CrossRef]
32. Jerrett, M.; Donaire-Gonzalez, D.; Popoola, O.; Jones, R.; Cohen, R.C.; Almanza, E.; de Nazelle, A.; Mead, I.; Carrasco-Turigas, G.; Cole-Hunter, T.; et al. Validating Novel Air Pollution Sensors to Improve Exposure Estimates for Epidemiological Analyses and Citizen Science. *Environ. Res.* **2017**, *158*, 286–294. [CrossRef] [PubMed]
33. deSouza, P. A Nairobi Experiment in Using Low-cost Air Quality Monitors. *Clean Air J.* **2017**, *27*, 12–42. [CrossRef]
34. D'Alvia, L.; Palermo, E.; Del Prete, Z. Validation and application of a novel solution for environmental monitoring: A three month study at "Minerva Medica" archaeological site in Rome. *Measurement* **2018**, *129*, 31–36. [CrossRef]
35. Lewis, A.C.; Lee, J.D.; Edwards, P.M.; Shaw, M.D.; Evans, M.J.; Moller, S.J.; Smith, K.R.; Buckley, J.W.; Ellis, M.; Gillot, S.R.; et al. Evaluating the Performance of Low-cost Chemical Sensors for Air Pollution Research. *Faraday Discuss.* **2016**, *189*, 85–103. [CrossRef]
36. Christakis, I.; Syropoulou, P.; Papadakis, N.; Stavrakas, I. On the correction of low-cost NO_2, O_3 and PM sensors. In Proceedings of the Eighth International Conference on Environmental Management, Engineering, Planning & Economics, Thessaloniki, Greece, 20–24 July 2021; pp. 301–310.

37. Christakis, I.; Moutzouris, K.; Tsakiridis, O.; Stavrakas, I. Barometric Pressure as a correction factor for low-cost particulate matter sensors. In *IOP Conference Series: Earth and Environmental Science*; IOP Publishing: Athens, Greece, 2022; Volume 1123, No. 1.
38. Christakis, I.; Hloupis, G.; Tsakiridis, O.; Stavrakas, I. Integrated open source air quality monitoring platform. In Proceedings of the 11th International Conference on Modern Circuits and Systems Technologies (MOCAST), Bremen, Germany, 8–10 June 2022.
39. Williams, R.; Kilaru, V.; Snyder, E.; Kaufman, A.; Dye, T.; Rutter, A.; Russell, A. *EPA Citizen Scientist Air Monitoring Tool Kit Air Sensor Guidebook*; U.S. Environmental Protection Agency: Washington, DC, USA, 2014.
40. Gerboles, M.; Spinelle, L.; Borowiak, A. *Measuring Air Pollution with Low-Cost Sensors. Thoughts on the Quality of Data Measured by Sensors*; European Commission: Ispra, Italy, 2017.
41. Borrego, C.; Costa, A.M.; Ginja, J.; Amorim, M.; Coutinho, M.; Karatzas, K.; Sioumis, T.; Katsifarakis, N.; Konstantinidis, K.; De Vito, S.; et al. Assessment of Air Quality Microsensors versus Reference Methods: The EuNetAir Joint Exercise. *Atmos. Environ.* **2016**, *147*, 246–263. [CrossRef]
42. Mueller, M.; Meyer, J.; Hueglin, C. Design of an Ozone and Nitrogen Dioxide Sensor Unit and Its Long-Term Operation within a Sensor Network in the City of Zurich. *Atmos. Meas. Tech.* **2017**, *10*, 3783–3799. [CrossRef]
43. Samad, A.; Obando Nuñez, D.R.; Solis Castillo, G.C.; Laquai, B.; Vogt, U. Effect of Relative Humidity and Air Temperature on the Results Obtained from Low-Cost Gas Sensors for Ambient Air Quality Measurements. *Sensors* **2020**, *20*, 5175. [CrossRef] [PubMed]
44. Karagulian, F.; Barbiere, M.; Kotsev, A.; Spinelle, L.; Gerboles, M.; Lagler, F.; Redon, N.; Crunaire, S.; Borowiak, A. Review of the Performance of Low-Cost Sensors for Air Quality Monitoring. *Atmosphere* **2019**, *10*, 506. [CrossRef]
45. Pang, X.; Shaw, M.D.; Gillot, S.; Lewis, A.C. The Impacts of Water Vapour and Co-Pollutants on the Performance of Electrochemical Gas Sensors Used for Air Quality Monitoring. *Sens. Actuators B Chem.* **2018**, *266*, 674–684. [CrossRef]
46. Solis, G. Test and Analysis of Key Factors that Can Affect the Reliability of Results Obtained from Low-cost Sensors for Outdoor Air Quality Measurements. Master's Thesis, University of Stuttgart, Stuttgart, Germany, 2019.
47. Bigi, A.; Mueller, M.; Grange, S.K.; Ghermandi, G.; Hueglin, C. Performance of NO, NO_2 low-cost sensors and three calibration approaches within a real world application. *Atmos. Meas. Tech.* **2018**, *11*, 3717–3735. [CrossRef]
48. Williams, D.E. Electrochemical sensors for environmental gas analysis. *Curr. Opin. Electrochem.* **2020**, *22*, 145–153. [CrossRef]
49. Farquhar, A.K.; Henshaw, G.S.; Williams, D.E. Understanding and correcting unwanted influences on the signal from electrochemical gas sensors. *ACS Sens.* **2021**, *6*, 1295–1304. [CrossRef]
50. Wei, P.; Ning, Z.; Ye, S.; Sun, L.; Yang, F.; Wong, K.C.; Westerdahl, D.; Louie, P.K. Impact Analysis of Temperature and Humidity Conditions on Electrochemical Sensor Response in Ambient Air Quality Monitoring. *Sensors* **2018**, *18*, 59. [CrossRef] [PubMed]
51. Castell, N.; Dauge, F.R.; Schneider, P.; Vogt, M.; Lerner, U.; Fishbain, B.; Broday, D.; Bartonova, A. Can commercial low-cost sensor platforms contribute to air quality monitoring and exposure estimates? *Environ. Int.* **2017**, *99*, 293–302. [CrossRef]
52. Mead, M.I.; Popoola, O.A.M.; Stewart, G.B.; Landshoff, P.; Calleja, M.; Hayes, M.; Baldovi, J.J.; McLeod, M.W.; Hodgson, T.F.; Dicks, J.; et al. The use of electrochemical sensors for monitoring urban air quality in low-cost, high-density networks. *Atmos. Environ.* **2013**, *70*, 186–203. [CrossRef]
53. Pang, X.; Shaw, M.D.; Lewis, A.C.; Carpenter, L.J.; Batchellier, T. Electrochemical ozone sensors: A miniaturised alternative for ozone measurements in laboratory experiments and air-quality monitoring. *Sens. Actuators B Chem.* **2017**, *240*, 829–837. [CrossRef]
54. Sun, L.; Westerdahl, D.; Ning, Z. Development and Evaluation of a Novel and Cost-Effective Approach for Low-cost NO_2 Sensor Drift Correction. *Sensors* **2017**, *17*, 1916. [CrossRef]
55. Piedrahita, R.; Xiang, Y.; Masson, N.; Ortega, J.; Collier, A.; Jiang, Y.; Li, K.; Dick, R.P.; Lv, Q.; Hannigan, M.; et al. The next generation of low-cost personal air quality sensors for quantitative exposure monitoring. *Atmos. Meas. Tech.* **2014**, *7*, 3325–3336. [CrossRef]
56. Gonzalez, A.; Boies, A.; Swason, J.; Kittelson, D. Field calibration of low-cost air pollution sensors. *Atmos. Meas. Tech. Discuss.* **2019**, 1–17, *preprint*.
57. Munir, S.; Mayfield, M.; Coca, D.; Jubb, S.A. Structuring an integrated air quality monitoring network in large urban areas—Discussing the purpose, criteria and deployment strategy. *Atmos. Environ. X* **2019**, 2, 100027. [CrossRef]
58. Hagan, D.H.; Isaacman-VanWertz, G.; Franklin, J.P.; Wallace, L.M.M.; Kocar, B.D.; Heald, C.L.; Kroll, J.H. Calibration and assessment of electrochemical air quality sensors by co-location with regulatory-grade instruments. *Atmos. Meas. Tech.* **2018**, *11*, 315–328. [CrossRef]
59. Malings, C.; Tanzer, R.; Hauryliuk, A.; Kumar, S.P.; Zimmerman, N.; Kara, L.B.; Presto, A.A.; Subramanian, R. Development of a general calibration model and long-term performance evaluation of low-cost sensors for air pollutant gas monitoring. *Atmos. Meas. Tech.* **2019**, *12*, 903–920. [CrossRef]
60. Spinelle, L.; Gerboles, M.; Villani, M.G.; Aleixandre, M.; Bonavitacola, F. Field calibration of a cluster of low-cost available sensors for air quality monitoring. Part A: Ozone and nitrogen dioxide. *Sens. Actuators B Chem.* **2015**, *215*, 249–257. [CrossRef]
61. Margaritis, D.; Keramydas, C.; Papachristos, I.; Lambropoulou, D. Calibration of low-cost gas sensors for air quality monitoring. *Aerosol Air Qual. Res.* **2021**, *21*, 210073. [CrossRef]
62. Mijling, B.; Jiang, Q.; De Jonge, D.; Bocconi, S. Field calibration of electrochemical NO_2 sensors in a citizen science context. *Atmos. Meas. Tech.* **2018**, *11*, 1297–1312. [CrossRef]
63. Gao, H.; Dai, B.; Miao, H.; Yang, X.; Barroso, R.J.D.; Walayat, H. A novel gapg approach to automatic property generation for formal verification: The gan perspective. *ACM Trans. Multimed. Comput. Commun. Appl.* **2023**, *19*, 1–22. [CrossRef]

64. Gao, H.; Qiu, B.; Duran Barroso, R.J.; Hussain, W.; Xu, Y.; Wang, X. TSMAE: A Novel Anomaly Detection Approach for Internet of Things Time Series Data Using Memory-Augmented Autoencoder. *IEEE Trans. Netw. Sci. Eng.* **2022**. *Early Access*. [CrossRef]
65. Kumar, A.; Kim, H.; Hancke, G.P. Environmental monitoring systems: A review. *IEEE Sens. J.* **2012**, *13*, 1329–1339. [CrossRef]
66. Guth, U.; Vonau, W.; Oelßner, W. Gas Sensors. In *Environmental Analysis by Electrochemical Sensors and Biosensors*; Moretto, L., Kalcher, K., Eds.; Nanostructure Science and Technology; Springer: New York, NY, USA, 2014; pp. 569–580.
67. Afshar-Mohajer, N.; Zuidema, C.; Sousan, S.; Hallett, L.; Tatum, M.; Rule, A.M.; Geb, T.; Peters, T.M.; Koehler, K. Evaluation of low-cost electro-chemical sensors for environmental monitoring of ozone, nitrogen dioxide, and carbon monoxide. *J. Occup. Environ. Hyg.* **2018**, *15*, 87–98. [CrossRef]
68. Li, J.; Hauryliuk, A.; Malings, C.; Eilenberg, S.R.; Subramanian, R.; Presto, A.A. Characterizing the Aging of Alphasense NO_2 Sensors in Long-Term Field Deployments. *ACS Sens.* **2021**, *6*, 2952–2959. [CrossRef] [PubMed]
69. Migos, T.; Christakis, I.; Moutzouris, K.; Stavrakas, I. On the Evaluation of Low-cost PM Sensors for Air Quality Estimation. In Proceedings of the 8th International Conference on Modern Circuits and Systems Technologies (MOCAST), Thessaloniki, Greece, 13–15 May 2019.
70. Air Pollution Measurement Data. Ministry of Environment & Energy, Greece. Available online: https://ypen.gov.gr/perivallon/poiotita-tis-atmosfairas/dedomena-metriseon-atmosfairikis-rypansis/ (accessed on 10 March 2023).
71. Larssen, S.; Sluyter, R.; Helmis, C. *Criteria for EUROAIRNET—The EEA Air Quality Monitoring and Information Network*; Technical report EEA; 12/1999 European Environment Agency: Copenhagen, Denmark, 1999.
72. Blanchard, C.L.; Carr, E.L.; Collins, J.F.; Smith, T.B.; Lehrman, D.E.; Michaels, H.M. Spatial representativeness and scales of transport during the 1995 integrated monitoring study in California's San Joaquin Valley. *Atmos. Environ.* **1999**, *33*, 4775–4786. [CrossRef]
73. Spangl, W.; Schneider, J.; Moosmann, L.; Nagl, C. *Representativeness and Classification of Air Quality Monitoring Stations*; Umweltbundesamt: Dessau-Roßlau, Germany, 2007.
74. Janssen, S.; Dumont, G.; Fierens, F.; Deutsch, F.; Maiheu, B.; Celis, D.; Trimpeneers, E.; Mensink, C. Land use to characterize spatial representativeness of air quality monitoring stations and its relevance for model validation. *Atmos. Environ.* **2012**, *59*, 492–500. [CrossRef]
75. Henne, S.; Brunner, D.; Folini, D.; Solberg, S.; Klausen, J.; Buchmann, B. Assessment of parameters describing representativeness of air quality in-situ measurement sites. *Atmos. Chem. Phys.* **2010**, *10*, 3561–3581. [CrossRef]
76. Christakis, I.; Hloupis, G.; Stavrakas, I.; Tsakiridis, O. Low-cost sensor implementation and evaluation for measuring NO_2 and O_3 pollutants. In Proceedings of the 2020 9th International Conference on Modern Circuits and Systems Technologies (MOCAST), Bremen, Germany, 7–9 September 2020.
77. Alphasense UK, Gas Sensors & Air Quality Monitors. Available online: https://www.alphasense.com/wp-content/uploads/2022/09/Alphasense_OX-B431_datasheet.pdf (accessed on 10 March 2023).
78. Alphasense UK, Gas Sensors & Air Quality Monitors. Available online: https://www.alphasense.com/wp-content/uploads/2022/09/Alphasense_NO2-B43F_datasheet.pdf (accessed on 10 March 2023).
79. Luftqualität—Stadt Zürich. Available online: https://zueriluft.ch/makezurich/AAN803.pdf (accessed on 10 March 2023).
80. Meteo.gr. Available online: http://penteli.meteo.gr/stations/athens/NOAAPRYR.TXT (accessed on 17 March 2023).

electronics

MDPI

Article

AoI-Bounded Scheduling for Industrial Wireless Sensor Networks

Chenggen Pu [1,2,*], Han Yang [2,3], Ping Wang [2,3] and Changjie Dong [2]

1 School of Computer Science and Technology, Chongqing University of Posts and Telecommunications, Chongqing 400065, China
2 Institute of Industrial Internet, Chongqing University of Posts and Telecommunications, Chongqing 401120, China
3 Key Laboratory of Industrial Internet of Things and Networked Control, Ministry of Education, Chongqing University of Posts and Telecommunications, Chongqing 400065, China
* Correspondence: mentospcg@163.com

Abstract: Age of information (AoI) is an emerging network metric that measures information freshness from an application layer perspective. It can evaluate the timeliness of information in industrial wireless sensor networks (IWSNs). Previous research has primarily focused on minimizing the long-term average AoI of the entire system. However, in practical industrial applications, optimizing the average AoI does not guarantee that the peak AoI of each data packet is within a bounded interval. If the AoI of certain packets exceeds the predetermined threshold, it can have a significant impact on the stability of the industrial control system. Therefore, this paper studies the scheduling problem subject to a hard AoI performance requirement in IWSNs. First, we propose a low-complexity AoI-bounded scheduling algorithm for IWSNs that guarantees that the AoI of each packet is within a bounded interval. Then, we analyze the schedulability conditions of the algorithm and propose a method to decrease the peak AoI of nodes with higher AoI requirements. Finally, we present a numerical example that illustrates the proposed algorithm step by step. The results demonstrate the effectiveness of our algorithm, which can guarantee bounded AoI intervals (BAIs) for all nodes.

Keywords: age of information (AoI); industrial wireless sensor networks; peak AoI; scheduling

Citation: Pu, C.; Yang, H.; Wang, P.; Dong, C. AoI-Bounded Scheduling for Industrial Wireless Sensor Networks. *Electronics* **2023**, *12*, 1499. https://doi.org/10.3390/electronics12061499

Academic Editors: Dionisis Kandris and Eleftherios Anastasiadis

Received: 20 February 2023
Revised: 17 March 2023
Accepted: 20 March 2023
Published: 22 March 2023

1. Introduction

Industrial wireless sensor networks (IWSNs) have emerged as a key technology enabling the deployment of Industry 4.0 for their flexibility, lack of wiring, low cost, and easy deployment characteristics [1]. IWSNs are capable of delivering time-sensitive periodic data flows generated by field devices to the gateway timely and reliably [2]. They have been widely deployed in industrial process automation applications, such as digital twin and remote state estimation [3]. These applications typically use data driven models, and the data inputted into the model should be fresh enough to accurately characterize physical objects [4]. An emerging application layer performance metric, age of information (AoI), has recently been proposed to measure the freshness of data [5]. AoI is defined as the amount of time that has elapsed since the most recent update was generated at the source and successfully received at the destination [6]. Compared with traditional performance indicators (such as delay), AoI not only considers the time spent by data packets in wireless link transmission but also considers the transmission interval specified by the network scheduler. A small AoI implies that there exist fresh data available at the destination. AoI comprehensively characterizes the freshness of data packets, which can be applied to the performance evaluation of IWSNs.

Industrial applications require a high degree of data freshness, as many industrial processes are dynamic and rapidly changing [7]. Timely and accurate information is critical in making decisions that can impact the efficiency, safety, and overall performance of industrial operations [8]. For instance, in industrial process control systems, real-time data from sensors are essential to adjust process parameters, prevent equipment failures, and optimize production. In predictive maintenance, fresh data from sensors can be used to monitor the condition of machinery and predict potential failures before they occur, allowing for proactive maintenance and reducing downtime. If the age of information (AoI) for a certain amount of data falls below its corresponding threshold, it can result in significant damage to industrial production.

Due to the real-time nature of industrial applications, the IWSN standards (e.g., ISA100.11a and WirelessHART [9]) adopt time/frequency division multiple access (TDMA/FDMA) as the medium access method to achieve collision-free transmissions [2]. A network scheduler in the gateway assigns time slots and channels to transmit a flow to the destination, with multiple time slots in a superframe repeating cyclically [10]. Over the past decade, researchers have proposed various IWSN scheduling algorithms [11–16], focusing on minimizing transmission latency, energy consumption optimization, avoiding conflicts, etc. Some studies have proposed using AoI as a metric to measure the freshness of data in WSN. For instance, the authors in [4] analyze the long-term average AoI in WSN, and they formulate the AoI minimization problem subject to energy and time constraints. In [17,18], the authors consider minimizing the average AoI in energy constrained scenarios, especially the energy-harvesting WSN. Meanwhile, the authors of [19] derive the worst case average AoI and average peak AoI of data packets in WSN with the MAC layer based on a carrier sense multiple access with collision avoidance (CSMA/CA) method. However, optimizing AoI from the application layer perspective with hard performance requirements in IWSNs has rarely been considered.

Numerous studies have investigated AoI optimization problems in wireless networks [20], mainly focused on optimizing the performance of the whole system, taking into account different types of queue models, packet generation/arrival processes, queue capacities, wireless channel models, etc. For example, the authors in [5] discussed the minimum AoI for various single-service queue models under the first-come-first-served queue discipline. In [21], the authors derived an expression for the long-term average AoI of multi-service queue models. The authors of [22] investigated the problem of minimizing the AoI in a network subjected to various interference constraints and experiencing time-varying channels. The authors of [23] examined the optimal sampling and updating processes for IoT devices in a real-time monitoring system to minimize the long-term average AoI. In summary, most studies on AoI optimization scheduling based on queue theory aim to target a weighted-sum long-term average AoI. There are also some works focused on reducing the violation probability, where the peak AoI exceeds a given age constraint [24–26]. However, the reduction of violation probability still cannot meet the deterministic requirements of industrial applications. The authors of [27] proposed a scheduling algorithm with the constraint that each source in the system has a maximum AoI threshold. They assumed that the time is divided into slots and each source node collects a new sample at the beginning of each slot. Nonetheless, it is still challenging for sensor nodes in IWSNs to sample data at each time slot due to computing capability and energy constraints.

In industrial applications, the timely delivery of sampled data from the source to the destination is critical, where there is a hard performance requirement for the AoI metric per data packet. It warrants attention that optimizing the average AoI does not guarantee a bounded peak AoI for each data packet; in industrial control systems, if one or a certain packet's AoI exceeds the predetermined threshold, it can seriously affect the stability of the industrial control system. In view of this, we propose an AoI-bounded scheduling algorithm for IWSNs that ensures that the AoI of all data packets sent by each node in the network is within a bounded interval, thus ensuring that the peak AoI of all nodes is bounded, which is crucial for ensuring the stability of the system. The main contributions of this work can be summarized as follows.

- We propose a low-complexity AoI scheduling algorithm for IWSNs that ensures that each packet's AoI is within a bounded interval, instead of optimizing the network's long-term average AoI. To the best of our knowledge, this is the first work in IWSNs that guarantees that the AoI of each data packet is within a bounded interval, which meets the high real-time demands of industrial applications.
- We analyze the schedulability conditions of the network and propose a method for reducing the peak AoI of nodes with higher AoI requirements by allocating more time slots to those nodes.
- We provide a numerical example to demonstrate the algorithm step by step, and the results show the effectiveness of our algorithm.

The rest of this paper is organized as follows. In Section 2, the system model and problem statement are presented by means of a comparison of the AoI evolution process for data packets at different transmission intervals, and Section 3 presents our AoI-bounded scheduling algorithm and analyzes the bounded AoI intervals (BAIs) of nodes. The performance of the proposed algorithm is evaluated and discussed in Section 4. The conclusions are presented in Section 5.

2. System Model and Problem Statement

We consider a data collection scenario in a time-slotted IWSN consisting of one sink node and N source nodes with a single hop. Each source node N_i collects data periodically according to its own sampling period T_i and sends packets to the sink through the wireless channel. We assume that the wireless channel is error-free, allowing us to ignore the underlying communication channel and simplify the scheduling policy design. The sampling period of nodes in an IWSN usually varies significantly due to differences in sensor type or data update rates required by the industrial application. We assume that each sensor node adopts a single-packet queue model due to the low-power and low-cost characteristics of IWSNs. In this model, the older packet is dropped from the queue when a new packet is generated. Therefore, to ensure the freshness and continuity of sampled data, the data packet in the queue must be transmitted to the sink node before the next new periodic data packet generation. The main notations used throughout this paper are summarized in Table 1.

Table 1. List of key notations.

Notation	Description
i	Index for node
t	Time slot number
$X_i(t)$	Indicator function that is equal to 1 when the node i transmits the packet in time slot t, and $X_i(t) = 0$ otherwise
$G_i(t)$	Data generation time
$A_i(t)$	The AoI of source node i at time slot t
T_i	Sampling period of node i
Δ_i^p	Peak AoI of the of node i
I_i	Transmission interval time of node i
U_{min}	Minimum transmission units
α_i	Transmission interval coefficient of node i
P_t	The duration of a superframe

The IEEE 802.15.4 is commonly adopted in IWSNs as the physical and MAC layer fundamental techniques [2]. Assuming that the system is synchronized, time is divided into equal-length time slots. Let $X_i(t) \in \{0,1\}$ be the indicator function that is equal to 1 when the node i transmits the packet in time slot t, and $X_i(t) = 0$ otherwise. It should be noted that interference may arise when multiple nodes transmit packets during the same time slot, and therefore, at most, one packet can be transmitted in the one slot, since we have

$$\sum_{i=1}^{N} X_i(t) \leq 1. \tag{1}$$

Similar to [27,28], we assume that a node will wait until its send time slot to transmit a data packet that it has generated, instead of sending it immediately. Each source node can send data to the sink node and receive an acknowledgment message within a single time slot. The scheduler cyclically schedules each source node through a superframe with a duration of P_t. The packet sent by a source node comprises the data and the data generation time, denoted by $G_i(t)$. The AoI of source node i at time t is represented by $A_i(t)$. When the sink successfully receives a new packet, $A_i(t)$ is updated to the difference between the current time slot t and $G_i(t)$. In other cases, $A_i(t)$ increases linearly. The update process of $A_i(t)$ can be expressed as follows:

$$A_i(t) = \begin{cases} t - G_i(t) + 1 & \text{if node } i \text{ update} \\ A_i(t-1) + 1 & \text{others} \end{cases} . \quad (2)$$

Figure 1 presents a comparison of the AoI evolution process of packets under different transmission intervals with the same sampling period ($T_i = 7$) of a node. The upper part of each subgraph illustrates the data sampling events and data packet transmission events in time slots, while the lower part depicts the AoI evolution process according to different generation and delivery sequences of data packets. The sampling event of the source node occurs at time slot $s_i(k), s_i(k+1), \cdots$, and the sink node receives the corresponding data packet at time slot $r_i(k), r_i(k+1), \cdots$. We define the peak AoI of the k-th packet of node i as $\Delta_i^p(k)$, $\forall k > 0$. The time between packet generation and sink reception is referred to as the system time $D_i(k)$, which is equal to $r(k) - s(k)$. We refer to the transmission interval time as I_i, which equals $r(k+1) - r(k)$.

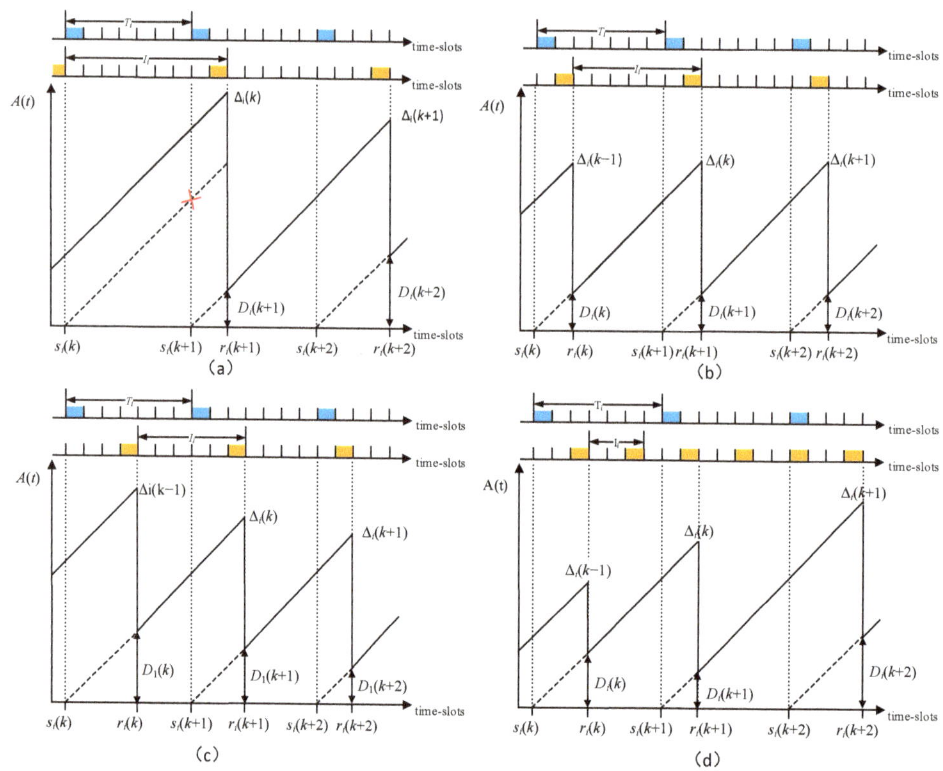

Figure 1. Evolution of node's AoI with respect to different transmission intervals. (**a**) The change of node's AoI when $I_i > T_i$; (**b**) the change of node's AoI when $I_i = T_i$ under ideal conditions; (**c**) the change of node's AoI when $\frac{T_i}{2} < I_i < T_i$; (**d**) the change of node's AoI when $I_i < \frac{T_i}{2}$.

The source node in our system employs a single packet queue model, which means that any data packet not sent from the queue will be dropped upon the generation of a new data packet. As shown in Figure 1a, where $I_i = 9$ is greater than T_i, the packet k was not sent to the sink node before packet $k+1$ was generated, resulting in the packet k being dropped. The red cross in Figure 1a indicates the moment that the data packet was dropped. To avoid packet drops, it is crucial that I_i must be less than or equal to T_i. Ideally, the transmission interval for each source node would be equal to its own sampling period, with the time interval between the packet generation time slot and the transmission time slot kept as short as possible. As shown in Figure 1b, in this scenario, the AoI of the node changes periodically with the peak AoI being equal to $D_i(k) + I_i$ time slots (i.e., one time slot plus seven time slots equals eight time slots). However, it is impractical to ensure that the transmission interval of each node is equal to its own sampling period in a scheduling since the sampling period of nodes in an IWSN usually varies significantly. In Figure 1c,d, it can be clearly observed that a smaller transmission interval does not necessarily improve AoI performance but can instead cause a waste of time slots.

Based on (2), it can be inferred that in the absence of new message arrivals, the AoI of a node shows a linear growth with a slope of 1. From Figure 1, it can be observed that the peak AoI $\Delta_i^p(k)$ of the k-th packet of node i is calculated as the sum of the system time of the current data packet and its transmission interval. As such, we can conclude that

$$\Delta_i^p(k) = D_i(k) + \left[\frac{T_i}{I_i}\right] \times I_i, \tag{3}$$

where $[\cdot]$ is the floor function. After the scheduler completes the network scheduling, the network executes the scheduling table repeatedly until the network parameters change. During the execution of the scheduling table, the value of I_i remains fixed. Here, the peak AoI of a node is determined by system time D_i. Figure 1 also depicts that $\Delta_i(k)$ is equivalent to the transmission period added to the system time of the subsequent data packet and is expressed as

$$\Delta_i^p(k) = D_i(k+1) + T_i. \tag{4}$$

By substituting (4) into (3), we have the update process of $D_i(k)$ as

$$D_i(k+1) = \begin{cases} D_i(k) + \left\lfloor\frac{T_i}{I_i}\right\rfloor \times I_i - T_i, \text{ if } \Delta_i^p(k) > T_i \\ D_i(k) + \left\lceil\frac{T_i}{I_i}\right\rceil \times I_i - T_i, \text{ others} \end{cases}. \tag{5}$$

The value of $D_i(k)$ changes periodically, which leads to the value of $\Delta_i(k)$, as stated in (4), being within a specific period. However, it is an intractable problem to determine the transmission interval I_i of nodes. If the value of I_i is less than or equal to T_i, it guarantees that there will be at least one time slot between two consecutive sampling slots. On the other hand, if I_i is greater than T_i, there will be no send time slot between two consecutive sampling time slots of a node. This consequently results in the node being unable to send the current data before the next sampling data are generated, ultimately leading to discarding the existing data. According to (4) and (5), when $I_i = T_i$, the node's peak AoI will be a specific value, and the duration of the superframe can be the least common multiple of T_i. Therefore, it is challenging to determine the length of the superframe due to the considerable variance in the nodes' sampling period. However, the value of $\frac{1}{I_i}$ indicates the proportion of the slot occupied by node i in a superframe. Thus, $\sum_{i=1}^{N} \frac{1}{I_i} \leq 1$ is necessary for the network to satisfy the scheduling feasibility. Therefore, we must choose an appropriate I_i for the node according to T_i.

3. AoI-Bounded Scheduling Algorithm

This section first presents the AoI-bounded scheduling algorithm for IWSNs. Then, we analyze the BAI of nodes with different sampling periods under this algorithm. Finally, we propose a method for improving the BAI in the proposed algorithm for nodes with higher AoI requirements.

3.1. Scheduling Algorithm

The algorithm primarily divides the superframe into multiple minimum transmission units (U_{min}) with the same length. Each node's I_i is an integer multiple of U_{min} and less than its T_i, guaranteeing that each node sends the current data before the next sampled data are generated and that the network's schedulability is satisfied.

As U_{min} is the algorithm's basic scheduling unit in a superframe, it is necessary to ensure that the U_{min} is less than or equal to the minimum sampling period in all nodes to prevent data packets from being dropped. However, since I_i is an integer multiple of U_{min}, a larger U_{min} can cause a superframe to have more available time slots. We take the minimum T_i among all nodes as the U_{min}, which can be obtained as

$$U_{min} = min\{T_i, \forall i \in \mathbf{N}\}. \tag{6}$$

U_{min} is taken as the least common factor of the transmission interval I_i for each node. I_i represents the length of the interval between two adjacent transmission time slots of node i within one superframe. The I_i of node i can be obtained as

$$I_i(k) = \alpha_i \times U_{min}, \tag{7}$$

and

$$\alpha_i = 2^{\lfloor \log_2 (\frac{T_i}{U_{min}}) \rfloor}. \tag{8}$$

α_i is the transmission interval coefficient (TIC) of node i, defined as a power of two, i.e., 2^n. $\lfloor \cdot \rfloor$ is a floor function. According to (8), $I_i < T_i$.

In addition to considering periodic data transmission, aperiodic data cannot be ignored. We reserve σ time slots for aperiodic data packets in each U_{min}; this means that there are $U_{min} - \sigma$ time slots that can be allocated to periodic data flows in each U_{min}. The general structure of the superframe defined by the proposed scheduling algorithm is shown in Figure 2. When considering network scheduling feasibility, it is crucial to ensure that the time slots allocated for periodic nodes as well as the reserved time slots of aperiodic nodes should not exceed the length of U_{min}, as constrained by condition (9):

$$\sum_{i=1}^{N} \frac{1}{\alpha_i} + \sigma \le U_{min}. \tag{9}$$

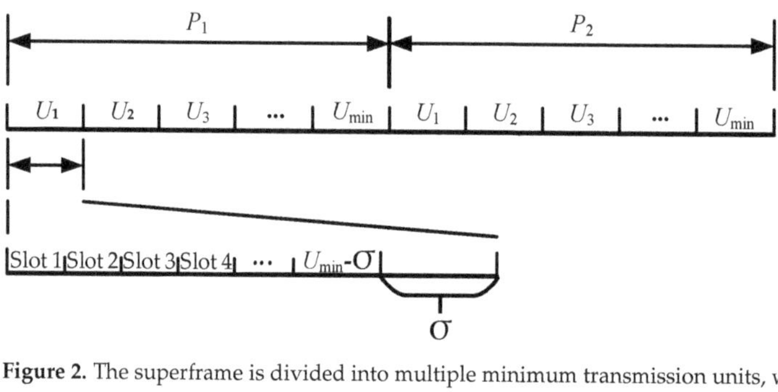

Figure 2. The superframe is divided into multiple minimum transmission units, where the length of each transmission unit is equal to the minimum sampling period in all nodes, and σ time slots are reserved for aperiodic data at the end of each transmission unit.

Under the condition of guaranteeing network scheduling feasibility, by (9), each node's transmission intervals I_i are multiples of each other. The duration of a superframe P_t is the least common multiple of all nodes I_i. Thus, P_t can be obtained as

$$P_t = max(\alpha_i) \times U_{min}, i \in [1, N]. \tag{10}$$

Finally, dedicated time slots are assigned to each node. In this step, priority is given to nodes with a smaller T_i to determine their initial scheduling time slot (IST). We allocate the first unused time slot t from 1 to U_{k_i} as the IST of node N_i. After determining the IST, the corresponding time slots of N_i in the remaining time slots of the superframe are determined accordingly, i.e., $IST + I_i \times m$, $m = \left[1, \cdots, \frac{\alpha_{max}}{\alpha_i} - 1\right]$. Take node i with its $\alpha_i = 2$ and the maximum α of the network (i.e., $k_{max} = 8$) as an example. If there exists a time slot $\mathrm{t} \in [1, 2 \times U_{min}]$ with $\sum_{i=1}^{N} X_i(\mathrm{t}) = 0$, we allocate the time slot t as the IST of node i, and the corresponding time slots in the remaining superframe can be determined as $t + m \times I_i, \mathrm{m} \in \{1, 2, 3\}$. The key steps of the proposed algorithm are given in Algorithm 1. The time is mainly consumed in Step 5 of Algorithm 1, in which we need to find the first unused time slot for each node as its IST. Therefore, the time complexity of Algorithm 1 is $O(n^2)$, where n denotes the number of nodes in the networks.

Algorithm 1: AoI-bounded Scheduling

Input: N, T_i
Output: U_{min}, IST, P_t, I_i
Determine the length of U_{min} based on (6) // **step 1**
for i = 1,2,$\cdots$,N **do** // **step 2**
    Determine the transmission interval I_i of node N_i based on (7) and (8).
end
// Validate the scheduability of the network. // **step 3**
if $\sum_{i=1}^{N} \frac{1}{\alpha_i} + \sigma \leq U_{min}$ **then**
    Network is schedulable, go to Step 4;
else
    Indicates the network configuration is overloaded;
    Return;
end
Determine the duration of superframe P_t based on (10); // **step 4**
for i = 1,2,$\cdots$,N **do** // **step 5**
    //Assign dedicated time slots to each node
    Allocate the first unused time slot t from 1 to U_{k_i} as the IST of node N_i;
end
Return

3.2. BAI Analysis

In this section, we analyze the upper and lower bounds of the node AoI under the proposed scheduling algorithm, which in turn shows that the node AoI is in a bounded interval. The BAI of node i can be determined by the interval between the minimum peak AoI and the maximum peak AoI.

The worst-case scenario for scheduling occurs when the send time slot of a node overlaps with a sample time slot, causing the latest sampled value to be delayed until the next send time slot. This delay results in the node's AoI reaching its maximum value. In the worst-case scenario, the latest sampled data is generated after an elapsed time of T_i from the last packet generation. Therefore, the AoI of the node at this time is T_i. The next sending time of the node requires I_i time slots, and given that it takes one time slot to complete the transmission of the message, the maximum peak AoI Δ_i^{max} of the node can be expressed as follows:

$$\Delta_i^{max} = I_i + T_i + 1 \tag{11}$$

Figure 1c depicts the optimal scenario of the scheduling algorithm, wherein a node immediately transmits sampled data in the following time slot. This allows the node to promptly send its data to the sink node and results in the minimum AoI for the node. In this case, with an additional time slot accounting for data packet transmission, the minimum peak AoI ${\Delta_i}^{min}$ is given by

$$ {\Delta_i}^{min} = 1 + T_i. \tag{12} $$

The analysis above indicates that the maximum and minimum peak AoI of a node is associated with its sampling period and transmission interval. Once the network scheduling concludes, both the sampling and transmission intervals remain constant, resulting in the AoI of the node being confined within a bounded interval. The BAI of node i is between the maximum peak AoI ${\Delta_i}^{max}$ and the minimum peak AoI ${\Delta_i}^{min}$, i.e., $A_i(t) \in [I_i + T_i + 1, T_i + 1]$.

3.3. Peak AoI Decrease Method

From (7) and (11), we can determine that the maximum peak AoI is positively correlated with α_i and U_{min}. Improving the peak AoI by reducing U_{min} will reduce the available time slots and affect network scheduling feasibility. The most effective way to improve BAI is to reduce the peak AoI of node i by reducing α_i. After reducing α_i to $\widetilde{\alpha_i}$, the the node's $I_i(k)$ reduces accordingly, adding the node's $\frac{P_t}{U_{min}} \times \left(\frac{1}{\widetilde{\alpha_i}} - \frac{1}{\alpha_i}\right)$ transmission slot to the superframe. The improved ${\Delta_i}^{max}$ can be obtained as

$$ {\Delta_i}^{max} = \frac{I_i \times \widetilde{\alpha_i}}{\alpha_i} + T_i + 1. \tag{13} $$

Due to the addition of time slots for nodes, reducing the value of α_i must still satisfy constraint (8), ensuring network scheduling feasibility. Other than that, reducing the maximum peak AoI can also reduce the average AoI. According to [29], the average AoI Δ_i of source node i can be obtained as

$$ \begin{aligned} \Delta_i &= \frac{E[D_i T_i] + E\left[{T_i}^2\right]/2}{E[T_i]} \\ \Delta_i &= E[D_i] + \frac{T_i}{2} \end{aligned} \tag{14} $$

Considering that the data packet transmission needs one time slot, the exception of D_i can be obtained as

$$ E[D_i] = \frac{1 + I_i}{2} + 1. \tag{15} $$

Therefore, a decrease in α_i will decrease BAI and the average AoI Δ_i, accordingly.

4. Evaluation and Numerical Results

In this section, we evaluate the performance of the proposed algorithm by providing an example and presenting the results of simulations. We consider a network of 10 sensor nodes that periodically sample data and send them to the sink node, i.e., $N = 10$. The sampling period T_i of each sensor node is uniformly distributed at random integers between 2 and 50 time slots, i.e., $T_i \in [2, 50]$, as shown in Table 2. The duration of a time slot is defined as 10 ms. We reserve one time slot for aperiodic data in each U_{min}. Following the five key steps of the proposed algorithm in Section 4, we present a detailed example below to illustrate the values obtained from each step of the proposed algorithm.

In step 1, the length of the minimum transmission unit U_{min} is determined according to (6), which involves taking the minimum T_i among all nodes. Since the sampling period of node #7 is the smallest among all nodes, the U_{min} is set to seven time slots.

In step 2, the minimum scheduling unit and the sampling period of all nodes are given. Each node's I_i and α_i are calculated using Equations (7) and (8), respectively. The results are presented in Table 2.

In step 3, the primary task is to examine the schedulability of the network. According to (2), the sum of the reciprocal of coefficients for all nodes, adding the number of reserved time slots for aperiodic data packets, is calculated as $\sum_{i=1}^{N}\frac{1}{\alpha_i}+\sigma=6$, less than the U_{min}. This indicates that the proposed AoI scheduling algorithm can find a solution that schedules all the nodes under the current network parameters setting.

In step 4, the duration of the superframe is determined based on (10), i.e., $P_t = 28$. The network will repeat the whole schedule table every 28 time slots.

In step 5, the dedicated time slots for each source node are allocated. Following the allocation process, the lower sampling period nodes are prioritized. For any node, its IST is determined, and then the remaining time slots are allocated according to $I_i(k)$. Table 2 shows each node's IST, and Figure 3 shows each source node's send time slot in a superframe. We take node #2 as an example, where $\alpha_2 = 1$; we need to decide the IST of node #2 in U_1. Since time slot 1 is occupied by node #7, the IST of node 2 is set to time slot 2. According to (7), the transmission interval of node 2 is seven time slots. Thus, the time slots of node 2 in the superframe are 2, 9, 16, and 23.

Table 2. Parameters setting of the example network.

Node ID	T_i	α_i	$I_i(k)$	IST
1	28	4	28	9
2	10	1	7	2
3	15	2	14	4
4	38	4	28	12
5	17	2	14	5
6	20	2	14	6
7	7	1	7	1
8	29	4	28	10
9	35	4	28	11
10	14	2	14	3

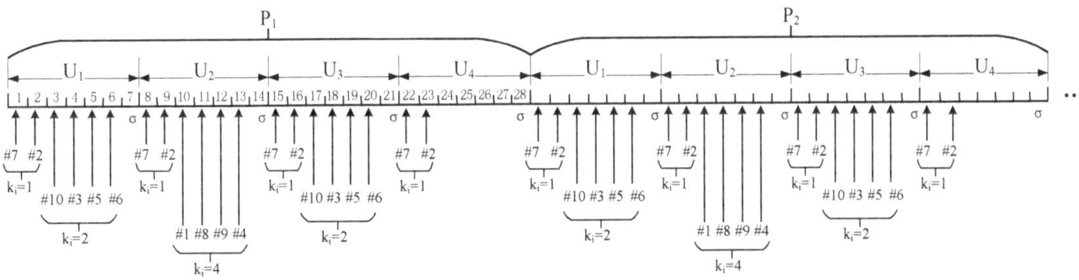

Figure 3. Illustration of the time slot allocation.

According to the parameters in Table 2, we evaluated the AoI performance of nodes in a discrete-event simulator that we built in a Python environment. To demonstrate the effectiveness of the proposed algorithm in ensuring AoI within a bounded interval, we selected three representative nodes and examined the changes in their AoI. These nodes include node #2 with $\alpha_2 = 1$, node #6 with $\alpha_6 = 2$, and node #9 with $\alpha_9 = 4$. We adopted the optimal greedy scheduling algorithm as a benchmark for comparison, in which each node transmits the data to the sink node in the next time slot after completing the sampling process, thereby obtaining the lower bound value of the peak AoI. It is crucial to note that it is impractical to apply the greedy scheduling strategy on every node in the network, given the large number of nodes and the possibility of significant variations in the sampling period of each node.

Figure 4 shows the real-time AoI, with the corresponding peak AoI and benchmark of the three representative nodes. Clearly, the AoI of all three nodes is below its corresponding peak AoI. The node's peak AoI changes periodically, guaranteeing each node's BAI. Under the proposed algorithm, since the peak AoI is positively correlated with I_i, it causes the maximum peak AoI of node 9 to be higher than the rest of the nodes, resulting in a greater BAI. The periodic variation of the AoI of nodes within the BAI interval can be attributed to the utilization of the floor function in determining the transmission interval of nodes in (8). As a result, the transmission period of nodes is shorter than the data generation period. Furthermore, the network employs a superframe-based periodic cycle scheduling method, which leads to the periodic variation of the time interval between node transmission slots and node sampling slots. In addition, we analyzed the AoI of all nodes and confirmed the BAI of all nodes and the peak AoI and average AoI of all ten sensor nodes, as shown in Figure 5.

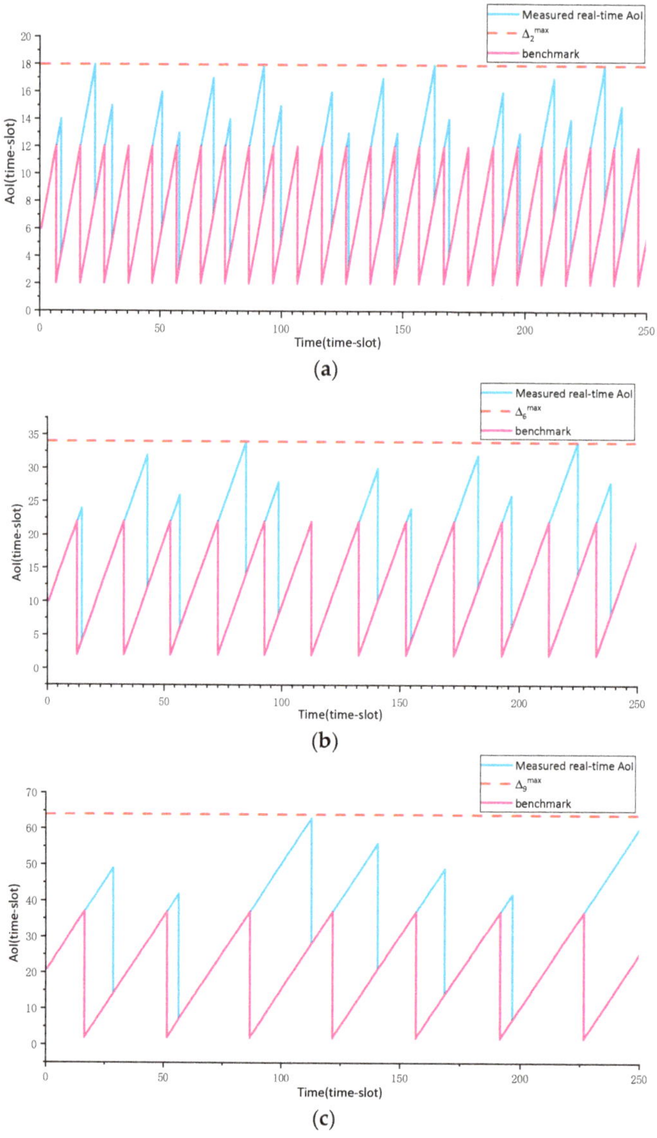

Figure 4. The real-time AoI, with the corresponding peak AoI and benchmark of three sample nodes: (**a**) node #2, (**b**) node #6, (**c**) node #9.

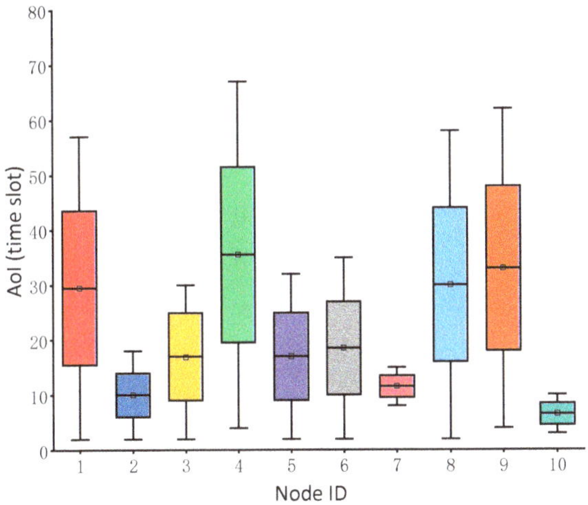

Figure 5. Boxplot of AoI for all ten sensor nodes.

In Figure 6, the effectiveness of reducing the peak AoI of a node by adjusting its TIC α_i is demonstrated, with node #9 used as an example. By adjusting the α_9 of node #9 from 4 to 2 and 1, a reduction in the peak AoI of the node was observed. Decreasing the TIC can increase the number of transmission slots allocated to nodes. However, the TIC cannot be arbitrary, as a coefficient that is too small would reduce the network's schedulability; the value of coefficient α_i must satisfy the constraint in (9).

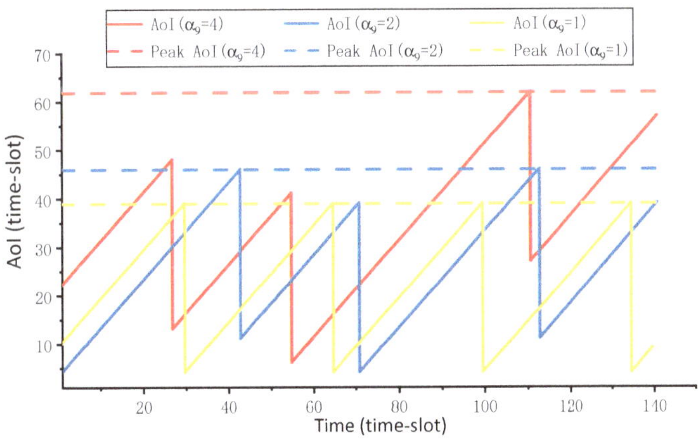

Figure 6. The AoI of node #9 with different α_9.

The boxplot of the AoI for node #9 with different α_9 is shown in Figure 7, which indicates that reducing the TIC α_9 leads to a decrease in the peak AoI, while the average AoI of the node also decreases accordingly.

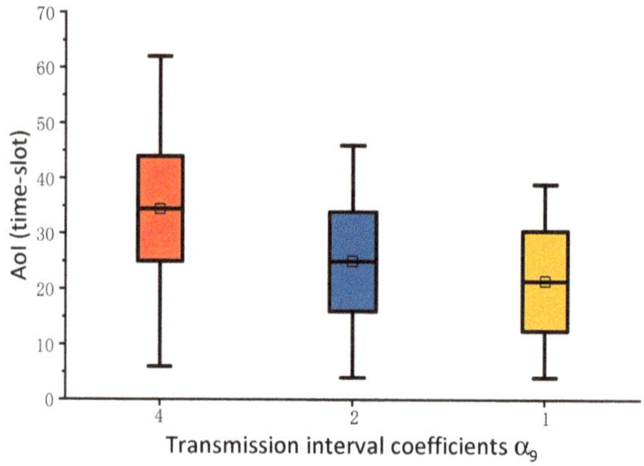

Figure 7. Boxplot of AoI for node #9 with different TIC α_9.

To evaluate the schedulability of the proposed algorithm, we conducted experiments to test its scheduling success rate under varying numbers of nodes and average sampling periods, as shown in Figure 8. The scheduling success rate is defined as the percentage of test cases for which the algorithm is able to find a feasible schedule [30]. The scheduling success rate exhibits a decreasing trend with an increase in the number of nodes, which can be attributed to the requirement for additional time slots as the number of nodes increases. Similarly, a decrease in the average time sampling period leads to a reduction in the scheduling success rate. This can be attributed to the fact that a smaller sampling period results in an increased number of packets being sent within a single superframe.

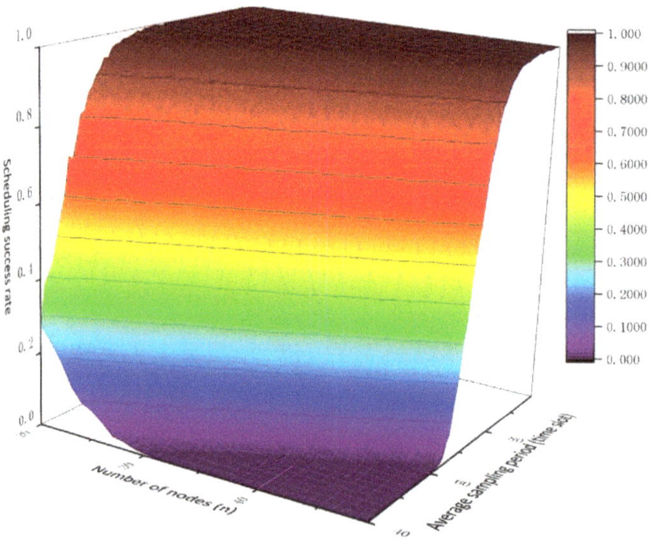

Figure 8. Schedulability analysis with respect to varying numbers of nodes and sampling periods.

5. Conclusions and Future Works

This paper proposed a scheduling algorithm guaranteeing each node's AoI within a bounded interval in an IWSN where the sensor nodes' sampling periods vary significantly, which is crucial for ensuring the stability of industrial systems. We determined the node's transmission interval and superframe length according to the node's sampling period to ensure network scheduling feasibility. Furthermore, we proposed a method to decrease the peak AoI by allocating more time slots for the nodes. A numerical example is given to illustrate the proposed algorithm step by step; the numerical results showed that the proposed algorithm could guarantee that the AoI of each node would be below the corresponding peak AoI.

In the future, the proposed algorithm is expected to be implemented in a real IWSN scheduler to test the AoI performance with real industrial data. Moreover, the algorithm can be extended to support multi-hop topology, while also taking into account lossy wireless channel models.

Author Contributions: Conceptualization, methodology, C.P. and H.Y.; software, validation, H.Y. and C.D.; writing—original draft preparation, H.Y.; writing—review and editing, C.P.; supervision, project administration, funding acquisition, P.W. All authors have read and agreed to the published version of the manuscript.

Funding: This research was supported in part by the National Key Research and Development Program of China, under Grant 2022YFE020527; in part by the Chongqing Talent Plan Project, China under Grant cstc2021ycjh-bgzxm0206; and in part by the Chongqing Municipal Federation of Trade Unions supporting the model worker's research project.

Conflicts of Interest: The authors declare no conflict of interest.

Abbreviations

The following acronyms are used in this manuscript:

AoI	Age of information
BAI	Bounded AoI intervals
CSMA/CA	Carrier sense multiple access with collision avoidance
TDMA/FDMA	Time/frequency division multiple access
IWSNs	Industrial wireless sensor networks
IST	Initial scheduling time slot
TIC	Transmission interval coefficient

References

1. Li, X.; Li, D.; Wan, J.; Vasilakos, A.V.; Lai, C.-F.; Wang, S. A Review of Industrial Wireless Networks in the Context of Industry 4.0. *Wirel. Netw* **2017**, *23*, 23–41. [CrossRef]
2. Wang, Q.; Jiang, J. Comparative Examination on Architecture and Protocol of Industrial Wireless Sensor Network Standards. *IEEE Commun. Surv. Tutor.* **2016**, *18*, 2197–2219. [CrossRef]
3. Majid, M.; Habib, S.; Javed, A.R.; Rizwan, M.; Srivastava, G.; Gadekallu, T.R.; Lin, J.C.-W. Applications of Wireless Sensor Networks and Internet of Things Frameworks in the Industry Revolution 4.0: A Systematic Literature Review. *Sensors* **2022**, *22*, 2087. [CrossRef]
4. Zhang, G.; Shen, C.; Shi, Q.; Ai, B.; Zhong, Z. AoI Minimization for WSN Data Collection with Periodic Updating Scheme. *IEEE Trans. Wirel. Commun.* **2022**, *22*, 32–46. [CrossRef]
5. Kaul, S.; Yates, R.; Gruteser, M. Real-Time Status: How Often Should One Update? In Proceedings of the 2012 Proceedings IEEE INFOCOM, Orlando, FL, USA, 25–30 March 2012; IEEE: Piscataway, NJ, USA, 2012; pp. 2731–2735.
6. Costa, M.; Codreanu, M.; Ephremides, A. On the Age of Information in Status Update Systems with Packet Management. *IEEE Trans. Inf. Theory* **2016**, *62*, 1897–1910. [CrossRef]
7. Vitturi, S.; Zunino, C.; Sauter, T. Industrial Communication Systems and Their Future Challenges: Next-Generation Ethernet, IIoT, and 5G. *Proc. IEEE* **2019**, *107*, 944–961. [CrossRef]
8. Qiu, T.; Chi, J.; Zhou, X.; Ning, Z.; Atiquzzaman, M.; Wu, D.O. Edge Computing in Industrial Internet of Things: Architecture, Advances and Challenges. *IEEE Commun. Surv. Tutor.* **2020**, *22*, 2462–2488. [CrossRef]
9. Petersen, S.; Carlsen, S. WirelessHART Versus ISA100.11a: The Format War Hits the Factory Floor. *IEEE Ind. Electron. Mag.* **2011**, *5*, 23–34. [CrossRef]
10. Yan, H.; Zhang, Y.; Pang, Z.; Xu, L.D. Superframe Planning and Access Latency of Slotted MAC for Industrial WSN in IoT Environment. *IEEE Trans. Ind. Inform.* **2014**, *10*, 1242–1251. [CrossRef]
11. Kharb, S.; Singhrova, A. A Survey on Network Formation and Scheduling Algorithms for Time Slotted Channel Hopping in Industrial Networks. *J. Netw. Comput. Appl.* **2019**, *126*, 59–87. [CrossRef]
12. Ding, Y.; Hong, S.H. CFP Scheduling for Real-Time Service and Energy Efficiency in the Industrial Applications of IEEE 802.15.4. *J. Commun. Netw.* **2013**, *15*, 87–101. [CrossRef]
13. Lin, F.; Dai, W.; Li, W.; Xu, Z.; Yuan, L. A Framework of Priority-Aware Packet Transmission Scheduling in Cluster-Based Industrial Wireless Sensor Networks. *IEEE Trans. Ind. Inform.* **2020**, *16*, 5596–5606. [CrossRef]

14. Mukherjee, M.; Shu, L.; Prasad, R.V.; Wang, D.; Hancke, G.P. Sleep Scheduling for Unbalanced Energy Harvesting in Industrial Wireless Sensor Networks. *IEEE Commun. Mag.* **2019**, *57*, 108–115. [CrossRef]
15. Abdalzaher, M.S.; Muta, O. Employing Game Theory and TDMA Protocol to Enhance Security and Manage Power Consumption in WSNs-Based Cognitive Radio. *IEEE Access* **2019**, *7*, 132923–132936. [CrossRef]
16. Elwekeil, M.; Abdalzaher, M.S.; Seddik, K. Prolonging Smart Grid Network Lifetime through Optimising Number of Sensor Nodes and Packet Length. *IET Commun.* **2019**, *13*, 2478–2484. [CrossRef]
17. Zhu, T.; Li, J.; Gao, H.; Li, Y.; Cai, Z. AoI Minimization Data Collection Scheduling for Battery-Free Wireless Sensor Networks. *IEEE Trans. Mob. Comput.* **2023**, *22*, 1343–1355. [CrossRef]
18. Hirosawa, N.; Iimori, H.; Ishibashi, K.; Abreu, G.T.F.D. Minimizing Age of Information in Energy Harvesting Wireless Sensor Networks. *IEEE Access* **2020**, *8*, 219934–219945. [CrossRef]
19. Moltafet, M.; Leinonen, M.; Codreanu, M. Worst Case Age of Information in Wireless Sensor Networks: A Multi-Access Channel. *IEEE Wirel. Commun. Lett.* **2020**, *9*, 321–325. [CrossRef]
20. Yates, R.D.; Sun, Y.; Brown, D.R.; Kaul, S.K.; Modiano, E.; Ulukus, S. Age of Information: An Introduction and Survey. *IEEE J. Sel. Areas Commun.* **2021**, *39*, 1183–1210. [CrossRef]
21. Kam, C.; Kompella, S.; Nguyen, G.D.; Ephremides, A. Effect of Message Transmission Path Diversity on Status Age. *IEEE Trans. Inform. Theory* **2016**, *62*, 1360–1374. [CrossRef]
22. Talak, R.; Karaman, S.; Modiano, E. Improving Age of Information in Wireless Networks with Perfect Channel State Information. *IEEE/ACM Trans. Netw.* **2020**, *28*, 1765–1778. [CrossRef]
23. Zhou, B.; Saad, W. Joint Status Sampling and Updating for Minimizing Age of Information in the Internet of Things. *IEEE Trans. Commun.* **2019**, *67*, 7468–7482. [CrossRef]
24. Hu, L.; Chen, Z.; Dong, Y.; Jia, Y.; Liang, L.; Wang, M. Status Update in IoT Networks: Age-of-Information Violation Probability and Optimal Update Rate. *IEEE Internet Things J.* **2021**, *8*, 11329–11344. [CrossRef]
25. Seo, J.-B.; Choi, J. On the Outage Probability of Peak Age-of-Information for D/G/1 Queuing Systems. *IEEE Commun. Lett.* **2019**, *23*, 1021–1024. [CrossRef]
26. Huang, W.; Li, X.; Liang, Y. Impact of Age Violation Probability on Neighbor Election-Based Distributed Slot Access in Wireless Ad Hoc Networks. *Electronics* **2023**, *12*, 351. [CrossRef]
27. Li, C.; Li, S.; Chen, Y.; Thomas Hou, Y.; Lou, W. AoI Scheduling with Maximum Thresholds. In Proceedings of the IEEE INFOCOM 2020—IEEE Conference on Computer Communications, Toronto, ON, Canada, 6–9 July 2020; pp. 436–445.
28. Li, C.; Li, S.; Chen, Y.; Hou, Y.T.; Lou, W. Minimizing Age of Information Under General Models for IoT Data Collection. *IEEE Trans. Netw. Sci. Eng.* **2020**, *7*, 2256–2270. [CrossRef]
29. Lin, W.; Li, L.; Yuan, J.; Han, Z.; Juntti, M.; Matsumoto, T. Age-of-Information in First-Come-First-Served Wireless Communications: Upper Bound and Performance Optimization. *IEEE Trans. Veh. Technol.* **2022**, *71*, 9501–9515. [CrossRef]
30. Jin, X.; Guan, N.; Xia, C.; Wang, J.; Zeng, P. Packet Aggregation Real-Time Scheduling for Large-Scale WIA-PA Industrial Wireless Sensor Networks. *ACM Trans. Embed. Comput. Syst.* **2018**, *17*, 1–19. [CrossRef]

MDPI AG
Grosspeteranlage 5
4052 Basel
Switzerland
Tel.: +41 61 683 77 34

Electronics Editorial Office
E-mail: electronics@mdpi.com
www.mdpi.com/journal/electronics

www.ingramcontent.com/pod-product-compliance
Lightning Source LLC
Chambersburg PA
CBHW061453210926
51850CB00008B/1698
9783725815135